Food Pollution

Gene Marine / Judith Van Allen

Holt, Rinehart and Winston

NEW YORK CHICAGO SAN FRANCISCO

Published simultaneously in Canada by Holt, Rinehart and Winston of Canada, Limited.

ISBN: 0-03-085070-3

Library of Congress Catalog Card Number: 73-117266

First Edition

Printed in the United States of America

for
DR. JACQUELINE VERRETT
and her colleagues
who are also trying
to tell the truth

Contents

Preface

We are not food freaks. On the contrary.

The pronoun "we," of course, reflects the dual authorship of the book; but the reader may experience some confusion, or at least a disturbing jar, at the occasional use of "I." When that happens, "I" am Gene Marine, and it happens because in the final version I occasionally draw from my own experience for an example, or have an observation that may not be shared. Those few sentences are the only ones for which we do not both share full responsibility.

As an example, it doesn't mean quite the same thing for both of us to say that we are not food freaks. *I* probably violate in my own life every tenet of this book, in addition to which I smoke cigarettes, go back and forth constantly between coffee and beer, and eat a decent meal perhaps once every other day. I buy for myself all the things we advise against buying. To make it even worse, I prefer hydrogenated peanut butter and I seem to be alone in the world in that I am even relatively uncomplaining about ordinary white bread.

Judith is far less cavalier about food, and devotes normal attention to balancing her diet (and sometimes mine) and to overcoming at least the standard American deficiencies. And she detests white bread.

Neither of us, however, devotes nearly the effort to intelligent nutrition that this book suggests should be devoted. The difference is that we understand what we're doing. If I want to ruin my body, that's my business; but to do it out of ignorance is to be cheated of choice in one of the most important of all areas of decision.

The facts in this book are facts, no matter how the writers live. We present those facts as reporters. There *are* food faddists (we find them boring) and food quacks (we find them contemptible).

But they are not nearly so dangerous as the malefactors of greed who determine—with their control of government agencies, of institutions of learning, and of advertising channels—what the unfortunate public should consider "normal."

If you know what this book tells you, and choose white bread and poison-riddled food, you are an adult making a free choice. If you eat it because you don't know any better, or cannot afford better, you are a victim.

And if you knowingly feed it to your children, *they* are the victims.

Many of the people who helped us with this book have asked to remain anonymous; to name others would be to put them into a position where they might be blamed for telling us things that came, in fact, from other sources. Because there *was* a lot of help, this is a difficult constriction; but it is a necessary one. You know who you are, and we thank you.

We can thank publicly, for his formal assistance, Larry Cornell of the San Francisco office of the Food and Drug Administration; he was extremely helpful in providing "straight information," and is in no way involved, even indirectly, in our criticisms of the FDA.

Since that agency *does* get irrationally sticky sometimes, we might also mention that we have never met the woman to whom we have dedicated the book, that she doesn't know about it, and that we hope the dedication doesn't cost her her job.

It is fascinating, incidentally, how many of the dedicated scientists in many disciplines concerned with food are women. You will find feminine names again and again in this book, not because we chose to do it that way but because it is that way. We know that it is partly because discrimination forces women with intelligence into certain "acceptable" fields, but we're glad they're there anyway.

Some of the concepts and perspectives involved in our presentation derive from conversations with people professionally accustomed to ecological and evolutionary thinking; George Child, Dr. Clarence Gordon, and Dr. Bernard Patten come quickly to mind. A particular debt is due to Dr. Charles Wurster for his brilliance of exposition and for his warm willingness to argue the merits of his particular approach to solving many of the questions we pose. None of those people will agree with all of what we write; but it would make a lot less sense without their contributions and challenges to our thinking.

Dr. Tom Bodenheimer, a good friend, came up with some timely aid to the section on genetics (but don't blame him, please, for any errors). The patience and understanding of Elizabeth Marine literally made the book's completion possible. As always, encouragement, guidance and assistance far beyond the merely professional earned for Cyrilly Abels our deep gratitude.

There must also be some very patient and thoughtful people at Holt, Rinehart and Winston. One is our apparently ulcer-proof editor, Robin Kyriakis. We can only hope it's worth her trouble.

Finally, we are grateful to the Louis N. Rabinowitz Foundation for a transfusion that kept the patient alive.

GENE MARINE

JUDITH VAN ALLEN

BERKELEY
SEPTEMBER, 1971

Food Pollution

CHAPTER ONE

The Painful Sight of Truth

The human race grows up, as individual children do, slowly and painfully.

Illusions, and the security they bring, fight hard against the assault of fact; what has always been true, it sometimes seems, must go on being true, or what will there be to cling to, what assurances can we have of order and meaning?

In the past two years, Americans have learned much about ecology, about the relationships among life forms and environment; about, if you will, the delicate balance that exists between man and the planet on which he lives. And much of what we have learned has been difficult, because for many of us it contradicts everything we have learned before. We grow up with assumptions—things are "natural" and "right" because they are *there*, and even more than that, they wouldn't be there if they weren't all right.

Towns and cities exist, and dams. The Colorado Desert is transformed almost magically into the startlingly green and blooming Imperial Valley. There are sprays to kill bugs, and leopard coats on well-to-do women; automobiles make the far places nearer for almost everyone, and great wide roads make them nearer yet; there are showers to take and toilets to flush, electric lights for reading and radios to fill the silences of the days, wood to build houses and steel to build skyscrapers.

It has always been all right, because the grownups didn't see anything wrong with it, and they wouldn't do it anyway if it weren't all right. And then, slowly, painfully, we grow up, and we look around.

The towns are built on flood plains, and sprawling across the precious agricultural land. The dams that, whatever else they do, hold back the floods that periodically renew that land, are silting up, and perhaps destroying fish life and other resources as well. Irrigation brings salinity to ruin fertile land, and drains poisons into water. DDT, far from where it was used, causes pelicans and falcons and petrels to lay omelets instead of eggs and wipes out whole species, of whose ecological importance we may never be sure. Leopards are disappearing. The automobile is an engine of invisible death, and its roads ruin the land and threaten the sources of oxygen itself. Sewage pours into the river and ocean, the plants that generate the electricity and manufacture the steel fill the air and water with death, the redwoods are in danger, and the cities are jammed-together horrors of filth and pollution.

Gradually, it dawns on us. Things are not all right just because we are told, over and over again, that they are. The feeling at first opens a great void before us—we have always believed, it has always been so—but that feeling, once the initial shock is over, is maturity, for an individual or for the human race.

Reporters confront this void early, or they are doomed to be poor practitioners of their profession. They come, as painfully as anyone else who undergoes the disintegration of illusions, to recognize that the condemned man may be innocent, that the public villain may be a hero, that the widely hailed beautification project is a carefully designed vehicle for someone's private profit. Above all, they come to learn, and not without sadness, that public statements are as likely to be lies as truth, no matter who makes them—baseball managers, movie actresses, or the President of the United States.

Somehow, the effect is sadder (though to reporters, perhaps, no longer surprising) when the false and misleading statements, the carefully adjusted half-truths, the condescending insults applied to opponents, come from persons who by their very positions are somehow supposed to represent the "public." Hawk or dove, we do not expect the President of the United States to tell the whole truth and nothing but the truth about a subject so diplomatically complex as the war in Vietnam; our cynicism is implanted far too deeply by now for that. But something in us still wishes it could expect the truth from the head of the Food and Drug Administration when he's talking about cyclamates.

In writing about food, then, we must confront the void twice—once in dealing with what is said about it, once in dealing with the food itself. For as we have grown up never questioning electric lights or automobiles or leopard coats, so to an even greater degree we have grown up accepting the food we eat. Vaguely, we know that there are government agencies with rules that do something to protect us. Even more vaguely, we assume, somehow, that it must be all right or they wouldn't let the food be sold to us.

Even when there is a flurry about cyclamates (what the hell is a "cyclamate," anyway?), or perhaps about MSG in baby food, it somehow seems an aberration. There must have been some new and intricate experiment that could not have been made before, some new danger that could not have been determined or predicted. And it must be that the government protectors then immediately jumped in on our side.

But that is not how it is, and it may even be that the publicity about cyclamates and MSG is more harmful than helpful, because it misleads us. The facts go much deeper. We cannot—in fact we must not—take it for granted that *anything* on the grocery shelf is healthy, FDA approval or not. We cannot—we must not—even assume that it is not poisonous. We are being poisoned. The only questions are how, and how much, and what can we do about it. And while we're at it, we had better also ask why.

For just as we must think of ourselves as living within ecosystems, we must think, for the sake of our own health, of ecosystems living within us. There are, for example, tiny life forms that live within our intestines that are necessary to our health and well-being. Not too long ago, the medical use of some broad-spectrum antibiotics was quickly discontinued for humans when it was learned that (although no one had known it in advance) they incidentally, as a side effect, killed those tiny life forms, and left an ecological niche to be filled by other, less beneficial, relatives.

Another reporter, William Longgood, looked at some aspects of this same subject more than a deçade ago, and he published a quotation from Dr. Edward J. Ryan, editor of *The Dental Digest:*

> Every time a natural substance is removed from a food, every time an adulterant is added to a food, the balance in nature is disturbed. . . . The chemical and cellular processes within the body cells cannot react to the passing whims of chemists without disturbance in function. It took thousands of years for the body to adjust itself to changing environmental conditions. When these conditions are suddenly altered by the actions of men, the cells cannot make the adjustment—disease is the result.*

* —Throughout the book, reference to quotations is by the author's name only, unless further identification is necessary to avoid confusion. The complete information is in the bibliography at the end of the book. Quotations from press sources are identified in footnotes.

We have, in other words, an "inner" ecology as well as an "outer" one. In our outer ecosystem, we have built dams and housing projects and power plants, not only without thought for the ecological consequences, but without even considering that there *are* ecological consequences. We simply did it without knowing what we were doing, because it was someone's misguided whim or, more often, because somebody made a buck.

Two years ago, I wrote a book about that. Paying, perhaps, too little attention to the profiteers behind the villainy, I was severely critical of Engineers, meaning not necessarily those who follow the profession of engineer, but those who are trapped in what I called an "engineering mentality," a way of taking one small problem at a time and solving it without looking at overall consequences. An Engineer, I said, will build a dam, and then, on being told that it blocks fish from swimming upstream, will build a fish ladder. And if the fish ladder causes another problem, he will design a further solution for that problem as well.

Those problems of our outer ecology are much better understood now, and, by whatever name we call it, many more of us understand how the engineering mentality works on the external environment. But just as we have violated our outer ecology with DDT and untreated sewage and automobile exhaust, so are we violating our inner ecology; for we have gimmicked and tortured and needled the contents of virtually everything we take into our bodies. And we have done it for the same two reasons, the whim of a tinkerer or the profit-hunger of an uncaring corporation executive, camouflaging it all with false doubletalk about health.

What follows will be called in some circles "controversial." It is not. I am reminded of a good friend and superb editor who, during the paranoid McCarthy era of the 1950s, was refused the use of a hall for a talk because he was "controversial." When he finally spoke in another hall, he pointed out that the refusal had not been because he was controversial but because he was on the wrong side of the controversy.

In this case, not even that is true. There is and will be noise and argument, but it is not controversy. Controversy assumes that there is a defensible opinion on either side. Experimenting blindly on human beings, in the face of clearly understood scientific principles, is not "controversial." It is damnable, the work of late-movie Mad Scientists portrayed perhaps by George Zucco or John Carradine—or less frivolously, the work of some never-to-be-forgotten Germans within the memories of some of us now alive.

They will defend themselves, however, they and the people who pay them and the people who are supposed to stop them but don't. As Dr. Ryan also noted:

> Anyone who speaks up against food adulteration in any of the many forms is subject to "name calling." The most common epithets are "food faddist" or "food fakir." If you object to [any of the things you will read about in the rest of this book], you are offending . . . some of our largest and most influential corporations. . . . [W]e can be certain that the public-relations counselors will go to work to change the situation—even if that requires a bit of character assassination.

We will be accused, too, of "emotionalism"—another favorite—and some people with apparently important scientific credentials will lend their names to such accusations. Reporters can only repeat the painfully learned fact that no one is too prominent to lie. But facts are facts: no matter what your opinion of the laws of gravity, when you walk off a cliff you fall. Antigravity belts are the property of Mad Scientists.

And the men who pay them.

CHAPTER TWO

. . . and Half a Bag of Peanuts

I have a son who is good at a number of things, among them baseball and eating, which makes it fun to take him out to see the Giants when the exchequer is in reasonable shape. I don't know whether I'll be able to do that comfortably ever again.

Soft drinks I won't even mention; they come later. But we do like to have at least a couple of hot dogs, while Old Father drinks a few beers. And sometimes, to keep it from being too awful, we'll take along something a little more nourishing, like maybe a nice fresh orange, or I'll get him to drink something a little healthier, like the chocolate milk they sell in little containers.

Everybody seems to know that it is now permitted to add up to 15 percent chicken to a hot dog without saying so, but according to the Stevens people—concessionaires at Candlestick Park, four other ball parks, 40 horse and dog tracks and a whole slew of arenas—*they* don't do it. Their hot dogs, they insist, are 92 percent meat (beef and pork in unknown proportions but no chicken), 3 percent corn syrup, 2 percent salt, 1.8 percent dextrose (call it sugar), 1 percent flavoring (otherwise unspecified), and a tiny bit of sodium nitrite.

The flavoring, whatever it is, does nothing whatever for you except to make the hot dog taste better, although 92 percent beef and pork ought to make it taste pretty good. Sodium nitrite is a preservative and a color fixer, so it doesn't do anything for you either, except maybe to fool you about how old the hot dog is. The Stevens people don't say anything about water, which is a good part of the hot dogs you're likely to buy in the store. Store dogs often have monosodium glutamate, too, which may be what Stevens uses for "flavoring." It is also called MSG and it is probably not good for you.

Store hot dogs are also likely to contain nonfat dry milk solids, which you can get by drinking milk but which are really used in a hot dog as cheap filler, and they may contain cereal, also as a filler (there is a football stadium in which the hot dogs have so much cereal that they're called "the Breakfast of Champions"). But even assuming that the Stevens people are telling the truth, it doesn't stop there.

It's extremely likely that before the beef and pork were ever ground up to put into the hot dogs, they were well laced with the broad-spectrum antibiotics fed to cattle and hogs and poultry, like chlor- and oxytetracycline and aureomycin. They aren't there to do you any good either. Finally—or probably not finally, but it's all I'm going to mention here—there's diethylstilbestrol.

Diethylstilbestrol is a synthetic estrogen—you might call it an "artificial female sex hormone"—used in cattle raising. The sole purpose in slipping it to animals, as is the case with almost everything we'll talk about that goes into the food you eat, is so that somebody can make an extra buck: in this case, it gets castrated male animals a little fatter on less feed. It is also known, in sufficient amounts (quite small amounts), to be carcinogenic in both men and women—more so to men, though it does affect women and relates particularly to breast cancers.

This is one of those places—there will be a lot of them as we go on—in which a fast-talking Mad Scientist, or his public relations man, can confuse you. He can tell you, quite truthfully, that government regulations allow the use of diethylstilbestrol only if *no* residue of the chemical remains in the product that reaches the consumer. He can also tell you, quite truthfully, that if any residue *is* found by a government inspector when the animal is slaughtered, the animal is condemned.

That's quite correct. But it's also interesting that when the government found residues in some chickens on the order of 45 parts per *billion*, they outlawed its use in chicken entirely. And it's interesting that government inspectors do in fact find diethylstilbestrol in one out of about every 200 cattle slaughtered. Not only that, but the average amount found is *higher* than 45 parts per billion, and frequently runs as high as 100 parts per billion—more than twice the concentration that was enough to have its use banned completely in poultry. And finally, what most people don't know is that not all, and not nearly all, cattle are inspected when they're slaughtered; the government doesn't have the manpower.

So you know, and the cattle raisers know, and the meat packers know, and the Mad Scientists know, and the FDA knows, that a lot of diethylstilbestrol is getting through to the public, rules or no rules. And if it should have anything to do with anybody's getting cancer, that person got cancer in order to put money in somebody else's pocket.

Rather a lot of money. The weight gains in cattle are believed to average out at about 20 percent, which at this writing means that the cattleman gets another $10 in profit—not much until you start adding the cattle together and find out that the additional profit to the industry is about $300 million a year. And of course, the Mad Scientists who *make* the stuff don't want any tighter regulation either; one of them at Eli Lilly and Company went so far as to say that without diethylstilbestrol "there simply wouldn't be enough beef to feed everyone," which is good old public relations scare-tactic nonsense.[1]

Okay, maybe the kid and I didn't get any diethylstilbestrol in our hot dogs this time out. But we raise cattle and pack meat in California, and of the 10 billion frankfurters eaten in the United States (it's a 1966 figure), one-third were sold within the state where they were manufactured, which means no federal inspection at all. So it's hard to tell.

Let's just settle for the MSG and the sodium nitrite, and the vegetable dye they put on the casing, which is made from something called annatto seeds. There's a story about that, too. Until 1964 they used a different dye (you understand this is just to make the hot dogs pretty; it has nothing to do with food), a coal-tar dye known officially as FD&C Red No. 4. We don't know for a fact that they give colors numbers like that to keep you from figuring out what they're made of, but we get that feeling.

Anyway, it turned out that FD&C Red No. 4 did strange and terrible things to animal bladders, which no one had bothered to learn, or to try to learn, until people had eaten billions of hot dogs. The government finally outlawed it. But there then descended on Washington a voluble lobby of cherry growers. Animal bladders or no animal bladders, FD&C Red No. 4 is now reauthorized for use on maraschino cherries, where its only function, again, is to make them pretty.

Our hot dogs, of course, come on buns. Buns are bread, which richly deserves a whole section of its own, so we'll just mention the use of butylated hydroxytoluene or butylated hydroxyanisole to keep the shortening from oxidizing, and the use as shortening of hydrogenated vegetable oils, the kind in which essential fatty acids are destroyed. Plus the mono- and diglycerides as emulsifiers, the chlorine dioxide used to age the flour, and the potassium bromate and iodate used to make the dough elastic and stable. Never mind for the moment what they did to the flour in the first place.

Happily our baseball park buns are white. In some supermarkets you can find packaged buns that are yellow and look richer. That's the same bun, with a coal-tar dye added to make it look yellow so you'll think of eggs and butter and things. That's all that it does except insofar as it may be poisoning you.

Chocolate milk (whether or not it contains synthetic Vitamin A, which is not necessarily good either) is certainly loaded with thickeners. They're either alginates, or carragheen, or carboxymethylcellulose ("carragheen" is a lovely Irish word, being the name of a substance derived from Irish moss). The antibiotics from the cow are also in the milk, which you might remember when you're reading about antibiotics later and watching the kids have an evening snack.

The orange, of course, is probably not only covered but impregnated with pesticide residues (no, you *can't* always wash them off; some of them go right through the peel into the orange itself). More than that, orange growers don't wait around any more for their oranges to ripen on the tree; they pick them when they want to, and run them all through a dye bath (this one is called Citrus Red No. 2) to make them pretty for you in the store, and then they shine the outsides with mineral oil or carnauba wax.

Carnauba wax is the stuff I would polish the car with if I ever got around to polishing the car. Mineral oil, except for specific reasons under a doctor's direction, is not so good for you either; it carries off fat-soluble vitamins so that you don't get the vitamins from any food eaten at about the same time. Of course no one eats orange peel much, but it takes very little mineral oil. It gets on your hands and then on the orange, and some people peel oranges at least partly with their mouths.

The kid doesn't drink beer. If he reads this he may never start. But I do, probably more than I should, which may explain the poorly concealed glee with which my wife handed me the list of chemicals used in beer manufacturing:

Sodium sulfite, sodium and potassium bisulfite, polymixin B, diethyl pyrocarbonate (like diethylstilbestrol in beef, this is supposed to be all gone by the time the beer gets to me), n-heptyl paraben (n-heptyl ester of *p*-hydroxybenzoic acid), synthetic ascorbic acid, dextrin, peptones (sounds like either a vocal group or the first baseman for the Cubs), methylcellulose, lactic acid, potassium sulfate, magnesium sulfate, diammonium orthophosphate, potassium chloride, and hydrochloric acid.

The old fashioned I might have had when I got home from the ball game doesn't seem so tempting now, with the FD&C No. 4 all over the cherry—not to mention the sodium benzoate with which it was preserved, the calcium hydroxide that improves its texture, the sulfur dioxide with which it was bleached, and the artificial flavoring it contains—and the stuff on the outside of the orange peel, plus the pesticide residues on and in both fruits, plus the sugar (more to come on that), plus dimethyl polysiloxane, sodium o-phenylphenate and, would you believe it, ammonia.

Note that we haven't offered you a meal, just a hot dog and an orange, some chocolate milk, a couple of beers, and a refreshing drink. Note, too, that *everything* you eat now contains residues of DDT and/or some of its near relatives—DDD, DDE, dieldrin, endrin, aldrin, and the other chlorinated hydrocarbons. Hot dogs being very fatty, it's worth noting that DDT residues store especially well in fat.

This book started out to be about intentional food *additives* in the ordinary-language sense, not about pesticide residues or other *contaminants*, and that's where we're going to try to keep the emphasis. But pesticide residues are there, and as we'll try to explain, you can't separate them in fact as easily as words make it seem at first. In point of literal fact, you can't separate them at all. Your body is one system, and it takes the consequences of everything you put into it, whether you can help it or not.

And incidentally, if you're already beginning to get that what-the-hell-*can*-I-eat feeling, it won't cheer you to know that the answer at the moment is almost literally nothing. However, we will try to offer some alternatives as we go along; and until the problem is laid out in all its ecological complexity, we won't be able to effect the turnaround that is absolutely necessary in American thinking about food if in fact we are going to survive at all.

The Mad Scientists like to call themselves "food technicians" (that phrase in itself should chill you, but they actually have a trade magazine called *Food Technology*, and it is an exact indication of their madness that they will read this sentence with brows furrowed in puzzlement over what we find wrong with the words), and we cannot improve on Longgood's list of their tools:

. . . dyes, bleaches, emulsifiers, antioxidants, preservatives, flavors, buffers, noxious sprays, acidifiers, alkalizers, deodorants, moisteners, drying agents, gases, extenders, thickeners, disinfectants, defoliants, fungicides, neutralizers, sweeteners, anticaking and antifoaming agents, conditioners, curers, hydrolizers, hydrogenators, maturers, fortifiers, and many others.

In public they talk about their work almost entirely in terms of adding things to food to make it healthier. In fact, they almost never do this, although, very occasionally, they do put in something nutritious to replace something they've already taken out. What they usually do is add something that is intended to fool you or cheat you or both.

They can fix up food so that when it is going bad you won't be able to tell by looking at it or smelling it, or sometimes even tasting it—but that doesn't mean that the chemical changes aren't taking place in the food. They can cover up for a lack of sanitation in preparation. They can provide cheaper substitutes for ingredients which have more nutritive value. They can make things look nutritious or appetizing which are not nutritious and ought not to look appetizing.

And they can do these things, and dozens of others whose function is to manipulate, con, and ultimately take more money from the consumer, with almost no concern about whether they are killing you while they're at it. The tests required for food additives are minimal at best; hundreds have never been tested at all, for anything. Some were never intended to be food additives in the first place. They were by-products of some industrial process unrelated to food, and became food additives when the house Mad Scientist was put to work to find a use for the waste so the company could make more money.

An example of how closely the Mad Scientist is related to the engineering mentality mentioned in the previous chapter is in a recent release from Science Service, which distributes scientific news items to newspapers:

> Addition of phosphates to the diet of children was found by Dr. Robert Harris, chairman of the Department of Nutrition and Food Sciences at MIT, to reduce the amount of tooth decay. Other researchers discovered that phosphates combined with additions of fluorides produced the best results in both preventing and reducing the number of caries—particularly where children eat pre-sweetened cereals and do not take care of their teeth regularly.[2]

They put the stuff in to rot kids' teeth, and then they put something else in so they'll only rot half as much. And they get angry because we call them Mad Scientists!

Let's get one thing straight, and very early in the game. Take any one food additive—calcium cyclamate, or hydrochloric acid, or butylated hydroxytoluene, or anything you like. Salt or sugar, if you'd rather. And remember this rule:

No argument about how much of that additive is safe makes any sense whatever.

You can talk about how much of an additive is unsafe. If a certain amount of it over a certain period of time kills people, then it's unsafe.

But safe, no, because we just don't know. It's like asking how much phosphate detergent you can put into a lake before the lake is polluted. Even assuming that you can agree on what you mean by "polluted," the answer is that *it depends on what else you put into the lake.*

You have only one stomach, one liver, one intestinal tract. Everything you eat goes in there and more or less mixes up together. What's the effect of mixing a trace amount of diethylstilbestrol, an uncertain amount of artificial ascorbic acid, a bit of sodium nitrite, some gastric juices, and half a bag of peanuts? Nobody knows. Nobody could ever test the possible permutations of food additive and food contaminant combinations that might be in any given American stomach at any given time (not that the rest of the world is immune; it's just a bigger book).

It is a familiar believe-it-or-not item of conversation that common table salt is a combination of two basic elements which, taken separately in the same amounts, will kill you immediately. Why may not two, or three, or ten "safe" substances, each chemically complex, combine to kill you just as dead but more slowly, over ten or 20 years?

Just as the blind construction of a few canals can ruin the Everglades through a series of effects no one could foresee, so, too, can we ruin our inner ecologies in ways that no argument about any single additive can justify. Never mind that many of us also drink booze, smoke cigarettes, take aspirin (I do all three), swallow tranquilizers and pep pills and sleeping pills and digestive aids and meaningless cold remedies and even, occasionally, a genuinely necessary medicine. With or without professional help, at least most of us could quit doing most of those, and doubtless it would help our inner ecologies quite a bit.

But in the first place, we cannot stop eating; and in the second place, those are all things we do to ourselves, and we can at least say that if we are killing ourselves, we're doing it deliberately and personally. We cannot, however, control the constant ingestion of chemicals into our bodies through our foods, and what is worse, we cannot control the ingestion into our children's bodies either.

To justify the statement that we are being poisoned, it is enough merely to argue that it is not the safety of each individual additive that must be considered and determined, but the interactions among them. There is, however, more to it than that.

For we are facing not only a chemical and physiological and ecological fact, but a social one: the Mad Scientists took over the food business (and "business" is not a word chosen loosely) before much of what is now known about disease and toxicity was understood. Social patterns have their own inertia; on a large scale, we can know that we are doing something wrong—as we know that segregated schools and the war in Southeast Asia are wrong—and be almost paralyzed by the sheer complexity of doing something about it.

When chemistry was called into action to serve the economic ends of the food industry, to make food prettier and to make it last longer on the shelf, so the manufacturer wouldn't have to take it back, and to make it cheaper to manufacture, chemists were not all villains rushing to poison their countrymen (and themselves). Allowing for a few scattered unscrupulous individuals, they did not and would not deliberately introduce a poison.

But their idea of a poison was something which made you fall down dead after you swallowed it. Only much more recently have we come to understand the mechanism of chronic toxicity: the build-up of a substance in the system to the point at which, perhaps after 20 or 30 years, its poisonous nature suddenly becomes clear (the importance of this concept even today may be clearer later, when we get to the subject of carcinogens).

If there is a positive statement to be made about the rush of nuclear tests in the 1950s, it is that through public discussion of strontium-90 and other by-products of testing, many of us have at last come to understand the concept of a slow and invisible poisoning over a period of time. It is essential to a discussion of food additives, even if we're talking about them only one at a time, just as a similar concept is essential to any other ecological discussion.

Nor, at the time, did chemists and food processors understand what has been called "the totality of toxicity"—the dangers inherent in the interactions among substances whose individual effects, even, are unknown. As we said before, it all goes into the same stomach, but it wasn't understood that the interaction, too, can take place over time.

Thus any simple argument about the safety "in small amounts" of any particular additive, no matter how prominent the voice that intones it, cannot make any scientific sense at all. Your body (or your child's body) does not take in, digest, deal with, distribute, and/or expel one substance at a time, ever. And, of course, this is the principal reason why we cannot simply "separate out" pesticide residues and other contaminants from our discussion.

Nothing better illustrates the importance of thinking ecologically about what goes into our bodies (and no aspect of the subject is less well understood among even conscientious laymen) than a look, however brief, at vitamins.

The fantastic boom in vitamin pills that began during World War II (Americans spent a quarter of a billion dollars on them in 1944) has never really died down, and nutritionists (at least nutritionists below the very top of their field) have never really quit saying that they may not be doing us any good, but they probably don't hurt anything. The fact is that nobody knows that at all, and the truth may be just the opposite, for reasons that can only be called ecological.

For one thing, natural vitamins and synthetic vitamins are not the same thing—and the man who invented the word "vitamin," Casimir Funk, said early in the game that the synthetic ones are both less effective and more toxic than the natural ones. It's true that we don't usually think of vitamins as "toxic" (don't look it up for an esoteric meaning: it simply means "poisonous," though not necessarily "immediately fatal"), but they are. Takahashi demonstrated the toxicity of Vitamin A in 1922; Spies and Brehm showed damage to the circulatory system and the kidneys by overdoses of Vitamin D in the 1930s; Mills reported the toxic effect of overdoses of thiamine (Vitamin B_1) in 1941.[3]

But, again, there is more to it than that. It was known for years that scurvy, the famous disease of seamen in the days when fresh fruit and vegetables couldn't be carried on slow sea voyages, was a Vitamin C deficiency; but it turned out that synthetic ascorbic acid (Vitamin C), while it helped, didn't help nearly so much as lemon juice. A great many more years went by before Szent-Györgyi of Hungary demonstrated that it is the interaction in lemon juice of Vitamins C and P (citrin) that is involved in treating scurvy properly, or preventing it. Obviously we landlubbers either drink lemonade or get our citrin elsewhere.

Vitamin D, it has been found, has an intricate relation to the body's balance between calcium and phosphorus, and—in animals at least—hypervitaminosis A (the toxic results of an overdose of Vitamin A) only takes place when the animals aren't getting enough of Vitamins B and C! Restoring some vitamin inadequacies without treating others may make you worse off than ever, as Dr. Agnes Fay Morgan reported in medical language in the 1940s:

> Attention should be given to the possible danger of the administration of large amounts of certain vitamins such as nicotinic acid [nicotinic acid, also called niacin, is a B-complex vitamin] to persons subsisting on diets having multiple deficiencies. Fortification of foods with those vitamins such as thiamine or nicotinic acid which are available in large quantities may precipitate conditions worse than the subacute deficiency state produced by the usual diet balanced in its inadequacies. Improvement in all directions equally is essential.[4]

That wasn't just an opinion; that was a report on a lengthy and conclusive experiment. And it was nearly 30 years ago. But we're still doing it all the time; take a look at the fine print on your bread wrapper.

Real, live human beings almost never get a deficiency in only one nutrient. If there's an essential kind of food missing from your diet, it'll make you deficient in more than one nutrient, and adding only one of them can make you sicker than ever. Some have to be in balance, so that adding one automatically makes you deficient in the other.

And vitamins, of course, have their catalytic functions; interactions involving other substances take place in your body when the vitamins are there, and don't take place when they aren't. They don't exist by themselves in a separate, closed system. Each of these details affects a thousand others in your body. You are, again, a single system—and there *is* a way it's all supposed to work.

One of the things it's supposed to do is reproduce. One of our favorites of the myriad posters to be seen around Berkeley (communicatively as well as aesthetically) is a large photograph of a woman's breast, stamped "Unfit for human consumption." Small type at the bottom explains that DDT concentrations in breast milk have been found to be higher than the concentrations allowed in milk transported across state lines. In other words, as biologist Charles Wurster frequently tells audiences, you couldn't take it across the line in any other container.

But, as proved to be the case with strontium-90, there can be other genetic effects as well. These are things which, despite recent successes in biological theory, we still know very little about.

It can be demonstrated that radiation causes mutations—changes in babies through changes in genes, which are then perpetuated—and that virtually all mutations are deleterious. Most mutations, whatever causes them, turn out to be miscarriages (sometimes so early in the game that the potential mother never knows that she was pregnant), but some are subtle and can be passed on for several generations. In some way—a lowering of intelligence, a reduction in physical vigor, weak eyesight—they render the genetic line weaker. In an otherwise healthy society, the weak line eventually dies out through evolutionary selection; but when, as with strontium-90 and other sources of radiation, the *cause* of the mutations is spread throughout the population, the effect is simply to make the entire species weaker.

Because of this long-range effect, biologists, for shorthand purposes, often speak of *mutations* and *birth defects* separately, although in fact birth defects *are* mutations only, in a way, more so. And now there are prominent scientists (Dr. Joshua Lederberg, professor of genetics at Stanford and a Nobel Laureate, is one) who are seriously concerned about food additives, with their unknown effects both singly and in interaction, as a possible source of birth defects and mutations.

Most of what little research has been done on food additives—"little" in relation to what should have been and needs to be done—has centered, as we said, on their immediate effects. Some researchers have concerned themselves with the fact that cancer may take 20 years to develop, and the only really strong law we have on food additives (the Delaney Amendment) is concerned with cancer.

But there is some evidence of chromosome breakage caused by both cyclamates and MSG (chromosomes are the strings of genes through which the mechanism of reproduction is carried out). Dr. Lederberg is among those who think that we should be at least as concerned about genetic effects as we are about effects on the individual, perhaps more so.

Is it possible? Can we, in fact, do anything to keep the pollution of our food from violating our inner ecology at least to the point at which we can be sure that it won't make us sick and, through our bodies, affect the bodies of our children yet unborn?

Not, certainly, until we change our ideas about what we mean by being healthy.

CHAPTER THREE

Look, Ma, No Cavities!

One writer tells a story of a veterinarian who was also a deer hunter, one who every year hunted farther and farther from civilization in his search for game. Finally, one year, he found himself high in the mountains, farther from other humans and their works than he had ever been in his life and farther than most of us will ever get—and there he found, shot, and dissected a deer.

"I was startled by what I'd found," he said, and after a moment explained why. "Until then I had never seen the organs of a healthy animal."

Your doctor probably never has, either.

In a book that is not about food at all, but about "the semantics of personal adjustment," a speech teacher, Wendell Johnson, once described a number of different meanings of "normal." In a *medical* sense, he wrote:

> . . . you are normal if you are not sick, generally speaking. The medically normal individual has a postnasal drip, perhaps, is rather susceptible to common colds, has a certain amount of dandruff, does not possess a highly satisfactory set of teeth, tends to be short of breath and to tire rather easily, but still he is not bedridden and he is able to hold down a job with sufficient competence to stay on the pay roll. . . . Medically, then, in a practical sense he is normal—but he may not be "strong" and he may die of "heart failure" before it would seem that he should. . . . When all is said and done, whether or not you are normal medically depends on how many doctors there are.

Johnson uses other methods to measure normality—legal, social, the equation of "normal" with "average"—but then compares the attitude of a skilled mechanic toward a relatively simpler mechanism, the automobile:

> [H]e examines your car, its driver, and the conditions under which it is used, and then he says, in effect, "It is constructed in such and such a way, its driver possesses such and such a degree of skill, it is used on paved roads, and is stored in a dry garage. Very well, it should, accordingly, start easily, accelerate and brake according to standard, do eighty miles an hour, go twenty miles on a gallon of gas, etc. If it isn't doing these things it isn't performing normally."

You, too. The difference is that the very first automobile appeared fewer than 90 years ago, and its "standard" has "evolved" for only a very short time. Yours has developed over somewhere between 100,000 and 1,000,000 years, and is consequently much more refined—and much more to be trusted as the best possible development. In fact, by now they are the only possible standard.

How, then, did we get from a standard refined over thousands of decades to postnasal drip, dandruff, "heart failure," and not highly satisfactory sets of teeth?

An old Shoshone chief, converted to Christianity and a life as a successful orchard grower, may have known the answer. "At the beginning," he once said, "God gave a cup to every race, and out of this cup they drank their life. Yes, in those days we partook of the strength of the desert. We knew nothing about the contents of tin cans. But now our cup is broken, and now it is all over with us!"

Is this nonsense, this Rousseauesque back-to-nature talk? Of course it is, if what it means is that we Americans, beginning tomorrow morning, must tear down the railroads and roads, spread across and divide the land, and tear our livings, each of us, from the earth by our own efforts in tiny groups. Half of us would be dead in a month (and while even that might be acceptable from the point of view of the species, we are against it because we would be among them).

But we need not give up comfort, nor even the contents of tin cans, to hear what the old Shoshone chief was saying—that our eating, like everything else we do, evolved in a harmony with nature which is now disrupted, and that that harmony can serve us as the ideal against which to measure our practice. We are hipped on "progress," and it is difficult for us even to entertain the idea that we might be healthier without Health and eat better food without Nutrition. But it is a possibility.

Major-General Sir Robert McCarrison, in 1939 physician to the king of England, thought so, and his remarkable work in the early part of this century remains for skeptics to explain away and for the Mad Scientists to discredit if they can.

High in the Himalayas, in the far north of the disputed land of Kashmir near the border with the thrusting eastern finger of Afghanistan, Rakaposhi rears more than 25,000 feet toward the sun. To the south of the mountain, less than 2 dozen miles on the map but over such territory that few would attempt the walk, lies Gilgit, where early in the 1900s young Dr. McCarrison was assigned. North of the mountain and 60 miles from Gilgit was and is the land of a strange, fair-skinned, deep-chested people known as the Dards or, more often, the Hunzas, who claim to be descendants of the soldiers of Alexander.

They may have been, at one time, but that was not what set the Hunzas apart, for the nearby Nagiris are descended from Hunzas, and the Hunzas themselves have intermarried with Kasmiris, Ishkomanis, and the people of the nearby Ghizr. What made the Hunzas unusual was that, except for an occasional case of eye inflammation caused by badly ventilated fires in their traditional huts, virtually nothing was ever wrong with them.

Dr. McCarrison's job included the medical care of the natives, but he saw the Hunzas only when he chose to visit their lands or when they came into Gilgit, walking the 60 miles, doing their trading, and immediately setting out to walk back. They were as intelligent as they were physically healthy; they were of notably gentle manner, given to wit and urbanity; and, though they were numerically few and their neighbors warlike, they were rarely attacked—because they always won. Around them, in the meantime, their neighbors seemed to suffer from the same illnesses and problems that all of us, in our mistaken superiority, associate with "primitive" peoples.

Dr. McCarrison wondered why. "I never saw," he later wrote of the Hunzas, "a case of asthenic dyspepsia, of gastric or duodenal ulcer, of appendicitis, of mucous colitis, or cancer . . . . Among these people the abdomen oversensitive to nerve impressions, to fatigue, anxiety, or cold was unknown." Their sanitation practices were no different from those of their neighbors; their climate was equivalent. The only difference that Dr. McCarrison could find was in their diet.

Initially, that was hard to believe. They were so poor that they couldn't keep dogs. They ate grains, vegetables, and fruits, along with goat's milk and the butter they made from it. On feast days they ate goat meat, but feast days were rare.

Yet there was, McCarrison finally noticed, a careful balance in the agriculture the Hunzas practiced on terraced mountain fields. Their soil was irrigated, and renewed, by water brought in aqueducts from the glaciers around them, water which brought its silt with it and carried the minerals ground out over the centuries by the glacial action. Their traditions demanded that all waste—human, animal, vegetable—went back to the soil. There were no parasites, no plant diseases. And their meager diet appeared to contain everything that a diet ought to contain.

Whether the methods and the diet of the Hunzas evolved "by accident" is not important (although there is ample evidence of sophisticated nutritional knowledge among "primitive" peoples). What is important is that it evolved, and the Hunzas with it, who are a far, far healthier people physically and temperamentally than you or I.

McCarrison did not stop with observation. Using rats, he set up a series of nutritional experiments that are still classic, still basic, still compelling, and still all but ignored by American medicine (the many misunderstandings about drawing conclusions for humans from animal experiments are discussed later; for the moment take our word for it that the experiments were valid, and that no one seriously disputes their validity on any scientific ground). Carefully studying the diets of the various tribes in the Gilgit area, McCarrison fed similar diets to groups of rats.

The "Hunza rats" were probably the healthiest laboratory rats that ever lived. They grew rapidly, were apparently never ill, had healthy offspring, mated with enthusiasm. They were killed at the age of 27 months—an age considered equivalent to about 55 years in a human—and autopsies were performed. McCarrison could find nothing whatever wrong with their organs. And during all their lifetimes, the doctor recorded, the "Hunza rats" were gentle, affectionate, and playful.

The contrast with the other rats was so startling as to be fantastic. They contracted a series of diseases too long to list on a single page of this book, including diseases of almost every part of the body: respiratory system, digestive system, urinary passage, womb and ovaries, skin, hair, blood, lymphatic system, heart, nervous system, you name it. They were snarling and vicious, and had to be kept apart at times lest they kill each other.

What is even more remarkable: the rats fed on Mahratta or Bengali or Madrassie diets (for ultimately McCarrison included diets from more urban parts of India in his experiment) developed precisely the diseases of the people from those regions, down to close percentages, and even seemed to adopt their tempers and behavioral characteristics!

All this without medicine, without food supplements, without any human help at all. The experiments were carefully controlled, and the only difference between one group of rats and the other was their diet.

Dr. McCarrison's experiments date from the 1920s. In 1949, Dr. Elmer M. Nelson, then in charge of nutrition for our own Food and Drug Administration, was quoted in *The Washington Post* as saying in a court hearing:

> It is wholly unscientific to state that a well-fed body is more able to resist disease than a less well-fed body. My overall opinion is that there hasn't been enough experimentation to prove dietary deficiencies made one more susceptible to disease.[1]

There is something about the idea that diet can prevent disease that seems to frighten doctors. Possibly, if Dr. McCarrison's India experiments, however famous, were an isolated fact, even Dr. Nelson might never have heard of them. He might even never have heard of McCarrison's later experiment, in which he compared in the same manner various Indian diets (including the Hunza) with a sample lower-class diet in Great Britain, with the same results.

But Dr. Nelson should have read McCarrison's *Studies in Deficiency Diseases,* a classic work, just as he should have read Dr. Weston Price's more recent (1939) *Nutrition and Physical Degeneracy.* Perhaps especially the latter, for Dr. Price was American.

Weston Price was a dentist, and apparently a fairly wealthy one by the time he was 60, in the 1920s in Cleveland. He was interested, understandably enough, in the connection between food and dental health—and consequently health generally, since the connection was as well-known then as it is now. While another 60-year-old might have pursued his search in a library, Price decided to tour the world and look at teeth (and his wife promptly decided that if he could do it, she could too).

Price wondered about American Indians, and studied them first on reservations, where he found the same dental problems he had found in Cleveland schoolchildren. Then, however, he managed to find two sizable groups who still lived very much as they had before the white man came, with one exception. One group still hunted and trapped in the old way, maintaining all the old rigors and traditions; but they accepted one shortcut in the process, trading their pelts to white men for white man's food. The second group, in the far northwest of Canada, maintained their old traditions with no contact with white men at all.

In the first group, Dr. Price found the same old dental troubles: not only bad teeth but underdeveloped jaws that caused the teeth to grow in crookedly (a problem that *we* call "normal"). In the second group, the isolated group, Dr. Price could find not a single malformed jaw—and when he examined 2,464 teeth in Indians of all ages, he could find only four teeth that showed any sign whatever of previous tooth decay.

All right. An accident, perhaps. But the old man and his wife had barely started. A short time later they were on the Tonga Islands, looking at natives whose teeth, strangely, showed no present decay but showed marked evidence of decay at some time—some particular time—in the past. Dr. Price stayed long enough to be sure that the explanation he ultimately evolved was correct: for a brief period, a rise in the world price of copra had brought unexpected affluence to the isolated islanders, and they had been able to purchase from the merchant ships American canned goods, which to them were a luxury. When the price dropped to normal, the canned goods stopped, and the islanders went back to their native diet and not only did tooth decay stop, but the decay already there began to heal itself!

The Prices went to the far North to study Eskimos, and even found a village in which most of the inhabitants had never eaten white men's food, but a few hardy travelers had. Only those few had tooth decay. Price studied the Seminoles who lived in proud isolation from the white man in Florida. He studied coastal and highland Indian tribes in Peru. He went to Australia and New Zealand to look at the teeth of aborigines and Maoris.

In Switzerland he found the isolated peasants of the Loetschental Valley, and in East Africa the Masai; he examined fishermen in the Outer Hebrides and Melanesians in New Caledonia. With his wife (and what a hardy pair they must have been!) he traveled by airplane and small boat, horseback and afoot. They took countless photographs and brought back innumerable diet samples for laboratory study.

And in Florida and in Peru, the Prices studied ancient skulls; of 50 mummified Chimús (pre-Inca dwellers of Peru) they found that only four showed any sign of tooth decay.

In every case, diet and only diet was the difference between virtually perfect teeth and teeth no better than ours. Among all the peoples he visited and all the skulls he studied, of those who existed on a native diet without intrusion by white men's foods, the percentage of diseased teeth was never higher than 4 percent. Of those who ate of the fruits of civilization, the lowest proportion of bad teeth was 13 percent, the highest 70.9 percent.[2]

Of course the Prices analyzed the diets, and among the enormous variety of foods they found that among those groups without diseased teeth the total intake of vitamins and minerals was almost exactly identical, and there was, of course, an ample supply of protein and other nutrients. Nor is that all; measured against the American government's "minimum daily requirements," he found that his natives' intake of such minerals as calcium, phosphorus, iron, and magnesium was much higher.

Not only good teeth came from the native diets the Prices found. That was the sexagenarian dentist's principal concern, but he was startled to find, further, that cancer, tuberculosis, high blood pressure, kidney and heart diseases, poliomyelitis, cerebral palsy, and other degenerative diseases were virtually nonexistent (there are, of course, many "primitive" peoples with inadequate diets, who do *not* have perfect teeth nor such excellent health; that doesn't change the point).

Keep your eye on the word "degenerative." Neither McCarrison nor Price wrote that the people they studied never suffered from any disease. The diseases that were almost completely absent were the diseases we associate with aging, principally those listed in the paragraph above. Even their name, degenerative diseases, suggests that we have already made up our minds that they involve a kind of wearing out, an inevitable collapse of the system.

But how do we know, aside from Price's natives and the Hunza rats with their perfect organs, that good health is related to what we eat?

To talk about that depends on studying some statistics (studying them, not just looking at them). The most frequent reply to any suggestion that diet may be important to health and growth is very simple and very deceptive. It goes, in its popular form, like this: if Eskimos who have never had any contact with white men's civilization are so damned much healthier, how come their life expectancy is only about 45 and ours is over 70?

That argument, and other pseudostatistical arguments like it, are fed to us by the Mad Scientists and their PR boys all the time. They wouldn't hold up for a moment if so many of us weren't afraid of, or so easily razzle-dazzled by, anything that is said rapidly and that sounds like mathematics. With apologies to the statisticians in the audience, we're going to try to blow away some of that statistical fog so that you'll be able to do the same when they start telling you that poison is good for you, and hiding behind numbers while they do it.

INTERLUDE

In Mournful Numbers

This is a sort of verbal time out, to talk about numbers and not about food, because numbers games are among the principal ways in which the public relations men of the food business (and of some parts of the doctor business, too) try to con you into thinking you're healthy.[1]

Among other things, they like to talk about America's high life expectancy and its low death rates. Neither of these has nearly so much to do with our national health as you might think; the fact that the life expectancy at birth of an Eskimo is about 45 and that of an American around 70 seems important, especially when some prominent figure utters a couple of mumbo-jumboish mathematical-sounding sentences that are quoted in magazines, newspapers, or on television. Or maybe even in a book review.

We will go on using the phrase "life expectancy," and related phrases like "life expectancy at birth," but there are a great many people who don't know what the figures mean or how they're determined, so we'd better start from the beginning.

One rule first, and no matter how much it may shake up your head, it must be mastered and kept firmly in mind: *"Life expectancy" has nothing at all to do with how long anybody can expect to live.*

Let's create a make-believe world. We'll use some percentages that would be pretty wild in the real world (for instance, we'll assume, for some examples, a 30 percent infant mortality rate, which in the real world would be far too high for any known culture), but the principles will be exactly the same as those used today in America, and we'll use the same phrases and the same assumptions that are used in real life.

Let's say that last year we lived in a very small town. In that small place, during the year, ten people died and ten babies were born.

Since we have a 30 percent infant mortality rate, three of the people who died last year were less than one year old; that's what "infant mortality" means to a statistician. For convenience's sake, we'll say those three died at one year of age. So that leaves seven of the ten who died last year. Let's say that two of them died by accident, being 21 and 34 years of age, respectively. And the last five of them—*that's half*—were 80 when they died.

Now we take the ages at which all ten of them died, and we add them together: $1 + 1 + 1 + 21 + 34 + 80 + 80 + 80 + 80 + 80$. Then we divide the result by ten. What we have then, of course, is the average age of death of everybody who died last year. It comes out to 45.8.

Of the ten babies who were *born* last year, then, we say that their average life expectancy at birth is 45.8 years.

That is all that "life expectancy at birth" means. It is sometimes referred to as "the average lifetime of Americans (or whomever) today"; that's all that *that* means too. It's simply the average age of death of everybody who died last year. It has really nothing to do with the ten babies who were born at all. For all we know, a sudden epidemic could kill them all tomorrow.

In real America, instead of adding together ten ages and dividing by ten, you'd have to add together 1,851,323 ages (for 1967, the last reliable year for death figures) and divide by 1,851,323. And what you'd get would be 70.1, the life expectancy at birth of Americans as announced some time afterward, when somebody got the counting and dividing done.

You'll notice that in our little make-believe world, although the life expectancy at birth is 45.8, half the population lived to be 80. You can see that there is something about the health of that little world that "life expectancy" figures don't tell us. In real America, with a much higher life expectancy figure, half the population does *not* live to be eighty.

Now let's suppose that at some time in the past, in our make-believe world, we had conquered infant mortality, so that those three babies didn't die last year. That leaves us with only seven dead people last year, and we can do the same thing. We add $21 + 34 + 80 + 80 + 80 + 80 + 80$, and this time we divide by seven, and we find that life expectancy at birth is now exactly 65. Life expectancy has gone up 19.2 years—an increase of 42 percent. Surely, a cause for joy!

Well, yes, of course; the babies are alive. But notice that *for seven out of ten people in the population, there is no difference in their lives at all!*

Statistically, you can now say lots of things about the babies who didn't die. Statistically, you can say that they will live to be 65. Statistically, you can also say that at least one of them will live to be 80 (but that means that at least one other won't live to be 65). Statistically, you can say all sorts of things, but they don't measure anything but statistical aggregates. And they aren't about any real babies; they're all statements about last year's dead people.

We can make a lot of other things clear with our model of ten dead people last year (let's take away our medical miracle now, and go back to the first example). A lot of people, for instance, read the perfectly accurate statement that life expectancy in America (real America) was 49.2 in 1900 and is 70.1 now, and they figure that on the average they have 20.9 more years to live.

But it doesn't work that way. The figure is for life expectancy *at birth.* It doesn't work the same way at other ages.

In our model, for instance, suppose you're 25. To get *your* life expectancy figure, you count, among those who died last year, only those who had reached 25 before they died. You ignore the three dead babies and the person who died at 21. You count the other six, one who died at 34, and five who died at 80; you add the 34 and the five 80s, divide by six, and you get 72.5—the average age of death of all those who died after reaching 25. Since you're already 25, your "expectancy" on the average is to live to 72.5, so you have 47.5 years to go.

Now if we change to our second example and cure the babies, so that they didn't die and only seven people did, what's your life expectancy at age 25?

It's exactly the same. We've done nothing for you at all.

And that is very nearly what has happened in real-world America since 1900. Life expectancy *at birth* has gone up 20.9 years as of 1968. But life expectancy at 25 has gone up only 8.8 years, and life expectancy at 45 has gone up only 4.8 years.

Obviously, not so many people die when they are very young; but the prospects for adults aren't a hell of a lot better than they ever were. That is, as you'll see in a few minutes, why this statistical seminar is so important when we're talking about food.

A third example from make-believe world: let's say there's no infant mortality, but this time we again have ten people dead last year. We'll stay with two deaths by accident at 21 and 34, but let's say that for everyone else, cancer strikes half the population suddenly and fatally at 55 and the other half at 65. We have to add together 21, 34, four 55s, and four 65s. Dividing by ten, we come out with a life expectancy (at birth) of 53.5 years.

Keep in mind example two, when with only seven people dead, we had a life expectancy of 65. In that model, we had our two accidents, and five people who lived to be 80. Now we're down to 53.5 because of cancer. Let us construct a fourth example by all contributing to the Make-Believe World Cancer Society, which educates everybody on the seven signs of cancer and subsidizes a crash research program. We succeed in coming up with medical treatment which will keep cancer patients alive for 15 years.

In our fourth example, then, keeping the same two accidents, we have one death at 21, one at 34, four at 70 (after 15 years of cancer), and four at 80 (after 15 years of cancer). Our figure for life expectancy at birth is now 65.5—the highest we've ever had.

But instead of having (as in example two) more than half the population living in perfect health until they're 80, we now have 80 percent of the population living with cancer for 15 years, and only 40 percent of the population reaching 80 at all! Life expectancy is higher in a much less healthy society (and life expectancy at age 25, incidentally, has dropped from 47.5 to 45.4).

So you see, it doesn't really prove anything to say that the life expectancy of Eskimos is a lot lower than ours. It means, probably, that infant mortality is high among Eskimos, and maternal mortality as well (they usually go together, and mothers are often young as total populations go). It means that infectious diseases are probably still widespread. It means that without fast American transportation networks, people who have accidents are in more trouble; a hundred years ago, an American who broke his leg in a remote mountain pass might well die of gangrene, but it's much rarer today. Children, of course, are notorious for having a lot of accidents, and so among Eskimos more children probably die accidental deaths. And when people of all ages get sick in any way whatever, they are less often kept alive by skilled (and expensive) medical care.

All of these things are regrettable and should of course be overcome if possible; but they do not prove that the lower life expectancy of Eskimos means that their bodies are less healthy. The opposite may be the case.

One argument is enough to demonstrate, really, that life expectancy at birth is not a measure of a population's health. Suppose, if you can imagine it, that 90 percent of our population were stricken with an incurable, crippling disease, but that the other 10 percent were somehow able to keep taking care of them, so that for a whole year nobody died. In that year, we would have the highest life expectancy figure of any nation in the world's history—and the sickest population.

All of this has really been intended simply to explain, and to try to make as clear as possible, three statements:

1. "Life expectancy" has nothing to do with life expectancy; it's just an average of last year's death ages.
2. "Life expectancy at birth" and "life expectancy at age 25" (or any other age) are not the same thing.
3. Life expectancy figures do not in any way measure the health of a population—only its lives and deaths. To measure health you need more data.

Now let's leave make-believe world and come to the real United States of America, and take a look at some real figures derived in 1968 from deaths in 1967, and from earlier years.

We've already noted that for 25-year-olds, the life expectancy figure has gone up only 8.8 years since 1900, and for 45-year-olds it's gone up only 4.8 years. That means that despite 68 years of progress—years that have seen pneumonia and tuberculosis all but wiped out, sulfa drugs, antibiotics, and other "wonder drugs" introduced, and pacemakers and intensive care units developed, years during which remarkable advances have been made not only in surgery and endocrinology and biological research but in sanitation and immunization and public health methods—the *average* remaining life span for *adults* in America has lengthened hardly at all.

In fact, the figure is exactly the same for 25-year-olds in 1968 as it was in 1960; likewise for 35-year-olds; and 45- and 55-year-olds have each gained exactly one-tenth of a year! To make it even more interesting, since 1960 the life expectancy for 15-year-olds has gone *down* one-tenth of a year.

Sure, we have one of the highest life expectancy figures in the world—at birth. But even 15 years ago, before nearly so much medical care had been exported to less fortunate parts of the world as is now the case, Dr. Norman Jolliffe of the New York City Department of Health could write:

> Although in America today life expectancy at birth is near the best of any civilized country in the world . . . at the age of 40 life expectancy is near the bottom.[2]

Please don't get the idea that we want to abolish the continuing efforts to reduce infant mortality. Of course we all want babies to live, and we're delighted to learn, for instance, that while in 1916 10 percent of all babies born alive in America died during their first year, today the figure in America is down to 2.2 percent (still only twenty-third among the world's nations, however). Maternal deaths from live births, too, have dropped, from 622 in 1916 (for every 100,000 births) to 83 in 1960. Nor do we want to see the return of infectious diseases, whose death toll has dropped from 676 per 100,000 population in 1900 to only 10 percent of that by 1949, and is still dropping.

All that we want is for Americans to understand that these are the improvements which are primarily responsible for the dramatic-sounding rise in that misleading "life expectancy at birth" figure—and that it has nothing to do with any major gains against the degenerative diseases: heart disease, cancer, arteriosclerosis, vascular lesions (strokes). Those four, together, account for 68.7 percent of the deaths in the United States in 1967.

And those are the diseases that Price and, particularly, McCarrison said were absent in their subjects, and the absence of which they could not explain except on dietary grounds. Suppose—just suppose—that they were right, and suppose that we could magically go back, say, 200 years and eat properly, while doing everything else the same. What do you suppose our life expectancy would be then?

There are other statistical fallacies that often come up, one of which has to do with the idea that death rates are somehow a measure of health. The death rate is simply the count of how many people die in a given year for every 100,000 people in the population. A high death rate keeps the population down, a low one (unless infant mortality is very high) lets the population increase. But the rate doesn't measure what people die of, nor at what age, nor whether they're healthy while they're alive. In any case, the death rate in the United States has not significantly changed in the last 20 years; it has been hovering around 9.5 deaths per 100,000 ever since 1950.

One can be misled by noting that infant deaths were 6.9 percent of total deaths in 1957, 5.1 percent in 1965, and only 4.5 percent in 1966, but that isn't entirely because the infant mortality rate is going down. The infant mortality rate, the percentage of babies born live who died during their first year, did go down during those years, from 2.64 to 2.34 percent. But that doesn't account for the other figures. If you're counting the percentage of total deaths, then you have to ask whether the proportion of infants in the population remained constant. As it happened, it didn't. The proportion of infant deaths to total deaths dropped partly because the proportion of infants to the total population dropped.

That's the kind of thing you have to look out for when they start throwing statistics at you.

One other point about life expectancy figures—and one other way in which they don't, and can't, measure the actual life expectancies, even in average terms, of real babies being born right now: if the average age of death of the Americans who died in 1967 was 70.1, then obviously they were born an average of 70.1 years before, which means that they were reared, and spent much of their lives, away from strontium-90 in the atmosphere, away from today's kind of air pollution, away from our dirty water problems and crowded highways.[3] They also grew up—or many more of them did—in rural rather than urban settings; at their mother's breasts they got milk and not DDT; and they certainly ate more food in their food, and less poison.

The baby born today, on whom we falsely slap that life expectancy figure, has none of those breaks going for him. He (or she) has a much better chance of escaping infant mortality, but it remains to be seen what kind of health awaits him after that.

If radiation certainly, and many food additives possibly, have genetic effects, then our babies may have weaknesses as yet unmeasured, but built in. If they grow up in a nation in which it is virtually impossible, without special effort, to buy food without a lot of chemical gunk in it (and almost literally impossible to avoid pesticide residues), then it's extremely unlikely that their bodies, complex chemical engines that they are, can turn out the way that a million years of evolution should have assured they would.

Certainly today's child won't have to wait until he's an old man to feel the onset of degenerative disease, unless he's one of the increasingly few lucky ones. Adults don't have to wait now: think of the heart attacks you've read about in men and women (more often men) in their thirties and forties, and the "premature" cancer deaths—an interesting term which assumes that there is a "mature" age for dying of cancer.

And think about this: when you see a chart or a table telling you that cancer is the cause of something like 17 percent of the deaths in America, remember that under "cancer" goes "leukemia." More than 6 percent of all cancer deaths are leukemia deaths, and leukemia causes about one-half of *all* deaths of children under 15 in America. This is a degenerative disease, but obviously it has nothing to do with getting old and "degenerating."

Cancer figures can truly be misleading. In fact they lend themselves, even at best, to differing interpretations. In 1900, 64 of every 100,000 Americans died of cancer, or of what was diagnosed as cancer. In 1940, the figure was 147 per 100,000. In 1967, it was 364.5 per 100,000.

The argument often heard is that, with infant mortality going down, more people live long enough to get cancer, and that's why the cancer figures are up. Others argue that statistics are higher because diagnosis is better and that deaths are better reported.

But these arguments, while they're true as far as they go, probably don't account for that dramatic 600 percent increase. It's also true that of the people who live to be any given age, more of them develop cancer than did in 1900 or 1920. That might be true in numbers if it were only that more babies survive, because of course the population has increased. But it's also true in percentages. A man or woman who has reached, say, 45 is more likely to develop cancer now—and not only lung cancer (which may increase with cigarette use and/or air pollution).

If an increasing cancer death rate (overall) simply reflected a decrease in infant mortality—if there were more cancers only because more people live long enough to get them—then two things would happen in the figures. The percentage rate for any given age group would hold even, and the average age of death from cancer would stay about the same. Neither is the case: the first keeps going up, and the second keeps getting lower. Cancer is striking more often and sooner. The rise of leukemia among children accounts for part of the second change. But there are still other things to be noted about cancer statistics.

Everyone knows that deaths from lung cancer have been mounting sharply, and that antismoking campaigns have had little effect on the death rate. Certainly the connection between cigarette smoking and lung cancer has been firmly established by now, and we have no intention of arguing with it. But it certainly should be noted that lung cancer is up dramatically among nonsmokers as well.

It should be noted, too, that according to estimates by the American Cancer Society, there were 59,000 lung cancer deaths in 1969, but there were also 46,000 deaths from cancer of the colon and rectum, where there is no known connection with smoking. Air pollution, which has also increased in recent years and which is another possible contributor to lung cancer, has no known connection to colonic or rectal cancers either. In addition, 42,000 women died from cancer of the breast or uterus. Lung cancer is certainly important, but it's responsible for only one of four of the cancer deaths in America.

There is another thought about lung cancer and smoking that might be mentioned in passing, again without taking away from the importance of the connection. The continuing emphasis on smoking as a cause of cancer can mislead us into thinking of cancer *only* as a disease which comes about as a result of individual self-abuse. It may be that, in spite of what any one person may do, we are abusing ourselves as a nation, and that emphasizing smoking, whatever the good results of that emphasis, may be taking our eyes off the total meaning of the cancer figures.

An oddity in the cancer death figures concerns cancer of the stomach, the second leading cancer site only ten years ago. It is now only the fifth leading site and the mortality rate from stomach cancer has dropped 40 percent in 20 years. Nobody has the slightest idea why, which is an excellent indication that we understand our inner ecology even less well than we understand our outer environment.

Cancer death rates, of course, are not measures of the prevalence of cancer in our society; in fact, cancer is probably increasing much more sharply than the death rates show. For instance, lymphatic cancers killed 17,000 people in 1969, but with modern medical treatment, many patients with lymphatic cancers live for years longer than they could have only a few years ago. The death rate could stay the same, but there could be three times as many people with cancer, just as in our example from the make-believe world.

The American Cancer Society (with the highest of purposes—the intent to urge us all to be on the lookout for cancer signs, as of course we should) tells us that in 1969 one of three patients found to have cancer was saved, as opposed to fewer than one out of five in 1937. On examination, though, "saved" turned out to mean, not cured, but "still alive after five years of treatment." Most people would of course find that better than dying (we certainly would) but statistically it probably means that more people are alive now who have cancer, by percentage, than was the case in 1937.

It's another example of how our life expectancy figure may go up, without being an accurate indication of how healthy the people are who are alive. Being older or living longer, as a population, does *not* mean being healthier. And while most of us would rather be alive and sick at 60 than dead, we would all rather be alive and healthy than alive and sick.

Cancer still seems to be, to Americans, the most frightening thing that can happen to their bodies, but it kills only a third as many people as cardiovascular diseases do. Where in 1967, 364.5 people died of cancer for every 100,000 people in the population, the figure for heart disease was 935.7, and strokes added another 157.2.

Here, too, figures are on the way up, and here, too, it has been argued that the rise is due to better reporting, better diagnosis, and the drop in infant mortality.[4] But before 1920, coronary artery disease was hardly ever reported at all, and it wasn't because doctors didn't know what it was. Besides, in heart disease as in cancer, the average age of death has been dropping; and in heart disease as in cancer, more people who are stricken by heart attacks or strokes are kept alive by better remedial medical care.

Here, the same pattern emerges: at any given age, there is more heart disease than ever before, by percentage as well as numbers. It is not because of an improvement in the infant mortality rate that heart disease is now the leading cause of death, not only of the old, but of everyone over 45. And again: to stay alive after a heart attack, however well an individual may function under certain restrictions, is not to be healthy.

To put the whole argument simply: the dramatic increase in American life expectancy, so often cited to prove that we are a healthy nation, proves nothing of the sort. It proves only that infant and maternal mortality, and certain infectious diseases, have been lessened, and that all of us, and especially children, have a better chance of surviving serious accident (to the point where some aren't called "serious" any more).

The degenerative diseases, our "big killers," are bigger killers now than ever, in numbers that are far too great to be accounted for by better reporting or diagnosis, or by population increase, or simply by the fact that more of us live to be older. The degenerative disease rate is steadily climbing among adults, and more of us are being affected sooner. We are increasingly developing the diseases the Hunzas don't have.

These points become particularly important when the PR men, and the prominent spokesmen they trot out, argue, as they often do, that we are the best fed people in the world, and start leaning on "health" statistics to prove it. Our statistics don't make us healthy (if we were healthier, we'd need fewer hospital beds per capita, not more), and the actual pattern of our health-and-sickness statistics doesn't show us to be well-fed. We eat a lot, but that's not the same thing.

We cannot know—although such extended and careful experiments as McCarrison's should certainly make us think seriously about it—that our mushrooming habit of tinkering with what we eat is responsible for our relative lack of health, or for the growth in degenerative disease. But we can draw our statistical correlations, too; and more than that, as we go on to explore what we do know about food additives, we can see that there is a lot more to go on than the work of one doctor in India and one globetrotting sexagenarian dentist and his wife.

Already, of course, researchers are finding more and more reasons to seek relationships between diet and heart disease, although the results still engender controversy. This work has tended, however, to zero in on particular narrow "causes"—cholesterol levels, for example. While of course the importance of anything that helps anyone is certainly evident, we are really talking about something much more complex in inner-ecological terms.

Inevitably, despite any formal statement about the nature of statistics, someone somewhere will respond with an individual case: "My father grew up in Pittsburgh when it was smoky, retired to Los Angeles just as the smog got bad, has smoked two packs of cigarettes a day for 50 years, and has lived all his life on spaghetti, cheap wine, and Coca-Cola—and he's 90 years old and perfectly healthy. What about that?"

That makes him unusual, that's all. Statistics measure the characteristics of groups; they don't measure individuals. They say only that people who smoke two packs of cigarettes a day are more likely to get lung cancer than people who don't smoke—and even then, they don't make the statement about any one person, only about percentages of people in large groups.

For reasons we don't yet understand in a great many cases, some people aren't affected by things that do affect most people. In fact, an unusually thorough medical examination within the past year resulted in my being told that my lungs are in excellent condition despite the fact that I've been smoking steadily for years; shortly before that, my own brother, with about the same smoking pattern—and he probably eats better—underwent surgery for lung cancer and nearly died (he is one of those who is alive, and keeping the life-expectancy figure high, though he is hardly "healthy"). There is simply no way to predict the individual case.

But if it should prove, say, that there is a 60 percent chance that good diet, starting now, might make your child less susceptible to a heart attack at 45 or cancer at 50, do you want to take the chance?

CHAPTER FOUR

Ethylenediaminetetracetic Acid and Its 2,999 Friends

Nobody knows how many food additives there are. Partly, this is because nobody can agree on what a food additive is. Even the name is artificial, and deliberately so—to keep you from asking too many questions.

Among a welter of definitions, we've decided to go with the one used by the Food Protection Committee (FPC) of the National Research Council-National Academy of Sciences. The NRC-NAS is technically a private, nonprofit organization that promotes and finances scientific studies for the general good, and also advises the government on scientific matters. Actually, it's quasi-governmental; the "private" gimmick enables it to use both government funds and those donated by private sources.

You might assume that the Food Protection Committee of the NRC-NAS might be a committee to advise the government on food protection. Well, it is—but it was the food business' idea to set it up, and they put up all the money. *All* of it. That gives them a little something to say about who gets on the committee and who doesn't.

Because the legal definition of food additives is not only complex but downright fictional (see Chapter Six), the FPC chose one that makes a little more sense:

> A food additive is a substance or mixture of substances, other than a basic food stuff, which is present in food as a result of any aspect of production, processing, storage or packing. The term does not include chance contaminants.

Some people quarrel with that last sentence, because they say it excludes pesticide residues, or materials that "leak" into foods from their packages. The FPC, however, includes these things in their studies, since they result from aspects of production and packaging, respectively; we're going along with that. "Chance contaminants" thus means things like simple dirt that get on food accidentally but are covered by public health laws.

There are a lot of reasons why the number of food additives can't be given exactly. For one thing, particular additives go in and out of use. For another, there is the sloppy legal definition (sloppy, at least, for scientific purposes). And for a third, there are different ways of classifying additives, and even of describing them.

If additives are classified by an exact chemical description, the number goes up sharply. Many chemicals which are not quite exactly similar can be used almost interchangeably; there are, for instance, four distinct chemicals in the group we have all learned to call "cyclamates," which is why the word is plural. If another authority uses a somewhat broader classification (as several pesticides are sometimes grouped under the name "chlorinated hydrocarbons"), the number drops.

If additives are classified according to what they do, the same chemical is likely to turn up more than once. Thus alphatocopherol may be a nutrient additive or an antioxidant, or adipic acid may be a neutralizing agent or a flavoring agent.

Then there's the "GRAS list," never fixed and not published. The present law exempts from its testing and safety standards any additive which was "generally recognized as safe" at the time of the law's passage in 1958, or which is "generally recognized as safe" now—which is where "GRAS" comes from. It also exempts items which had previously been given specific approval ("prior sanction" is the trade phrase). But as will be seen, things go on and off the GRAS list all the time, and in fact there isn't any list that you can sit down and read, although a lot of additives are known to be GRAS.

We don't know the numbers, but we do know the amount, at least as of 1965. In that year, at least 661 million pounds of food additives were used in the United States. The *wholesale* value of those additives was at least $285 million. We're talking about big business.

If you're an average American (another statistical fiction), you ate three pounds of food additives that year, and probably an increasing amount every year since. Most of it, you were suckered into eating.

You and your dog. The various laws having to do with food additives cover all additives for human and animal food. Most of what goes into commercial animal food—that is, food for farm animals raised for the butcher—are vitamin and mineral supplements, and antibiotics. They're there to make or to keep the animals healthy, or just to make them bigger, so that the farmer/rancher can make more money (and you can pay more).

Pet foods, on the other hand, must meet certain legal requirements for nutrition, but many of the additives used in their processing are there solely to make the food look pretty to you. Your dog or your cat is colorblind, but we feed our pets, it seems, by looks.

If, for instance, you buy one of those premium-priced, television-advertised dog foods that generate a "gravy" when you put water on them, you might like to know that the "gravy" is a food-additive merchandising trick, and does nothing whatever for either the nutritional value or the taste of the dog food. It's described by color-additives expert James Noonan:

> An interesting use is made of color in the gravy type pet foods. This consists of extruded pet food to the surface of which is added a dry powder consisting of a gum, dry color (brown) and an inert material. When water is added to the food, the gum imparts viscosity to the liquid. This together with the dissolved color resembles a natural gravy.[1]

In other words, it's only there to con you. If you don't believe us, taste it before and after.

The difficulty of determining how many food additives are in use, and of what kind, is quickly demonstrated by a look at two publications: *Chemicals Used in Food Processing*, a 1965 list published by the aforementioned Food Protection Committee, and *Handbook of Food Additives*, a 1968 book the size of an encyclopedia volume, published by the Chemical Rubber Company and edited by Thomas E. Furia.

The FPC list—and remember that this is a quasi-official body—is based entirely on a questionnaire sent to manufacturers, which is a little like asking crabgrass manufacturers to help protect the lawn. The committee says that it thinks the list is relatively complete. It also says that it is not talking about how much of any additive is used. All it can be sure of is that some corporation somewhere says that it uses some additive somehow, or at least has permission to do so.

Sometimes the FPC list appears to confuse its methods of classification (as noted above, it can be done in a number of ways). It is possible, for instance, to discuss together all enzymes used as food additives—including papain (injected into cattle to tenderize meat), fungial proteases (used among other functions to make bread crumbs compressible), hemicellulases (used with other enzymes to clarify wines), and a number of others. But since we're sticking to a functional classification, we can break down the FPC list as follows:

Nutrient additives	116
Preservatives	30
Antioxidants	28
Surface active agents (emulsifiers, defoamers, etc.)	85
Buffers, acids, alkalis	60
Coloring agents	35
Flavoring agents (synthetic)	720
Flavoring agents ("natural")	357
Bleaching and maturing agents	24
Non-nutritive sweeteners (including cyclamates)	9
All others	158

That adds up to 1,622 (the list was made before cyclamates were banned), and before we go on, you might take a quick look at the breakdown. When the Mad Scientists, their bosses, and their PR men talk about food additives, you'll notice that they almost always talk about the nutrients: the vitamins and iron and so on. But on the FPC list, all the nutrients in use, in amounts however small, add up to fewer than 10 percent of the total, while more than 1,000 of the 1,622 additives listed are there only to make something taste better than it would ordinarily taste (or possibly to keep it from tasting bad when it *is* bad).[2]

That list of 1,622 additives includes only direct, intentional additives. It doesn't include anything like pesticide residues or materials that "migrate" from packaging. Furia's *CRC Handbook*, however, which used FPC lists among other sources, says that there are "2-3,000 chemicals added to food," and it actually lists more than 2,000, while not claiming that its listing is complete either. It excludes "pending petitions, FDA proposals, rejections of petitions and petitions withdrawn without prejudice" since "these represent, at best, a transient state or non-approved use." It also leaves out animal feed chemicals and some indirect additives.

These discrepancies may have something to do with the different definitions and uses of the term "food additive" and the changing nature of the GRAS list, but even leaving aside the "indirect additives" which are, in fact, included in the law, there are certainly very nearly 3,000 additives that we know about.

Of the more than 2,000 additives listed by Furia, about 1,400 are on the GRAS list. This doesn't quite mean that anybody can use them in any way he wants. The GRAS designation means that the chemical is "generally regarded as safe" *for a specific use.* Sometimes it's allowable only on a specific product, like the FD&C Red No. 4 that's now legal for maraschino cherries but not for hot dog casings. Sometimes there are limits on the amount that can be used, and sometimes an additive can be used only in the presence or absence of other substances. To cite just one example, cholic acid is GRAS for use as an emulsifying agent, in dried egg whites only, and only up to 0.1 percent.

Our classification system is similar to, but not the same as, the FPC's. By adding in the indirect additives which we feel we have to talk about, we wind up with 13 classes of additives (including one "miscellaneous"). Most of them are the subject of extended treatment later, but in brief the breakdown is as follows:

Nutrients

This is the big one that the food business likes to talk about. But there are a few things they rarely say.

One of them is that while almost all of those 8,000 items in your supermarket have food additives in them, only an extremely small number of them have added nutrients. If you count only items which are eaten by a large percentage of the public, the number is smaller yet.

Another is that a lot of what is added is put in because somewhere along the line, the food business took the natural nutrients out. So they aren't added nutrients—they're simply replacements of some of what the Mad Scientists took out in the first place.

A third thing they don't talk about is how few of the food additives in use are of any nutrient value whatever, and a fourth is the question of why any of them should be necessary. It's a good idea to keep in the front of your mind that the healthy Hunzas don't find it necessary to add Vitamin D to their goats' milk.

It's just as well that there are nutrients in only a few foods, because of the ease with which vitamin intake, and particularly the intake of B vitamins, can get out of balance in the diet. The entire B complex of vitamins, for instance, is taken out of white wheat flour, polished rice, and other refined cereals by the processors, but they only put three of the B vitamins back in. Such industry apologists as Dr. Frederick Stare are fond of referring to these as "key vitamins," but they aren't any more "key vitamins" than the ones that are missing.

Nor is it really true, even using the FPC list, that 116 out of 1,622 additives are separate sources of nutrients. For instance, among those 116 are 11 separate sources for manganese, where a given processor is likely to use only one. There are four sources listed for niacin, seven calcium salts, eight iron salts, and so on. In other words, if you think of manganese, niacin, calcium, and iron as being the additives, then there would be four items on the list instead of 30.

And most of the nutrients listed are not used in ordinary foods anyway. Regardless of the natural or synthetic chemical used as a source, what most Americans get out of nutrient additives are vitamins—A, B_1, B_2, B_6, C, D_2, D_3, niacin, and carotene (carotene is actually a "vitamin precursor," which is broken down in our bodies and produces Vitamin A)—and the minerals calcium, iron, phosphorus, and (in salt) iodine, the latter intended primarily to prevent goiter.

All the other nutrients, including amino acids as well as a number of the vitamin and mineral sources, are used either in animal feed or in specialty diet foods. Of those which do reach ordinary Americans, almost all are in bread, flour, and cereals, in milk (Vitamin D), in margarine (A and D), and in fruit drinks (Vitamin C).

The essential points here are that most nutrient additives are put in to replace something the Mad Scientists themselves took out; that very few food additives have any nutrient purpose; and that while almost all our foods contain additives, only a very few contain nutrient additives, so that even the small percentage of nutrients among the known additives is misleadingly large.

Preservatives and antioxidants

Already, classification gets difficult.

Looked at in another way, there are three kinds of "preservatives"—antimicrobials, antioxidants, and antibiotics. For other reasons, we're separating out the antibiotics, which also have other functions.

Further, some of the preservatives that are left also have other functions: ascorbic acid is sometimes a preservative but is also Vitamin C; sodium nitrite and sodium nitrate (check your lunch meat wrappers) are preservatives in some processed meat products, but they also serve to keep it a pretty color after it would otherwise have changed.

Health factors aside, the big thing about most preservatives is that while they do indeed preserve, they do so not for your benefit but for the benefit of the manufacturer, the processor, the wholesaler, and the retailer. The merchandising term is "shelf life." The longer the food processor can hold his product in storage without special care (like refrigeration), the better he can deal with fluctuations in orders, and the more profitably he can run his factory. The longer time he can allow to ship his product from his plant to the wholesaler, the cheaper the method he can use to send it. The longer it lasts after that (but before you buy it), the less has to be taken back, replaced on the shelves, and so on—and the older the food you buy. The propaganda from the industry never describes preservatives this way. They only talk about keeping food from spoiling, with the clear implication that it's going to spoil in your kitchen without the magic additives. The preservative additives are merchandising aids.

Good bread, reasonably well wrapped and put in a bread box, will easily keep for a week. The junky mixture of air and water that is the standard American white bread loaf will not. To disguise this fact, and not to do you any favors, things like sodium diacetate, monosodium phosphate, or the calcium and sodium salts of propionic acid have to be added.

Most of the stuff that you buy is bought with the intention of eating it right away, or very soon. The food you're going to keep for a while comes in cans or goes into the freezer. If the food was fresh when you got it, you wouldn't need preservatives in most of it at all. The preservatives are there so the stuff can sit on a shelf somewhere before you buy it.

The most common of the "antimicrobial" preservatives (those intended to inhibit, not prevent, bacteria or mold growth), besides those used in bread, are sorbic acid and some other chemicals used in syrup, cheese, and pie fillings; benzoic acid and sodium benzoate among the most commonly used of all, found in fruit juices, pickles, jam and jelly, margarine, and confectionery; and sulfur dioxide in dried fruit. A number of others are used in carbonated beverages. Diphenyl is used in the wrapping material of citrus fruits to inhibit mold growth and to fool you about how old the oranges are (waxes and coloring agents are also used for somewhat the same purpose).

Among the antioxidants are the two additives which, for some reason, a lot of people seem to notice on labels: butylated hydroxyanisole and butylated hydroxytoluene (the label often reads, "BHA [or BHT] added to retard spoilage"). These, along with lecithin, propyl gallate, and several others, are used to combat, in fatty foods, what most of us call rancidity.

Another little thing they don't tell you about additives is that most of these antioxidants are good at keeping oils and shortening from going so rancid so soon at room temperature. Oils will keep perfectly well if refrigerated, or even if kept in a cool cellar, but that doesn't do anybody any good when he wants to keep oils on the shelf of a store or a warehouse. Refrigeration would mean that they would have to spend money and possibly cut their profit margin.

This goes not only for oils but for crackers and cake mixes. The antioxidants are intended to make the stuff keep on the shelf, just as bleached, refined white flour will keep because the Mad Scientists have smashed out of it all the natural oils.

If you have to have an antioxidant, one of the better ones to have is alpha-tocopherol. All tocopherols are cleared for antioxidant use (they're on the GRAS list), and they're used in vegetable oils and sometimes in lard and other animal fats. While for a time it was believed that all tocopherols are Vitamin E sources (which is why you sometimes hear from behind-the-times doctors that it's hard to imagine a diet deficient in Vitamin E), it has turned out that only alpha-tocopherol is useful in this way. Since in fact it is quite easy to find a diet deficient in Vitamin E, the use of alpha-tocopherol as an antioxidant may be an unintended blessing.

Another use of antioxidants is to prevent browning in fruit, sometimes called "enzymatic browning" because it comes about when enzymes inside the fruit are exposed to air after the fruit is cut. Vitamin C, ascorbic acid, is an antioxidant widely used to combat this; without it, for instance, frozen peaches would change color. What it does in the package is just what is done at home when people dip cut fresh fruits into lemon juice to keep them from turning brown.

Besides peaches, there are "browning enzymes" in apples, apricots, bananas, cherries, pears, and potatoes. But, as the Manufacturing Chemists' Association lets slip in their propaganda brochure, "Food Additives," the use of antioxidants doesn't only apply to *cut* fruits:

If foods containing certain enzymes are cut, bruised or overmature, and exposed to air, the tissues darken and turn brown. . . . Antioxidants prevent or delay this enzymatic browning.

In other words, the fruit you buy may not be cut, but it may be "overmature," which means just plain too old, and thanks to antioxidants you may be conned into thinking it's fresher than it is.

Again leaving aside the fact that we don't know what we're doing to our inner ecology by ingesting all these chemicals in random combinations, the thing to remember about preservatives is that for the most part, they're there to help somebody make a buck, and the somebody isn't you. Once you've bought the food, almost all the preservatives are unnecessary.

Antibiotics and hormones

The principal use of antibiotics as "preservatives" is actually to extend the "fresh life" of perishable food, just as described above for other preservatives, but the food most often treated is poultry or fish. Frequently these foods are dipped in chlortetracycline or oxytetracycline, and some residue is permitted. Antibiotics can also have the effect, however, of causing animals to grow more *rapidly* (which of course profits the grower), as can hormone additives such as diethylstilbestrol, discussed in an earlier chapter, and phenylarsonic acid.

I.S. Siddique, a Tuskegee veterinarian, is one of the leading researchers on the effects of antibiotics in feeding or treating animals, and he believes that antibiotics in animal food very often leave residues in the meat, residues that can have undesirable effects in our intestines. He also worries a great deal about mastitis, an udder infection of cows which can be treated with penicillin or other antibiotics. Although it's illegal to sell the milk for several days after the antibiotic is used, the financial loss involved tempts too many dairymen to rush the process, and milk does go to market with antibiotic residues; in milk, they're serious enough to be considered a public health hazard.

The use of antibiotics in animal feed can be shown, by using some test results, to have no effect on the animal as food, but not every authority accepts the tests. Their argument is that the amounts fed, though small, add up—good old "totality of toxicity" again—to affect the intestinal flora and other internal bacteria of the animals, with a variety of results, mostly undesirable.

Any antibiotic residue in the food you eat can have at least one undesirable effect, no matter how small: it can help build up your own resistance to antibiotics so that when you're sick and need them as medicine, they won't work as well, and possibly not at all. And of course, some people get bad side effects from antibiotics anyway.

We mentioned a while back that some broad-spectrum antibiotics are rarely used in humans now, because they destroy desirable kinds of life in the intestines and leave their ecological niches open to less desirable kinds. More technically, the antibiotics alter the intestinal flora, lowering our ability to synthesize vitamins while enabling our bodies to develop bacteria which not only cause trouble but are resistant to antibiotics.

Among these bacteria are the poisonous, and sometimes deadly, salmonellae. Salmonellosis is the most common kind of food poisoning in America, and salmonellae also cause occasional cases of paratyphoid fever (one of the 1,200 species of salmonellae causes typhoid itself). Some experiments now seem to indicate that in foods treated with antibiotics, the growth of salmonellae, in a form resistant to antibiotics, is selectively encouraged. The most common carrier of salmonellae into the public diet in America, not incidentally, is meat.

Emulsifiers, surfactants, stabilizers, etc.

There are a lot of names for the things in this class ("surface acting agents," "thickeners," "gelling agents," and so on), but around the house we describe the whole group with the rather horrid word "uniformifiers."

Their principal function is to make things smooth. Though sometimes described as "improving" texture, what they actually do is come as close as the food processor can manage to taking away texture entirely. They're called "surface acting agents" or "surfactants" because what they do is act chemically on the surfaces of various substances so that they will be more permeable to each other.

They come in all forms: vegetable products like gum arabic, carragheen from Irish moss, pectin from fruits, and chemicals with names like hydroxypropylmethylcellulose, polyoxyethylenated sorbitan monostearate, or stearyl monoglyceridyl citrate. They go into the making of virtually textureless ice cream, process cheese, bread, cake, salad dressing, pie filling, margarine and other shortenings, and any other product which the manufacturer thinks you won't like if it looks a little bit natural.

This, too, is purely merchandising. Homemade ice cream has (or, for most of us, had) larger ice crystals, a "rougher" texture—and much better taste. In fact, if you think about the things that tend to be "uniformified" in American stores, and if you're old enough, you can almost always remember (or your parents can) that it used to taste better before the Mad Scientists made it smooth: peanut butter and bread are only obvious examples, and ask anyone who really likes cheese what he thinks of the process products so many of us are stuck with today.

There is also a possibility, which we'll describe later, that emulsifiers may emulsify your stomach.

Acids, alkalis, buffers, neutralizing agents

Other possible classifications are "acidulants" and, in general, "pH adjusting agents." These are all additives designed to make foods more or less acid, although some of them also serve as emulsifiers (in process cheese and chocolate), gelling agents (in jam and jelly), or flavorings (in confectionery).

They include a number of acids (acetic, citric, lactic, phosphoric), sodium acetate, pyrophosphates, hydroxides or carbonates of sodium, calcium, or magnesium, and a few other compounds. They put the acid taste in some soft drinks, jellies, and jams, or take it out of other products.

Used in preparing sour cream, butter and other dairy products, canned olives, tomato soup, and even some wines (Americans have abominable taste in wines anyway), they are sometimes there because some other, acid additive was used in processing and would leave a residue unless the acid was neutralized (in other words, they're the fish ladder the engineer builds after he finds out that his dam blocked off the fish). You'll also find them in process cheese, pickles, cured meat, prepared cereals, evaporated milk, and a lot of canned vegetables—anywhere the manufacturer thinks you might find the taste "too strong."

Colors

In case our hints about whose side the FPC is really on haven't penetrated, let us quote, *in full,* the section on food colors in the introduction to the Food Protection Committee's own list of *Chemicals Used in Food Processing:*

Food colors of both natural and synthetic origin are extensively used in processed foods, and they play a major role in increasing the acceptability and attractiveness of these products. However, the indiscriminate use of color can conceal damage or inferiority, or make the product appear better than it actually is. In view of these factors, food colors must be used with discretion. Classes of foods that are frequently colored include confectionery, bakery goods, soft drinks, and some dairy products, such as butter, cheese and ice cream. Natural colors used in foods include alkanet, annatto, carotene, chlorophyll, cochineal, saffron and turmeric.

Note the mild tone, the use of "attractiveness" as a major criterion, the seemingly short list of colored foods, the solemn bow to "discretion," and the innocuous little catalogue at the end. What "housewife" could be afraid of chlorophyll, or of saffron and turmeric?

Of course some "housewives" might be disturbed to learn that cochineal is simply a name for the dried bodies of a female cactus insect found in the Canary Islands. But much more important: would you ever guess from that paragraph that more than 90 percent of the food colors used in America are synthetic chemicals? Or that almost everything in the store has food coloring in or on it?

The indiscriminate use of color, we're told, "can conceal damage or inferiority, or make the product appear better than it actually is." But what other reason is there for putting fake coloring into food at all?

There is none. The color is there solely to con you into thinking that the product is healthier than it is, or will taste better than it does (or even that your dog will like it better). The *CRC Handbook* virtually admits this; after noting that manufacturers always try to make artificial orange-flavored drinks the same color as orange juice, the contributor on colors writes that they are used "to create new food products and to modify the color of established food products which show color shift as the result of manufacture or storage." In other words, they're disguising either its age or what they've done to it.

Colors, too, have their own section later in this book, but at least we can say that we must eat an awful lot of them. And some of us, who are old enough, have eaten an awful lot of poisonous ones.

In dry mixes, for instance (cake mixes, pancake mixes, etc.), a coloring, if you're going to use one (and everybody does), has to remain stable in both a dry and wet state; otherwise you won't be fooled into thinking that that's the real color. For a great many years, manufacturers accomplished this by using two synthetic dyes, FD&C Yellow No. 3 and FD&C Yellow No. 4. In the 1950s, virtually under orders from the Supreme Court to enforce the Food, Drug, and Cosmetic Act of 1938, the Food and Drug Administration banned both dyes. It took the manufacturers a while, but they managed to find substitutes which they now argue are "safe," just as they argued that the previous toxic dyes were "safe."

You will sometimes read that synthetic colors have to be certified by the FDA, while natural products which impart color (like caramel or the insect bodies) do not. This is almost, but not quite, accurate. The FDA uses the word "uncertified" instead of the word "natural," because some natural products, like beta-carotene, can be and are duplicated synthetically. The "certified" food colorings—which doesn't mean that they're better or that they're more desirable, but only that they are created completely in a laboratory—all have names like FD&C Red No. 2. The letters come from the Food, Drug, and Cosmetic Act, and many of the products are coal-tar dyes. Each belongs to one of four classes, based on their molecular structure, but that's getting a little beyond what we need to know about them.[3]

The uncertified colors, on the other hand, have names, ranging from chemical names like beta-apo-8′-carotenal to everyday names like saffron and beet powder. Certified and uncertified colors are sometimes used together, and to get some effects (chocolate cake mixes are an example), as many as four separate colors may be used.

Why color everything, including even some fresh fruit and vegetables? Ask them and they'll tell you: it's because *you* demand it. Press them and they'll haul out one of the oldest, but best, of PR tricks, and call out the psychologists (that trick is so good, in fact, that PR men often play it on their own employers).

And they're right, too. I remember white margarine, and the streaky mess we got when we tried to color it at home. Almost no one reared on butter could bring himself to eat it. But the butter wasn't always the *same* color. Nor was chocolate cake; nor were vegetables; nor were ice cream cones, at either end.

Which is why we have so difficult a task ahead of us. The food *is* colored with fake colors because we want it that way. It looks "wrong" now if it isn't—because the food business, using the products of its Mad Scientists, has successfully conditioned us that way, and that's the way they're going to keep it no matter what's happening inside our bodies (and, ironically, theirs).

Flavors

"If strawberry-flavored foods," say the Manufacturing Chemists, "were produced only from actual strawberries grown in this country, the entire annual supply would be used in a few months by a city the size of Syracuse, New York."

Actually, they wouldn't, because we grow most of them in California and we doubt whether they would all be shipped to Syracuse, New York. Aside from that, however, the chemists seem to be saying that somebody is eating one hell of a lot of phony strawberry flavoring, probably with a little FD&C Red on the side. The flavoring agent could be methyl cinnamate, or it may be an aliphatic aldehyde called C_{16}, which is also used for raspberry or cherry flavoring. So much for our taste buds.

Out of 1,622 additives, you'll recall, the FPC listed 1,077 designed only to add or enhance flavor, mostly phony flavor. That makes flavoring agents by far the largest single category of additives—and again, they'll tell you it's because you demand it.

Undoubtedly you would like to have flavor in your food—possibly even more than is there in the bland, smooth, emulsified junk that makes up so much of our diet. But a lot of us thought we were demanding the real flavor of the food. It could be said, perhaps, that you do demand peach flavor when you see something with the word "peach" in it. But according to them, if it turns out that the taste comes from phenylethyl isovalerate, why, obviously you're demanding food additives!

They can do lots of things with flavors. They can even blend them so that one flavor will be the first impression you will get, and thus the one you're likely to retain as an impression, even though it's not the predominant flavor in the product (that's called "top note" flavoring). All too often, what they can and do do is use artificial flavors to replace some of the taste taken out of the food in the processing. In any case, it's all designed to fool you.

Included under flavors are the substances some additive experts call "flavor potentiators," which is almost as horrid a phrase as "uniformifiers." They're also called "flavor enhancers," and what they mean is things like monosodium glutamate which are supposed to "bring out" whatever flavor is already there. There must be something wrong with the way it does it: by accident I was just looking at a package of Knorr Leek Soupmix, and found that it contains not one but *three* "flavor enhancers," monosodium glutamate, disodium inosinate, and disodium guanylate. That's especially interesting because a few years ago, when the Campbell Soup people first took over the European Knorr firm and began to import the soup mixes, the first thing they did was change the formulae to make the Knorr soups blander "for American tastes."

Aside from the fact that I can't stand the stuff, the appearance and success of "instant coffee" has had at least one bad effect: it has gotten us used to the idea of somehow having the "essence" of something around, so that adding it to water turns the water into a pungent and magical brew. It helps to add to the vague idea, which a lot of us have, that to add flavoring to a food product means to add something like instant coffee—a small and concentrated amount of the real thing.

But that's not how it works. Certainly the best way to give something a cinnamon flavor is to add a little cinnamon; but you can do it quite well by adding cinnamic aldehyde—a completely synthetic chemical. Even the similarity in name comes only from the similarity in taste; cinnamon has nothing to do with it at all.

Flavors, too, have their own section later on.

Bleaches, maturing agents, bread improvers, etc.

We don't call them "bread improvers." That's a trade designation.

This classification also includes starch modifiers, leavening agents, and a couple of other items, and will be dealt with at some length in the chapter on bread. Mostly, they are things used in bread and in flour, cake mixes, and related products, although there are bleaches used in cheese and in a few other products.

For instance, they think that you think that blue cheese is supposed to be a certain color (for one thing, they think you're too stupid to know that it's called "blue cheese" in the first place to con you into thinking that it's *bleu* cheese, which comes only from Denmark). So they bleach it to make it the color that they think that you think it should be.

Nonnutritive sweeteners

Just what they sound like. Four of them, now banned, are the cyclamates (or cyclohexylsulfamates)—calcium, magnesium, potassium, and sodium cyclamate. Three others are ammonium saccharin, calcium saccharin, and sodium saccharin (saccharin also has a longer chemical name: 2,3-dihydro-3-oxo-benzisosulfonazole). And two more are used for special dietary supplements: sorbitol, which also has many other uses, and xylitol.

Miscellaneous

We have to lump together a number of additives, with a number of uses, to keep the list of classifications from going on forever. Included in this group are humectants, used to prevent loss of moisture; glycerol, propylene glycol, sorbitol (also a sweetener), and others are used to keep the moisture in packages of shredded coconut, marshmallows, and some confectionery. Anticaking agents, like calcium phosphate or magnesium stearate, are used to do what it sounds like they do in salt, powdered sugar, garlic and onion salt, and other products.

Firming agents are used in a number of products—canned vegetables, canned fruits, pickles, seafood, and others—to keep the texture attractive to the purchaser. Those nice, firm stewed tomatoes in the can, or those nice, firm pickle chips in the jar, or those nice, firm canned peas don't look at all the way yours would look (try stewing tomatoes and see how firmly they hold together). That's because you're all out of alums, or calcium citrate, or magnesium chloride, or whatever.

Clarifying agents are used to make liquids clearer (vinegar, for instance), and clouding agents are used to make them less clear (some citrus-flavored drinks, for instance), to fool you about what they really look like. Curing agents are used in curing meat. There are foaming and whipping agents, and antifoaming agents (a chemical determines just how much head I get on my American beer). Solvents are added to some foods in order to keep *other* additives "properly" distributed. Coating agents (like shellac, silicone, or carnauba wax) are added to make things shine prettily (including chocolates and oranges).

"Propellants," or aerating agents, add carbonation to beverages, or make it possible for foods with creamy textures to come out of those cans with the pushbuttons on top. Most propellants in use are carbon dioxide or the nitrous oxides; propane and butane are approved, but they haven't taught us to like the odor or the taste yet. Most recent figures on propellant use, by the way, are from 1966, when 2.2 billion aerosol units were used (up from 1.3

billion in 1964), and their use, according to *Chemical and Engineering News*, "showed every sign of increasing."

Ethylenediaminetetracetic acid and other chemicals are used to "separate out of the way" traces of chemicals that might otherwise interfere with chemical processing. They're called "sequestrants," and are used in frozen fruits, soup bases, process cheese, evaporated milk, vinegar, soft drinks, and other products. They also serve to neutralize the taste of trace items that might detract from the blandness of whatever's being processed.

Radiation sources and products

Sources of radiation are considered "food additives" under American law. The idea of irradiating foods in order to preserve them has been around for a long time, but it hasn't been used much, and most of the uses approved at one time or another by the FDA have since been disapproved again.[4] Irradiation of potatoes does stop them from sprouting, and ham and bacon can be kept from spoiling—but there's a great deal of concern about what else may be happening besides.

The best that can be said is that a lot of the work is inconclusive, but that's just a government-agency word for the fact that in talking about irradiation, as in talking about so many other things, we do not for the most part know what we're doing.

What we do know is that ionizing radiation appears to destroy many of the essential ingredients in many kinds of food, and that when you irradiate something you change it. To call it "ionizing" radiation means that it adds electrons to, or subtracts electrons from, the atoms of the material being irradiated, and that in turn can cause all sorts of other changes (X-rays are ionizing radiation, and as such can cause mutations or, in some cases, sterility just by changing a few electrons around).

At least some authorities believe that chemical changes during or after irradiation may result in the production of new toxic substances. Particular types of chemical changes may result in the production of carcinogens. And since it is known that the irradiation of living animals (including people) can cause cancer, but it isn't known just how it works, there is some fear that the process of irradiating food and then eating it can, in complex ways, result in cancer in the consumer.[5]

There are just too many unanswered, and at the moment unanswerable, questions about irradiation. Fortunately, the FDA seems to agree, for the moment, and there is no big push from the food business or their Mad Scientists, partly because it's expensive going (although the army, for its own purposes, would like to see the process developed—but then they make nerve gas, too). A change, however, could come tomorrow, and we all ought to remember that at least until there's a lot more known about it, irradiating foods is simply not a good idea.

Pesticide residues

You know what they are. You may not know some of what we will have to say about them later, but they're poisons, and you're eating them.

Those are our classifications of additives. Aside from what we have said and will say about specific additives and the possible dangers they pose, this may be a good spot to repeat the general warning: we don't take in any one of them in isolation, in carefully controlled amounts, as though they were doctors' prescriptions. We take them in all at once, indiscriminately, without control, in odd combinations. There is no way for us to know what we're eating. Most of them have never been tested at all, and none has ever been—none could ever be—tested in all possible combinations.

We don't know what they're doing to us. What we do know is that our bodies took hundreds of thousands of years to adapt to being what they are, and that in a few decades we are drastically changing the chemical nature of what we put into them—a process that cannot by any stretch of the imagination make any ecological sense.

But that's progress.

CHAPTER FIVE

Progress Is Our Most Important Murphy

A "Murphy," not to keep the innocent among you in suspense, is a con game.

To be specific, a Murphy is a caper in which a male, A, makes contact with a male mark, B, and offers for a fee to take B to a prostitute. A collects the money and then, somewhere along the line, loses B. There is of course no prostitute; A has sold B something that doesn't exist and is never delivered.

The food business, for its own benefit, wants you to consume food additives. But they don't sell you food additives; they sell you a nonexistent prostitute named progress.

To start with, they insist that the reason you consume three pounds of food additives a year is because you simply can't bear to be without them. The Manufacturing Chemists' Association publishes a dandy little booklet, from which we quoted in the last chapter (in the section on flavors). Called simply "Food Additives," it is obviously designed to brainwash schoolchildren, and it oozes propaganda about the American need for variety—as for instance (with our italics):

> Iowans enjoy the corn, pork, and other good foods produced locally. But they also *want* oranges from Florida, blueberries from Michigan, and peaches from Georgia. People in Arizona like the traditional foods of the Southwest but they *want to try* Maine lobster and salmon from the Northwest. New Englanders have their favorite foods, too, but they *depend on* beef from the Midwest, cantaloupe from Colorado, and citrus fruits from the groves of Florida, California, and the Southwest.

"Wanting," in the middle of the paragraph, suddenly became "depending on," although New England was presumably settled and flourished without cantaloupe. But watch what happens to those "wants" in the hands of an expert—in this case J.F. Mahoney, who works for the Merck Chemical Company and who begins four successive paragraphs as follows (still our italics):

> The consumer *wants* foods that are nutritious. . . .
>
> The consumer *wants* foods that look and taste good and retain their desirable appearance and taste. . . .
>
> The consumer *wants* variety, novelty, and convenience in processed foods. . . .
>
> The housewife has many well-recognized *needs* for processed foods. . . .[1]

All those sentences are on one page, and "wants" have now noiselessly become "needs." By the next page, Mahoney is telling us what industry has to do about it:

> [I]t is . . . necessary to consider trends in consumer *demands* and what these are likely to imply in terms of future *needs* for food additives. Keep in mind that it takes 5 to 7 years to develop a new food additive to the point of commercial use.

Note what has happened. We have not only gone from "wants" to "needs" to "demands." We have also gone from oranges, which can be shipped to Iowa in refrigerator cars or trucks with no additives at all, to processed foods. Never mind that Iowans didn't want oranges in the first place until salesmen from California and Florida taught them to want them. A chapter could be written on the national advertising campaign, early in the twentieth century, to sell oranges to Americans as a "necessary" health food despite the fact that "navel" was considered too dirty a word to use in the advertising. But at least oranges are good for you except for the wax and mineral oil and pesticide residue and things.

Mahoney's paragraph on processed foods goes on to list, among others, freeze-dried foods, fish sticks, TV dinners, and frozen pizza. But according to him, these are all things that exist because somebody considered "trends in consumer demands" and, presumably, spent from five to seven years developing the necessary additives. Does anyone seriously believe that somehow, out of the blue, there came into being in America a spontaneous "consumer demand" for frozen pizza? And that the manufacturers of food additives have simply responded to this demand by giving you what you want?

Here is this poor suffering manufacturer trying to guess what the "housewife" is going to "demand" seven years from now. He spends anywhere from $30,000 up developing his new chemical additive for making the new product available, and it turns out that the "housewife" doesn't demand it after all. So of course he throws it all away and goes back to the old drawing board. No big advertising campaigns, no massive effort to see how many products can be developed to use the new additive. You didn't demand it, so he just loses out.

Sure he does.

Understand, frozen pizza is okay except that it's usually terrible pizza. But then I grew up in San Francisco and I've eaten a lot of *real* pizza. On the other hand, not until I was well into my adult years did I ever eat Maine lobster. It's good; I like it. But among the millions of families in the Greater San Francisco Bay Area, I doubt whether very many of them ever set up a demand to have it available at will in northern California.

It's here, so some of us (who can afford it) eat it. But we wonder how many of us would eat it—or any of the other foods from other parts of the country or the world, or the foods we eat canned or frozen that are out of season—if we knew what was done to it in order to get it to us, in such form so that it at least *looks* unspoiled and doesn't taste outright awful.

Because remember: very few of us in California know what Maine lobster ought to look like, and very few of us know what it ought to taste like. Iowans may for all I know eat a lot of strawberry ice cream, but unless they've been out there they don't know what strawberries taste like fresh from the ground of California's Pajaro Valley. To them, that aliphatic aldehyde is strawberry flavoring, and those chemically preserved, chemically colored, chemically taste-changed things in the frozen food bin are strawberries.

Still, that's the picture they want you to see. You, the consumer, the "housewife" in their terms, are in absolute charge (their stereotyped "housewife" or "homemaker"—female—is the only person who has anything to do with using or stocking the kitchen or who spends any time thinking about what to have for dinner). All they do is give you what you want, and make your life easy for you. In fact, you've never had it so good. Progress.

Refrigerator cars and fast shipment are one thing. Even canning is, in itself, relatively innocuous. But the whole point of the argument about how you've never had it so good is that it depends on your being sold more and more different and more complicated foods and combinations of foods. And that depends on more and more things being done to them in order to get you to eat them at all. After that, the only tricks left are to convince you that you demanded them, and to keep you from asking too many questions about how they got there.

But no matter what some expert, real or pseudo, may say in a magazine article, he can't say that he knows whether our growing load of food additives is safe or not. Under pressure, and asked the right questions about the cumulative toxicity of substances over time, and about the "totality of toxicity" from a number of substances used indiscriminately without regard to their possible interactions, he may even admit that he can't. That's when the "balance" argument comes out.

In 1960, the Food Protection Committee held a symposium on food additives, and one of the speakers was a vice president of DuPont, David Dawson. He talked about the high cost of proving that a food additive is "safe" (meaning only that it meets current government requirements), and said that it's too bad such a cost is necessary. It is necessary, Mr. Dawson went on ruefully, because some tiresome people want "to reduce the risk of any harmful effect of any chemical additive to foods, and preferably to make that risk approach zero."

That might seem like a pretty good thing to do, but not to Mr. Dawson. What he and his fellow industrialists have to do, he said, is "to convince the public that governmental controls and administration should not be concerned solely with the elimination of risk, but rather with a judgment of risk versus gain."

That's what they, and the Mad Scientists they hire, all like to say. But keep your eye on the pea: you and I are taking the risk, the risk of slow poisoning, hidden illness, maybe degenerative disease, while DuPont and the other manufacturers are doing the gaining.

To say so is to stand condemned of blocking progress, which is taken to be an end in itself and which always, we are told, entails some risk. But so does giving the Murphy man your money before you see the woman. Progress is only the most important product of those who make a lot of money at it, and the money comes from conning us and sometimes from poisoning us. The risk, on the other hand, is ours.

There is always an "expert" at hand to back up the manufacturers. This is not to denigrate experts; we have quoted several so far, and will quote more. Nor is it to appeal to that strain of antiintellectualism in America which likes to believe that any expert is *ipso facto* a phony. It is merely to say that experts can be phonies. There are fat consulting fees to be had in this as in any major industry; there are often conflicting interests; there are people who are simply flattered by the attentions of the rich. There are others who can be misled or are simply wrong. And not everyone whose title is somehow medical is necessarily an expert on nutrition, nor even on toxicology.

One of my favorite "expert criticisms" came at that same FPC symposium from B. Blackwell Smith, Jr., then head of the Medical College of Virginia, who, in possibly the longest sentence this side of William Faulkner, criticized

> . . . the professional alarmists, such as the poison pen pushers with literary royalties in mind; those unscrupulous and cynically provocative writers who thrive on sensationalism; those politicians who see here an almost perfect chance to appear in the noble role of protectors of the helpless and the weak; those who apparently take positive delight in shivering anticipation of dangers unknown and perhaps nonexistent; and some few dedicated but legalistically minded bureaucrats who firmly believe the public interest cannot possibly be protected save through ever more complicated and restrictive regulatory enactments.

Aside from the fact that we could use a few more politicians who would even pose as "protectors of the helpless and the weak," we find that all rather charming. We are especially taken with "poison pen pushers," a much better than average phrase. But since we definitely have literary royalties in mind, we might point out that one can earn them much more easily by writing on the other side. The pay is much better in public relations than in journalism.

More important, it is characteristic of all such diatribes that they are nonspecific. Beta-naphthylamine is associated with bladder cancer in human beings no matter how many glistening phrases are written about "poison pen pushers" or "shivering anticipation of dangers." Facts are facts.

Some argument about balance is, indeed, essential to this or to any other public question. But always, it is necessary to ask: what is being balanced against what? It is not enough to say that we must balance risk against "progress," or against the "fears of sensationalizers." We must first ask, what is the risk? and then, what is the "progress"?

Sometimes we find that the question doesn't make any sense at all. To meet growing public concern about the external environment, a California utility company (Pacific Gas and Electric) ran newspaper ads calling for "a balance between energy and ecology." The phrase makes absolutely no sense in the English language. You cannot *balance* ecology against something else; you can only change the ecology of a system or leave it alone.

And that is as true of inner ecology as of the ecology of a mudflat or a planet. If you change your own inner ecology at your own risk, then perhaps it makes some sense to talk of balancing that risk against some gain. But if you have no choice, if the ecological change is to be brought about through the food you have to eat (and have to feed your children), then you are balancing nothing. And if you don't know what's in the food and they won't tell you, then you certainly have no choice.

Another thing that Mr. Mahoney wrote we can agree with, although it's possible that he didn't mean exactly what we would mean. He called for more consumer education,

> to develop in the consumer a more informed knowledge concerning the proper and safe use of additives, which, in turn, will permit the food processor to take full advantage of the additives available to him to produce foods to meet consumer needs.

We're for that. It's just that we believe that "a more informed knowledge" may leave the "food processor" with fewer additives "available to him," and may result in the production of food that really does "meet consumer needs."

What we don't need is the kind of pseudo-authoritative "consumer education" we get from people like Dr. Frederick Stare. Stare is often, and correctly, identified as the head of the Department of Nutrition of Harvard's School of Public Health. But high position does not always mean detached objectivity. Stare, who also writes a newspaper column, is a member of the board of directors of the Continental Can Company, and almost invariably reflects the position of the food industry. In the middle of 1970, when breakfast cereals came under attack before a Senate subcommittee and the attack made newspaper headlines across America, Stare was in front of the committee almost immediately, testifying on behalf of Kellogg and the National Biscuit Company. Which should give you an idea.

Stare went so far, earlier in 1970, as to say that "food additives are far safer in actual use than the basic natural foods themselves."[2] And what are we to say about the passage that follows?

> Do you remember when there were no mixes, when no one could choke down the dry peanut butter after the jar had been opened a few days, when white bread and cereals lacked certain key vitamins

and the mineral iron, when water lacked fluoride and salt lacked iodide? When bread became moldy in a day?

I don't know where Dr. Stare grew up. Where I grew up, in the middle of a fairly poor neighborhood in a big city (and without a refrigerator or even an icebox), we had enough sense to mix peanut butter before we used it and to keep the jar tightly sealed. What is he trying to say we did—throw the jar away after one sandwich?

The bread my mother made had all those vitamins and minerals in it (not just "key" ones, which is another fiction) because the "food processors" hadn't ruined the flour by taking them all out—so that today they have to add a few, but only a few, in partial compensation.

And bread does not become moldy in a day and it never did, and we're sure Stare knows it. In fact, as I type this passage I have just finished a sandwich, made with store-bought bread kept in a kitchen breadbox for a week. It is labeled "No preservatives added." It hasn't even begun to dry out.[3] Stare is not giving us "consumer education," nor expert advice, nor even reasonable argument. He is giving us rhetorical public relations for the additives industry.

There are balances to be struck that can make sense. Much of what we now know about food preservation, for example, was discovered in research done during World War II, by or at the behest of the armed forces, in order better to feed troops engaged in a global war. We might well consider the risks involved in using some additives as relatively small alongside the risks involved in losing a war to Nazi Germany or imperialist Japan.

To take another example, there was some concern in the spring of 1970, after cyclamates were banned (more or less) in America, because one manufacturer donated a large amount of already prepared and canned diet drink, complete with cyclamates, to be fed to the hungry people of Laos. Assuming that it gets to the people instead of the black marketeers, it is certainly possible to speak meaningfully of balancing the possible dangers from cyclamates against the known effects of malnutrition and possible starvation.

In the same way, it is possible to understand the arguments of such organizations as the Southern Christian Leadership Conference who argue that concern over subjects like food additives is essentially a middle-class concern, of no importance to the people in Mississippi and elsewhere who are literally starving to death.

But the food processors are not getting rich by feeding the hungry in Mississippi (and Libby, McNeil and Libby would hardly have sent its generous gift of diet drink to Laos had cyclamates not been banned in America). They are getting rich by feeding the great unhungry majority of Americans, and what they are feeding us, *in toto*, is poison.

They don't really want us to think about it, as witness Dr. Stare, in that same *Life* article, blandly observing that "if you put cinnamon on your toast you are using an additive." That is simple word juggling. It is not the way the word "additive" is used in the trade. And Dr. Stare almost certainly knows that even the phrase "food additive" is as vague as it is because the industry made a determined effort to push that vague phrase into use.

You see, they used to call them "chemical additives," until they found out that an increasing number of Americans were getting concerned about the idea of adding a bunch of chemicals to their food. The PR men went into high gear. And, to nobody's surprise who knows how the agency works, they took the Food and Drug Administration with them.

The Manufacturing Chemists' Association, in their booklet, "Food Additives," reassures us:

> Too often, people without scientific background believe that the word "chemical" means something dangerous, or, at best, "unnatural." They fail to realize that everything in the world is essentially chemical—from the concrete and steel in the Empire State Building to the vitamins in our food. The average homemaker [here she is again] would look incredulous if she were told she had just fed her family triglyceride esters of palmitic, oleic, linoleic, and stearic acids. But when she becomes aware that the words are simply the chemists' terms for the chief components of cooking fats and shortenings, she begins to understand that foods are made up of many chemicals.

The tone, perhaps, may make clear how the PR men feel about that dumb broad, "the average homemaker," but aside from that, it can be argued that there is nothing in that paragraph that isn't true. What it doesn't say is that while no chemical may be "unnatural" (except in the sense that some of them, including a lot of food additives, do not occur anywhere in nature), a lot of them may very well be dangerous. Arsenic is a chemical too. To know that "normal" foods can be described in chemical terms does not mean that all "chemicals" are, or should be, foods.

Which, of course, is why the paragraph was written in the first place. PR men who write propaganda booklets know, as any writer knows, that the public makes subtler distinctions where words are concerned. When you talk to a friend about adding chemicals to food, your friend doesn't take you to mean that you're putting cinnamon on your toast. You're talking about putting in a substance that doesn't belong there, and the Mad Scientists and their PR men know it. That's why "chemical additives" have become "food additives."

They even like to point out—or they used to, before they were successful in putting across "food additives" so that we'd forget about "chemicals" entirely—that we're all made out of chemicals, which of course we are but which again is intended to confuse, rather than to clarify, the question. Another apologist for the system, Paul R. Cannon of the University of Chicago, has referred to "an excessively apprehensive state of 'chemophobia,' " which he defines as "an exaggerated fear of toxicity brought about by the widespread use of chemicals in food production and processing."

It is worth noting about Dr. Cannon (a member of the FPC) that he proudly defends the industry for adding some nutrients to some foods, and doing so "not for pecuniary gain." Apparently he believes that the manufacturers of these additives give them away, or that the processors pass them along to us without any profit markup. Apparently, too, he believes that the processors add them for some other reason than that they are either required to by law, or are convinced that it will help them to sell their products.

Fortunately for us, it costs a lot of money to test new additives—probably the only thing that protects us from having three times as many as we have. This, too, bothers our Mad Scientists. At a 1966 symposium, G.W. Ingle of Monsanto complained about restrictions on new packaging materials, and the cost of demonstrating their safety, which he said keeps many of them off the market. What he wanted, of course, was for the testing standards to be relaxed.[4]

That takes us back to Mr. Mahoney of Merck Chemical, who was at the same symposium:

> During the last ten years, food additive research in the United States has become a distinctly less attractive field of industrial opportunity. New knowledge of toxicity has increased the time and expense required for safety evaluation. Regulatory requirements have further prolonged the time required to develop a product, clear it, and get it to market.

This is a pretty blatant example of the industry's attitude: Mahoney's sole concern about our "new knowledge of toxicity" seems to be that it makes it tougher for the industry. He estimated that with that new knowledge "it takes about four years and $100,000 to $500,000 to establish safety and secure clearance of a new food additive in the United States," assuming that it passes the tests. And he went on:

> The solution to the problem of making industrial research on new food additives more attractive is not obvious. Certainly no one wants to cut expenses and effort on safety evaluation if this increases hazard to the consumer. Perhaps a more open and cooperative relationship with regulatory agencies could help eliminate unneeded testing and enable a better assessment of the likelihood of regulatory approval.

He doesn't say why "making industrial research on new food additives more attractive" should be considered a "problem" in the first place, although we can guess. But notice how, within a few lines of print, a reference to our "new knowledge of toxicity" has been married to a reference to "unneeded testing." In fact, if you translate that last quoted paragraph into English, it sounds suspiciously like, "We don't, of course, want to cut down on safety. All we want is for the regulatory agencies to do it our way, and eliminate tests we don't like."

Because of the testing cost, those additives now on the GRAS list—additives, remember, which don't have to be tested, and most of which have never been tested—are not likely to be replaced by new and possibly safer substitutes. But if Mr. Mahoney's figures are accurate, he tells us even more than that about the GRAS-list additives, more, perhaps, than he meant to say.

If it actually takes from $100,000 to $500,000 to establish the "safety" of a single additive, think of what we don't know about each of possibly a thousand or more additives on the GRAS list, which have never been tested. Perhaps if we did require tests on GRAS-list additives, we could get rid of a lot of them because it would cost too much to test them. Manufacturers might then have to use the decent storage procedures or provide the better quality foods that many additives are intended to replace.

The manufacturers are not, of course, interested in any tightening of regulations in the United States. On the contrary: the United States is one of the most liberal countries in the world in allowing the Mad Scientists to mix all kinds of gunk into our food, and the industry is doing its best to persuade the United Nations, which is working on overall world standards, to adopt as many American rules as possible. They want to be able to sell their additives in other countries as well.

Germany, for instance, does not allow antioxidants in animal feed; the United States does, and American manufacturers would love to sell their antioxidants in Germany. This is only one of dozens of examples, another being a major propaganda effort to convince "underdeveloped" countries that American food additives are going to solve their nutritional problems—and to keep them from learning that they may, in fact, aggravate them instead.

All of which you pay for. For those of us who are concerned about the genuinely hungry in America, and about those just above them who are on extremely tight budgets, it should be noted that the food industry forces them to pay for it too. They have even less choice about the food they buy than the rest of us do. What you may not know, though, is how much you pay for it.

The usual supermarket contains something like 8,000 food items, most of them new in our lifetimes. More than two-thirds of them were never seen in a store before World War II, and it's doubtful whether many, if any, appeared in response to a spontaneous "consumer demand." About 5,000 new items appear on the shelves each year, although most don't stick around very long, thanks partly to consumer *non*demand.

Almost all of these new items are processed foods, or formula foods: food items which are in some way artificial, ranging from a simple mixture of two kinds of fruit juice to the completely prepared and frozen four- or five-course meal. Nor are there, of course, only new combinations of foods that were formerly available separately. There are completely artificial puddings, cheese products produced in a laboratory, synthetic "whipped cream" (which is to whipped cream, by taste, what frozen pizza is to real pizza). You can make your own inventory on your next trip; simply count the number of things that you couldn't make at home, starting from scratch, even if you wanted to and had the time.

Obviously those of us who can afford it are willing to pay something for all this. I would rather write than spend the time making my own beer (even though I haven't yet written enough to escape, except occasionally, the bland sameness of the watery American beers). Convenience is a part of affluence, and sometimes, for working women without husbands for example, convenience is very near necessity.

But how many of us know that we pay *more than one-third of our food bill* just for convenience? At least that's what Mr.

Mahoney of Merck Chemical says, and we're willing to take his word because if he were going to con us about this, he'd use a lower figure, not a larger one:

> [I]t has been estimated that 35 percent of the retail cost of food is being paid by the consumer for convenience features. Examples of such developments are ready-to-eat breakfast cereals, instant foods, soup mixes, cake mixes, brown-and-serve rolls, freeze-dried foods, fish sticks, formulated salad dressings, etc. Consider also the trend toward compound foods: TV dinners, frozen pizza, vegetables with sauce, and many gourmet delicacies. The development of most of these products would not have been possible without the use of a variety of food additives.

Exactly. And the next time you're spending one-third of your supermarket budget for "convenience" (whether you want to or not), remember that even five years ago, the wholesale value of the additives alone was $285 million.

You *know*, when what you're buying is in a can, or when it comes from far away, or when it is an instant food or a frozen compound, that its convenience costs extra. You can decide for yourself whether to spend the additional money for that convenience, just as I know I have to pay somebody for making and bottling and transporting beer if I want to drink it.

There is no such choice available about the food additives you eat.

CHAPTER SIX

The Webs of Anacharsis

You will not, when you have finished this chapter, understand the GRAS list. That is because nobody understands the GRAS list—nobody, that is, except possibly some people who like to keep the whole thing vague so that food additives won't have to be subjected to expensive and possibly revealing tests.

Nor will you understand the complex of laws, regulations, trade association agreements, hidden correspondence, and off-the-cuff remarks that together control the use of assorted forms of glop in practically everything we eat. The best we may be able to do is to keep your head from spinning too much, and to let you know just how confused it all is.

Most of us have a fairly simple view of the whole thing. There are laws about pure foods, and there is a Food and Drug Administration, and we sort of trust them to keep things more or less safe. We believe that they must have the legal power to do that and in fact, they probably have, if they would read the words in the laws instead of trying to find ways to get around them.

But writing laws is not, and will never be, enough. Six hundred years before Christ, the philosopher Anacharsis wrote to Solon, "Written laws are like spiders' webs, and will, like them, only entangle and hold the poor and weak, while the rich and powerful will easily break through them." Things haven't changed much in the intervening millennia.

The Food and Drug Administration is full of wonderful people. They are dedicated, sincere, underpaid, overworked, and for all practical purposes bound and gagged. If some of them, after a while, fall into a bureaucratic routine, shifting papers from In Basket to Out Basket and doing their best not to be noticed until pension time, we shouldn't blame them. We don't pay them enough and we don't protect them when they try to protect us. If they try to do anything more zealous than what they're doing now, they may very well find themselves collecting unemployment insurance.

Near the top of the ladder, you run into another type: the politicians. It is their job, as it is the job of all politicians who are six or eight levels down from the top, to see that no waves are made. They are oil-pourers long before the waters are ever really troubled. They know, or they soon learn, that you and I are not big political contributors; that we do not have lobbyists in Washington; that we do not recruit "experts" to industry-front "professional organizations" and pay them off in corporate directorships. All we do is eat junk and die of degenerative diseases.

For some reason Americans seem constantly to have to relearn that the government's regulatory agencies belong to the men they're supposed to regulate. Right after the 1960 election, John F. Kennedy appointed James Wilson to report on the regulatory commissions, and the Wilson Report—widely reprinted as a paperback book—pointed out the connections in detail. Ten years later, Ralph Nader's study groups are doing it all over again. And for the FDA, the late Senator Estes Kefauver performed the service once in between.

Yet we forget what it means. It means, by simple logic, that the higher the position of an official in the FDA, the less we, his employers, can trust what he says. It means that the FDA does not do what the law clearly tells it to do. It doesn't even do as much of it as the budget allows. It means that the FDA's priorities are not the priorities you and we would like it to have, nor those the law requires it to have.

It also means that the FDA, which is partly a scientific organization, debases the meaning of "science." It does that every time it sits on the results of an experiment, every time it issues a bureaucratic or political "interpretation" of a scientific finding, every time it keeps a scientific discovery to itself.

It is the essence of science that its experiments are objective. That is, they must be conducted in such a way that another scientist 3,000 miles away, performing the same experiment with the same kind of equipment under the same controlled conditions, will get the same results. If he doesn't, then that proves that in one of the two cases, there was some "experimental error," and the results are suspect.

The care that goes into such experiments is difficult for the nonscientist to imagine. In testing FD&C Red No. 4, for instance, the experimenter didn't say he was testing it on rats; he said he tested it on the Osborne-Mendel and Sprague-Dawley strains of rats, two strains which are carefully bred especially for laboratory experiments so that differences among test animals will be minimized.

But nobody can check an experiment that hasn't been published anywhere, that no one has ever heard of. And so it is also the essence of science that communication must be open, results must be published, other scientists must be able to check experiments. Every time the FDA sits on the results of an experiment, or even on an educated conjecture which might lead to an experiment being done, this process is debased.

There are a lot of examples of this in the FDA—James S. Turner's *The Chemical Feast* describes many of them—but somebody about to leave for the supermarket may be even more disturbed by the ways in which the FDA "interprets" the laws it is supposed to enforce.[1]

The basic law is the Federal Food, Drug, and Cosmetic Act, passed in 1938 to go into effect in 1939 (that's why some books give one year and some the other). A few substances—butter is one—are covered in part by other, separate laws. In 1954 came the Miller Pesticides Amendment. In 1958 came the Food Additives Amendment. In 1960 came the Color Additives Amendment. And in 1968 another amendment covered the administration of drugs to food animals. Most of our concern is with the 1958 and 1960 amendments, and with what they do and don't say.

Remember that our overall definition of food additives is not the legal definition. Under the 1958 amendment, a lot of food additives, including most of those which are most common, are not "food additives" at all. But we'll try to keep it clear as we go.

The Food Additives Amendment *excludes* from its definition pesticides or their residues. It *excludes* color additives (later covered in another amendment). It *excludes* "any substance used in accordance with a sanction or approval granted prior to the enactment of this paragraph." Those last items we call food additives approved by prior sanction. More on those later. The amendment also *excludes* any substance which is:

> generally recognized, among experts qualified by scientific training and experience to evaluate its safety, as having been adequately shown through scientific procedures (or, in the case of a substance used in food prior to January 1, 1958, through either scientific procedures or experience based on common use in food) to be safe under the condition of its intended use[.]

Thence came the GRAS list. Notice in that paragraph that there are two different ways that a substance can be excluded from coverage by the amendment. If experts agree, not that it is safe, but that it has been shown to be safe through scientific procedures, then it isn't covered. On the other hand, if the substance was used in food before the law was passed, then you don't need the scientific procedures, but you still need the experts; in that case they can say that it has been shown to be safe through common use in food. The law does not otherwise say what an "expert" is.

The Congressional committee which held hearings on the Food Additives Amendment wrote a report to tell their fellow congressmen what it was about, and they made their intent quite clear:

> The purpose of the legislation is to protect the health of consumers by requiring manufacturers of food additives and food processors to pre-test any potentially unsafe substances which are to be added to food.[2]

But it did not require them to pretest anything if the FDA decided that the substance was "generally recognized as safe," or if the FDA had already given the substance sanction. The GRAS list, then, is exempt from the testing requirements. Except for one pesky clause.

Representative James Delaney, the New York Democrat who was chairman of the committee that reported the bill, wrote an amendment of his own into it which is now famous (or infamous) among the Mad Scientists as the Delaney Amendment, or the Delaney Clause. This part of the law says:

> [N]o additive shall be deemed to be safe if it is found to induce cancer when ingested by man or animal, or if it is found, after tests that are appropriate for the evaluation of the safety of food additives, to induce cancer in man or animal.

Although a nitpicker might raise some questions about exact interpretation, the clause is generally taken to mean that even if a substance is exempt by being GRAS, it is still covered by Delaney's exception. The reasoning is that if it is known to cause cancer, it can't be generally recognized as safe. When Robert Finch was HEW Secretary, he gave authority to this interpretation by using the Delaney Amendment as his reason for taking cyclamates off the GRAS list. He probably did this for other than legal reasons, but that belongs in Chapter Eight, where the Delaney Amendment (and cancer) are discussed much more fully.

Finally, the Food Additives Amendment makes another possible area of interpretation very clear. It says that before the Secretary of Health, Education, and Welfare can allow a food additive to be used (not counting the exemptions), there must be a regulation. And it says that "no such regulation shall issue if a fair evaluation of the data before the Secretary fails to establish that the proposed use of the food additive . . . will be safe."

That doesn't mean that he can okay it if they think maybe it might be safe. It doesn't mean that the secretary or the FDA or anyone else gets to balance (our) risk against (the manufacturer's) gain. It means that an additive *must* be shown to be safe before it can be allowed.

The word "regulation" is a reminder that you ought to know about two publications, both of which should be at any good library. One is *The Federal Register*. The other is the *Code of Federal Regulations*.

When Congress passes a law telling the FDA, or some other governmental agency, to make specific rules for something, the agency publishes those rules in *The Federal Register* as they're made. This gives interested parties the chance to complain, or to attend hearings, or to do whatever is appropriate under the law. When the regulations take on the force of law, they go into the *Code;* agencies keep it up to date day by day in loose-leaf form (or they're supposed to), but it's published in bound form once a year.

Title 21 of the *Code of Federal Regulations* has to do with the FDA, and comes in two volumes. Part 121 of Title 21 is the part that deals with food additives (as the term is used under the law), and a subpart of that, Part 121.101, contains a partial list of the substances regarded by the FDA as GRAS. Part 8 of Title 21 has to do with color additives. And so on. You'll find an abbreviated form of this in the footnotes of this book occasionally; if it says "8.303 CFR 21," that means Part 8.303 of Title 21 of the *Code of Federal Regulations*, which happens to be the regulation on caramel as a food coloring.

The point of all this is to make it easier to wend our way through the labyrinth of the laws about the stuff they put in what we eat. To do that, we can break down what we call food additives into the following categories:

1. "Food additives" within the meaning of the 1958 amendment,
2. Additives with prior sanction,
3. Additives on the GRAS list,
4. Color additives,
5. Pesticide residues.

The last two are covered by other laws and discussed in future chapters, but in general new color additives are handled pretty much the same way as the first category we've given. In a minute we will discuss the fact that they have their own watered-down version of the Delaney Amendment.

Items in the first category are legally new additives. If a manufacturer or processor wants to use one in food, he applies to the FDA. His application is supposed to describe in detail the proposed use (including the amounts he proposes to use), and it is supposed to include scientific data to show that the additive is safe.

The FDA doesn't test the substance itself; in general, it takes the applicant's word. It does, or it says it does, "evaluate" the test data it gets. It can, if it wants to, call in outside experts for their opinion. This is rare. If the FDA then decides that use of the additive is okay (sometimes with restrictions attached as to amounts or specific uses), it says so in an announcement in *The Federal Register*. As of that day, it can be used.

Anybody who doesn't like it can then file his objections (in quintuplicate), and the FDA can, if it wants to, pay attention to the objections and reverse the order. Or the objector may request a hearing, which he may or may not get. Finally, an objector can take the FDA to court, but that, of course, gets expensive. If the order stands past the required time for filing objections, or survives any complaint, it will go into the *Code of Federal Regulations* under the appropriate section and the additive is then completely approved and legal. The same process applies if the applicant wants to use an old additive in a new, previously unapproved way.

This is what the law says. It also says that the scientific data *must* establish that the additive is safe. In practice, it doesn't always work out that way. Take the part about consulting experts. Furia (and we should emphasize that his *CRC Handbook* is an industry-oriented source) says, for instance, that "the FDA relies on the opinions of an expert committee of this organization in reaching clearance decisions on flavor components." An expert committee of what organization? The Flavor Extract Manufacturers Association!

If that sounds a little like the cops asking Willie Sutton to help guard a bank, we agree. But much more important is that the FDA does not, in fact, require that even halfway decent scientific work be done to show the safety of a proposed new additive.

Of course you can't prove harmlessness. That's first-year science, or even first-year logic. But there are reasonable limits. It is one thing to talk about cinnamon on your toast. It is quite another thing to feed some complex chemical to chickens, and to have the baby chicks formed with their legs turned around backward (which happened in a cyclamate experiment some time before cyclamates were banned), or to find that a chemical which American millers used for decades to bleach flour turned out to cause running fits in dogs. Those things don't prove that the chemicals aren't safe for humans either, but they certainly raise enough questions to get them banned under the clear wording of the law, so that you and we don't have to be the chicks and dogs in the next series of tests.

Beyond that, we started early with a rule that no argument about how much of an additive is safe makes any sense whatever, and one reason for that rule is that additives aren't taken by themselves. There is no record that the FDA has ever asked a manufacturer to show that a proposed additive won't turn into a harmful substance in combination with any other known additives. The FDA does not normally require that substances be shown to be "safe" over long periods of time. Tests for birth defects are normally not required. In fact, the FDA can hardly be said to require that substances be shown to be safe at all.

Even though it cannot be shown, in either a scientific or a logical sense, that something is harmless (you can't prove a negative), it wouldn't be too difficult for a group of genuine scientific experts, not in the pay of or closely allied to the food business, to come up with some criterion that an additive would have to meet, some set of tests that every additive would have to go through. It is not a complicated part of science, and the rest of us wouldn't need doctorates in biochemistry to know whether the tests had, in fact, been conducted properly. Everything that is added to our food should have to go through such tests, and as new testing procedures are developed, every approved additive should have to be subjected to them.

But if the tests required for "new additives" are minimal, they are vastly protective compared to the standards applied to the rest of the junk we eat. Take the second category, additives approved by prior sanction.

Ordinary common sense will tell you that this category refers to additives which were authorized by the Food and Drug Administration before 1958, presumably on some sort of application by a manufacturer and presumably with at least some showing of safety. Ordinary common sense will tell you wrong. There are a few such additives, but most which have "prior sanction" got it in another way entirely. They were quietly given sanction under the section of the law having to do with food standards.

Under the basic 1938 law, the FDA has the right (on petition by a manufacturer or a trade association, usually) to formulate standards for a particular food or food product. Over the years, federal standards have been established for various cacao products (cocoa, chocolate, etc.); flours and related products; macaroni and noodle products; bakery products; milk and cream; cheeses, cheese foods, and related products; frozen desserts (ice cream and things like it); food flavorings (mostly vanilla and things like it); canned fruits, juices, vegetables, etc. (orange juice and orange drinks is a section by itself); jellies, jams, and related products; soft drinks; some shellfish; canned tuna; eggs and egg products; peanut butter; margarine; and a number of others.[3] Meat standards are set by the Department of Agriculture, but often *enforced* by the FDA.

For an example of a standard, take catsup (or ketchup, or even catchup, it says in the regulation). Once the FDA has established the regulation, you can't sell anything called "catsup" unless it meets the description in the regulation. You have to get some tomato juice and concentrate it. You add salt. You add one or more vinegars. You add spices or flavorings or both. You add onions or garlic or both. You add sugar, or dextrose, or corn sirup, or glucose sirup, or any mixture of the four (but there are rules about how you can mix them). And then, presumably, you put it into bottles the customer can't get it out of.

If you skip any one of those steps except the last, you cannot call it catsup or ketchup or catchup. If you put in anything else, you cannot call it any of those things either. You would be guilty of misbranding. The only label requirement, besides the word "catsup," is that if you get your concentrated tomato juice from some kinds of tomato by-products you have to say so (if your label doesn't say anything about "residual tomato material," then you're getting the best-grade concentrate or somebody's cheating). There is no law that says you can't put the ingredients on your label (Heinz does, for instance), but you don't have to.

Now take tomato paste. It starts with the same concentrated tomato juice (and there are the same label requirements about it), but you can then add, as optional ingredients, salt, spice, flavoring, or baking soda (the last is to neutralize tomato acids). You don't have to use them. But if you use spice, flavoring, or baking soda you have to say so on the label; you can say "spice" or "flavoring," or you can use the name of the particular product used. The salt is also optional, but you don't have to list it.[4]

Some of these standards have been around for a while; without really making much noise about it, the FDA has used its existing food standards as a medium for making a lot of food additives legal through "prior sanction."

For instance: the standards for flour and for bakery products allow, as an optional ingredient, dicalcium phosphate as a bleaching agent (if it's in flour, it has to be labeled "bleached"; if it's in bread, the label doesn't have to say anything about it). Some frozen desserts allow the presence of ammonium caseinate. Aluminum calcium silicate is listed in the standards for vanilla-vanillin products.

As a result, these additives are regarded as having been "cleared by prior sanction." They don't have to be put on the GRAS list, and they don't have to be tested by anybody. And they never were.

An additive can even be cleared by being approved for adding to an additive. Acetic acid, for instance, is considered to be safe because the FDA allows it to be used in making caramel, which is itself a color additive. Because it has so many other uses, acetic acid is also GRAS.

But most additives in use are GRAS, and how they get to be GRAS is something that strong men have turned gray trying to find out. Nobody even knows how many additives *are* GRAS. That sounds like an exaggeration, but it isn't. It may be that there is no one person or group anywhere in the world who could find out how many additives are GRAS, with a year of free time and a Ford Foundation grant.

The Introduction to the *CRC Handbook* has this to say about the 1958 Food Additives Amendment:

> It was a clarion announcement that our government had concluded that the use of chemicals is necessary and unavoidable for the provision of an adequate food supply. It was also clear that government subsidy of safety research on compounds for the market was removed. Now the total cost of proving the material safe must be borne by the manufacturer. Much of the $200,000 to $3,000,000 required for testing would have been spent by reliable companies with no legal compulsion. This cost is not unreasonable for successful new products by large companies, and is negligible when considered a means of retaining public confidence in the face of scare books.[5]

The writer may have believed that nonsense when he wrote it, but it has as much to do with the real world as military reporting by a South American junta. We're particularly fascinated by the news that the food business would have spent somewhere up to $3 million per additive anyway, especially since they hadn't made a move to spend *any* of it until the law was passed (the authors' source for that sentence, by the way, is the Manufacturing Chemists' Association's 1963 booklet). It is even more interesting in view of the frantic haste with which the industry descended on the FDA with pressure to get everything possible put on the "generally recognized as safe" list so they wouldn't have to test it, regardless of the state of toxicological knowledge.

It's difficult at this distance to say exactly what Congress had in mind when it made provision for a GRAS category, but they certainly didn't intend what they got. Probably they started out with the simple idea that it might be unreasonable to ask manufacturers to run thousands of dollars worth of tests on things like salt and pepper. But we doubt whether they were thinking of 3,6-octadiene-1-ol. Or even propane.

They got them, though. And they got them, step by step, like this.

Passage of the Food Additives Amendment meant, of course, that the FDA had to come up with some idea of what chemicals were "generally recognized as safe." So there were a lot of meetings at 200 C Street, and they finally came up with a list of 189 items. It's a pretty safe bet that they didn't bar the doors against food industry representatives while these decisions were being made; they're big cooperators down at FDA.

They also came up with a list of 900 scientists who, they decided, might be called "experts." Then they sent the list of 189 additives out to the 900 scientists, and asked them whether the 189 additives were safe.

That was the first evasion of the law. It doesn't say anything about whether the experts think the substances are safe. It says that the experts are supposed to decide whether they have been *shown by scientific procedures* to be safe, or in some cases by common use in food. The FDA ignored that distinction immediately, and has virtually ignored it ever since in its approach to the GRAS list.

They like to talk about how they sent the list to 900 experts for approval. They don't like to talk about the fact that only 350 of the 900 scientists even bothered to reply. And of the 350, 156 had adverse comments about at least one substance on the list, and not all the same substance by any means. There were objections about the cyclamates, about benzoic acid (sodium benzoate), about ammonium hydroxide, ammonium carbonate, nordihydroguaiaretic acid, Vitamin D, and several others. Seven items were ultimately deleted from the list of 189, but every one we've just named was designated GRAS. They sent it to 900 experts all right, but they ignored the replies.

"Ammonium hydroxide," said the FDA, "does not exist as such in the food or at most as a very limited amount." It is GRAS today, as a "miscellaneous and/or general purpose food additive," and is also cleared under the color additives sections for use in caramel. Nordihydroguaiaretic acid, after ten years, was taken off the list because it causes mesenteric cysts and kidney lesions in rats. Vitamin D, objected to by 12 different "experts" because of possible hypercalcemia, went on the list anyway, only to have the permissible amount sharply lowered six years later because, as you might guess, of possible hypercalcemia. And you know about cyclamates.

Turner tells an interesting story about how folic acid, also called pteroylglutamic acid, didn't make the GRAS list:

> [FDA Assistant Commissioner Kenneth Kirk] told how Dr. Barbara Moulton came into his office "hopping mad" about seeing folic acid on the proposed GRAS list. Later she testified before the Kefauver Committee about the inclusion of folic acid on the GRAS list. But what she didn't know, according to Kirk, was that after she left his office, "I took out a red pencil and crossed through folic acid and it never appeared on the list." This won Kirk a debating point before Congress but it is just one more illustration of the arbitrariness that governed selections for the GRAS list.

It is an illustration of rather more than that. For today, thanks to an application by a manufacturer, folic acid is again an approved food additive, cleared for use in dietary supplements in designated amounts. Dr. Moulton, a medical officer, quit in 1960 in disgust over the FDA's sloppy regulation of both drugs and food additives.

The important thing about that first, beginning GRAS list is that it was not compiled in accordance with the law. The communications that went out to the 900 scientists were obviously window-dressing to make it look as though "experts" had been consulted, as the law says they must be. But the experts weren't asked any questions about scientific procedures, nor about common use in foods; more than 60 percent of them didn't answer at all, and those who did object to some of the substances were ignored.

Incidentally, the *Code of Federal Regulations* says that a manufacturer can get into trouble if he hides from the FDA the fact that there is "a difference of opinion, among experts qualified by scientific training and experience," about anything on a food label.[6] But the FDA is apparently willing to allow itself to overlook such differences all the time, no matter what the law says.

Once the original list was out, the FDA abandoned even the pretense of scientific consultation, and simply started adding items to the GRAS list, usually by publishing an announcement in *The Federal Register*. Once in an extremely great while one has been taken off the GRAS list, using the same method. The result of those publications can be found today in the *Code*, in Part 121.101 of Title 21.

But that is not by any means all. Even the FDA obviously doesn't know how many chemicals have become GRAS by various means. 1961 testimony by an FDA commissioner said 718; three years later an FDA annual report said 575 (that number has since been repeated, although there are no publications during those three years indicating that anything like 143 items were taken off). The first figure we ever saw was 680, in a newspaper story early in 1970, and Turner found the same figure in a Philadelphia newspaper in October, 1969.

Having given up the use of the legally required criteria for the GRAS items it was designating by publication, the FDA naturally went on to giving up publication entirely. Some manufacturer or association would write to the FDA and say, "Our information leads us to believe that flatulated hyperglop is generally recognized as safe," and the FDA would write back and say, "Yep, we think so too." Flatulated hyperglop, without expert opinion or publication, would then become GRAS, though the fact wouldn't be printed in any official list.

The *CRC Handbook* is, in part, an attempt to define the legal status of as many additives as the editors could find out about. A random page-flipping through the listing section reveals these entries.

Cottonseed oil: Listed as GRAS by FDA in material submitted to the House Interstate Committee [*sic;* it must mean Interstate Commerce Committee] in 1958. GRAS cottonseed oil meeting U.S. Pharmacopoeia specifications in dietary supplements affirmed by FDA letter to trade association.

Peppermint oil: GRAS for use in dietary supplements affirmed by FDA letter to trade association.

Algin: . . . GRAS affirmed by FDA in communications with industry for flavoring use [algin also is given prior sanctions in the standards for cream cheese and frozen desserts].

These are some of the many items on the "hidden GRAS list." We know about them only because we stumbled across the right book, privately produced by people who took the trouble to look through House committee testimony or to find out about private letters to private manufacturers or their associations. From there it was only a short step for the FDA to letting the Mad Scientists make up their own GRAS list, which is just what they did. A trade publication neatly misstated the "law" as early as 1961:

> [T]he law confers no special authority on the FDA for deciding that an ingredient is GRAS. If experts who are qualified to evaluate the safety of a particular ingredient generally recognize it as safe under the conditions of its intended use, then according to the language of the law itself, . . . it is not subject to the provisions of the Additive Law.[7]

Again, the law has nothing to do with experts recognizing additives as safe; and as pliant as Congress sometimes is in the hands of sizable business interests, it's hard to believe that they intended, in effect, to put the fire laws in the hands of a committee of arsonists.

Still, that same *CRC Handbook*, in its listings, shows hundreds of items designated simply, "Deemed GRAS by FEMA." FEMA stands for Flavor Extract Manufacturers Association. In 1960 they simply drew up their own GRAS list, nearly a thousand items which they said were, according to their experts, generally recognized as safe. Even the FDA balked at that for a while, but sure enough, they went along. If there is a slight technical, legal difference between the FDA's GRAS list and the manufacturers' GRAS list, there is no difference whatever in practice.

And finally, in the newspapers for October 21, 1969, we have Deputy Commissioner of Food and Drugs Winton B. Rankin making this astonishing statement:

> The manufacturer is entitled to reach his own conclusions, based on his scientific evidence that a subject is, in fact, generally recognized as safe. And he is not required to come to us then and get the material added to the list.

Thus, finally, we have the manufacturers not only deciding for themselves what is GRAS, but not even bothering to tell the FDA what they're doing. That's what's happened to the law that is supposed to protect you in what you eat. And that's what's happened to the men who are supposed to enforce that law, on your behalf, against the Mad Scientists and their bosses.

It doesn't even mean anything, in fact, when the FDA takes something off the GRAS list. An Associated Press story printed on January 24, 1970, says that the FDA "gave food and drink manufacturers six months" to prove "a commonly used type of food additive" safe or quit using it.[8]

What had happened was this. Brominated vegetable oils have been GRAS since the beginning as stabilizers in soft drinks (their principal use was to keep citrus-flavored drinks looking a little cloudy, because the makers think that you think that they look "right" that way). But Canadian experiments showed that brominated vegetable oils cause heart damage in rats, along with retarded growth, difficulty in digesting fats, enlargement of the kidneys and liver, and spleen and thyroid damage.

Sweden, Britain, and Canada banned brominated vegetable oils entirely. In *The Federal Register* for January 22, 1970—the Canadian experiments were done in 1968—the FDA said that manufacturers would have to quit using brominated vegetable oils within 180 days, unless somebody filed a petition for them as a "new additive" with supporting evidence of their safety.

In other words, they took brominated vegetable oils off the GRAS list. The FEMA promptly filed a petition "proposing the issuance of a food additive regulation to provide for the safe use of brominated vegetable oil as a stabilizer in citrus and other fruit-flavored beverages."

You will recall that no such regulation can be issued, under the law, unless the information before the FDA definitely establishes that the additive is safe. In this case, those Canadian results showing heart, kidney, liver, and other damage are still there.

Nevertheless, on July 28, 1970, the FDA announced issuance of a regulation providing for "the interim use" of the brominated oils, at a stated level, "pending the outcome of additional toxicological studies." The final results of those studies—which are to be done *by the manufacturers*—are to be "submitted to the Food and Drug Administration not later than December 1, 1973."[9]

In the meantime, enjoy your orange soda. The FDA regulation is illegal as hell, but don't let that bother you.

As far as the GRAS list is concerned, the item about brominated vegetable oils simply means that taking something off the GRAS list does not in any sense mean taking it out of your food. Because people were worked up about cyclamates at the time, the AP happened to run a story on the brominated vegetable oil "ban" that was widely picked up. Six months later, when people were worked up about something else, there was no story on the fact that the same additive is still in the same soda water, and continues to be there.[10]

Of course, we older folks don't need these fresh items to bring joy to our dinner table. We can always sit around and think about the constant and widespread use, for 30 years, of nitrogen trichloride as a bleaching agent in flour. Also known by the trade name Agene, nitrogen trichloride was finally banned in America in 1949 after it proved to give fits to dogs. The public-spirited manufacturers, of course, pooh-poohed the whole thing. Those tests were done in England.

I don't know why the manufacturers aren't happy with the law, considering the way the FDA enforces it. Maybe they're afraid that so long as there's any law at all, some bright young Congressman might accidentally read it and start to wonder. In any case, the manufacturers' line has steadily been in favor of "administrative discretion" on the part of the FDA instead of "unnecessary and rigid" requirements in the law, and you can certainly see why they want that. A good example of what the food barons like is in the differences between the Delaney Clause, in the Food Additives Amendment, and the cancer clause in the Color Additives Amendment.

The first, you'll remember, simply says that you can't call any additive "safe" if it causes cancer when "ingested by man or animal," or if after appropriate tests it is shown "to induce cancer in man or animal." The FDA has used a little administrative discretion on this one, too, although it's not provided for in the law. They pay some attention to the first part, about ingestion, when they have to; and they ignore the second part, which is a little broader. Other reporters have written, and an FDA official confirmed to me privately, that the agency takes the clause to be inapplicable unless the substance is taken in through the animal's mouth. If it causes cancer when it's dripped on the skin, it's apparently still okay, although the *law* doesn't say so.

But the Color Additives Amendment cancer clause, which superficially looks something like the Delaney Clause, is really something else again. It leaves the Commissioner of Food and Drugs with a lot of that "administrative discretion" the manufacturers like, and under that discretion he has issued a couple of rather long regulations.[11]

One of them seems to say that the Commissioner will act toward color additives and cancer pretty much the way he's supposed to act toward other additives. It's a handy regulation for quoting to critics. But the other one says that "any person who will be adversely affected"—and a corporation is a "person" under the law—can ask that the matter be referred to an advisory committee. It also says that the Commissioner can use an advisory committee on his own if he wants to.

The Delaney Amendment doesn't leave any interpretive room for horsing around with advisory committees, but the Color Additives Amendment does. It is no accident. The Color Additives Amendment was passed two years later, and the food business was alert and ready to cut down a Delaney-type clause as far as it could.

The method is no accident either. Say "cancer" in the same room with a bottle of color additive and somebody will yell for the advisory committee. That committee, say the regulations, will be selected by the National Academy of Sciences, whose own Food Protection Committee is completely industry-financed and remarkably industry-oriented. The manufacturer can figure that he'll get a pretty friendly committee. Undoubtedly most appointees would not approve a blatantly obvious carcinogen. But things are rarely blatantly obvious in this field, and the borderline cases are apt to go in the direction of continued use, which is exactly what the laws are supposed to guard against.

The provisions in the law for setting food standards, described earlier, provide other examples of how a law originally designed to protect the public can be "enforced" in such a way as to fool us. The idea of food standards, of course, was to keep manufacturers from selling inferior products, and to set some kind of minimum quality rules so that the purchaser, confronted with something called "pasteurized process cheese food," could have some knowledge about the cheesy-tasting thing he was buying.

Also, as an additional safeguard against something being put over on the public, Congress ordered that, when the standards do leave a choice about an ingredient, the maker has to say on the label whether he used that ingredient or not. Anyway that's what they thought they said. But written laws are like spiders' webs.

The actual phrasing of that part of Section 401 of the Food, Drug, and Cosmetic Act is as follows:

In prescribing any definition and standard of identity for any food or class of food in which optional ingredients are permitted, the Secretary shall, for the purpose of promoting honesty and fair dealing in the interest of consumers, designate the optional ingredients which shall be named on the label.

With a great deal of cheering and not a little coaching from the food business on the sidelines, the FDA promptly interpreted this to mean that if an ingredient is not optional—if it's something that you have to use according to the standards—then you do not have to put it on the label. You can figure out how much that has to do with honesty, fair dealing, and the interest of consumers.

Certainly it's true that the name of chemicals on labels don't mean much to a lot of people (except that their food is full of chemicals, which is one of the things the food industry doesn't want you to think about). But there are some people, at least, who know something about, say, sodium hexametaphosphate. And a lot of them, like a lot of us, have the silly idea that you can tell what's in food by looking at the label.

Consider, then, that such a knowledgeable person may pick up something called "pasteurized process cheese food" and be confident that, if it contains sodium hexametaphosphate, it will say so. But beside it on the shelf may be another package, labeled "pasteurized process cheese," which says nothing whatever about sodium hexametaphosphate, but which contains the same amount of it by relative weight.

Sodium hexametaphosphate is one of a number of emulsifiers that can be used in either product. If you wanted to, you could use, say, calcium citrate instead. So in that sense it's an optional product. But that's not the way the FDA chooses to interpret the word "optional."

The trick is that according to the standards for "pasteurized process cheese food," the use of any emulsifier is optional. So if you use any emulsifier, you have to put its name on the label. But according to the standards for "pasteurized process cheese," you must use some emulsifier. According to the FDA, since the use of an emulsifier is required, you don't have to put on the label the name of the emulsifier you used. You don't even have to say that you used one.[12]

This is not what the law says the FDA is supposed to do. This is what the FDA by administrative discretion chooses to do. It can, if it wants to, order that all ingredients, optional or not, be listed on the label. In fact, it does not even order that all optional ingredients be listed.

You may already have gathered, from reading about brominated vegetable oils, that the people who make soft drinks have a pretty powerful lobby around 200 C Street. You have gathered correctly. Part 31.1 of Title 21 of the *Code* will make your head swim; it's the standard, if it can be called that, for "soda water," which includes just about everything we usually call "soft drinks" if they're carbonated.

For one thing, the standard allows the following ingredients, clearly described as "optional":

any one of 12 "nutritive sweeteners" or any combination thereof (forms of sugar),
any one of a vaguely described but quite large number of fruit juices or natural flavorings,
any natural color additive,
any one or more of eight acidifying agents,
any one or more of 32 buffering agents,
any one of 17 chemicals loosely describable as emulsifiers (including brominated vegetable oils),[13]
another additive that helps the emulsifier(s) emulsify,
any one or more of six foaming agents, and
one antifoaming agent.

All are "optional." But *none* of them has to appear on the label.

Optional ingredients you do have to list, if you use them, are artificial flavoring or coloring, any of the 21 permitted preservatives, and quinine (the latter can be used in the name of the product). And there is one more optional ingredient given: caffeine. Now you have to watch the magician's hands very closely.

Caffeine is quite clearly listed as an optional ingredient in paragraph (b), subparagraph (8). Paragraph (d) then says in part:

> Soda water that contains the optional ingredient caffeine as provided for in paragraph (b)(8) of this section . . . shall be labeled to show that fact. . . .

No exceptions are allowed. So if there's caffeine in your soft drink, you can tell it by looking at the label, right?

Wrong.

A previous paragraph tells you that soda water flavored to taste like grape is to be called "grape soda," etc., and then follows an exception to what various kinds of soda water must be called:

> If the soda water is one generally designated by a particular common name, for example, ginger ale, root beer, or sparkling water, that name may be used in lieu of the name prescribed . . . For the purposes of this section, a proprietary name that is commonly used by the public as the designation of a particular kind of soda water may likewise be used in lieu of the name prescribed . . .

A "proprietary name" is a brand name, and that section means that if the name is commonly used by the public, Coca-Cola can be called just Coca-Cola. We do not choose the example by accident.

Now we go way back to the first paragraph, where the mandatory standards for all soda water are set forth, and we find this (the italics are ours):

> Soda water designated by a name, including any proprietary name . . . , which includes the word "cola" or a designation as a "pepper" beverage that, for years, has become well known as being made with kola nut extract and/or other natural caffeine-containing extracts, and thus as a caffeine-containing drink, *shall* contain caffeine in a quantity not to exceed 0.02 percent by weight.

There is only one soft drink we've heard of that is designated "as a 'pepper' beverage" in its name—Dr. Pepper—and there aren't all that many colas. But observe the effect of this section. It is now the law that if your drink has been around for a while and if it's called a cola or a "pepper beverage," it *must* have caffeine in it. For those beverages, caffeine is not optional, and therefore it definitely does *not* have to be listed on the label. And these are, of course, virtually the only beverages which do contain caffeine.

More than that: read the standard carefully and you will note that it does not say that the caffeine in the drink has to be there naturally from the kola nuts or whatever. As a matter of fact, the caffeine, or some of it, is quite often added artificially to make sure it's at the maximum.

We wonder how many people actually know that Coca-Cola (the firm that lobbied this sentence into the regulation), Pepsi-Cola, Royal Crown Cola, Dr. Pepper, and other cola drinks, diet colas included, contain caffeine.[14] What about the children? But the point here is the demonstration of how far we have come from the intent of the law which provides for food standards.

Congress passed a law more than 30 years ago authorizing the setting of standards to protect the customer. The FDA immediately began interpreting the law to protect the food business. And by the time "the Coca-Cola Amendment" was written into the soda water standard, we had reached the point at which the industry, with complete disregard for both you and the law, was actually writing the standards itself.

The standards law, read in isolation, is beautiful. If someone from another country asked you how our foods are protected, and you read the law to him, he'd think we were great. But the FDA, or any other governmental agency, for that matter, can—if Congress doesn't rise up in total anger—do just about what it wants to do with any law that is passed. The best of laws can be twisted to do the opposite of what they're intended to do, and they will be, so long as a $125 billion food industry is there to protect itself against a disorganized public. The best Commissioner of Food and Drugs you can imagine, or the best HEW Secretary, couldn't change things for more than a year or two.

And the law is, to begin with, far from perfect. One of its cozy provisions says that a product which uses a distinctive name doesn't have to meet the standards for any product it might closely resemble. Thus the distinctively named product Miracle Whip does not have to meet the standards for mayonnaise (it's not a mayonnaise, it's a "whip"), and Velveeta and Chateau do not have to meet the standards for the various kinds of process cheese products.

"Distinctive name" products do (unlike standardized products) have to list all their ingredients on the label, so that it might seem that we are actually better off for this con game. But it's not necessarily so.

The standard for mayonnaise, for instance, says that no ingredient can be used which would "impart to the mayonnaise a color simulating the color imparted by egg yolk." A product that evades that standard can fake its color. It can also use oils that don't meet the quality description given in the mayonnaise standard. Miracle Whip may not do these things—we don't know—but about yolk-colored dyes, the label will only tell you whether a color is used (not what color it is), and it won't tell you anything about the oils at all.

The latest "distinctive name" product to be driven by the truckload through this legal loophole is Gatorade. You might think it would be a nonalcoholic beverage. It's not. It's a "thirst-quencher." They aren't just using the phrase in the ads; they're using it to avoid a food standard.

Thanks to the standards law, then, certainly half, and probably as many as two-thirds, of the chemicals that are added to our food do not have to be listed on any labels. People who may have certain illnesses or allergies, or who just want to know what they are eating (which is enough to make them "food faddists" to the Mad Scientists), are simply out of luck.

Even people who know about standards, and know where to find them, can be out of luck, as Turner shows in giving an example using a meat-product standard set by the Department of Agriculture:

> The Department . . . requires that Beef Stroganoff consist of 45 percent meat, the rest being noodles, garnish and sour cream. Lipton's, with government acquiescence, adds much less than 45 percent beef to its Stroganoff product. It packages the Stroganoff ingredients in three subpackages: one contains noodles, one garnish, the third beef, sour cream and soy protein meal. Forty-five percent of the third package is beef. When prepared, Lipton's Stroganoff contains approximately 20 percent beef rather than the 45 percent required by the food standard.

The FDA, which has enforcement authority, could require Lipton's to promote a little more "honesty and fair dealing" on its labels. But nobody's getting hurt except the people the laws are supposed to protect.

Most people do not know about food standards, or if they do, know only enough to assume that standards are there simply to protect quality. To a small extent, they do that; you have to use cow's milk to make ice cream, and there are some quantitative minimums you have to meet.

But mostly their existence fools you. We shop at the Berkeley Co-op, where there are home economists who function to aid the customers. A letter to one of them said that the writer wanted to avoid monosodium glutamate, and asked how. The reply was, "Just read the labels." Fine—but it won't work for mayonnaise, French dressing, or salad dressing, in which monosodium glutamate is included in the standards but not in the labeling requirements. It may be in there or it may not.

The Manufacturing Chemists' Association, in its propaganda booklet, tried to distort the historical truth to make the industry look good, and slipped into telling more of the truth than it realized:

> Since 1906, the efforts of food industry members to protect the food supply have been backed by federal legislation.

It's only that the industry, backed by the FDA, has been protecting the food from us—lest we force them to make it fit to eat.

CHAPTER SEVEN

"Is She Not Pure Gold, My Mistress?"

The "efforts of food industry members to protect the food supply" go back at least to 1879, when government investigators found lead in the powders used to color coffee berries. There was no law against it at the time, and nothing the government could do about it.

Government bulletins in 1887 warned of the widespread adulteration of milk, whiskey, lard, and spices (charcoal in pepper was a common stunt). In 1892 they found chloride of tin in molasses, dangerous aniline dyes in confectionery, and all sorts of adulterants in coffee—including even fake whole coffee beans, a lot of which turned out to have been fabricated out of things like sawdust.[1] Copper salts were used as a coloring in canned peas at least that long ago, and acids were added to other canned vegetables.

There were a number of attempts to pass some kind of pure food and drug law (the drug situation was pretty frightening, too, but that's another book), and most of them revolved around a remarkable man named Harvey Washington Wiley. Wiley, for many years head of the Division of Chemistry of the Department of Agriculture—ancestor of today's Food and Drug Administration—was a stubborn, irascible, tough old guy who was totally devoted to his concept of serving the public.

Reared as a fundamentalist in pre-Civil War Indiana, Wiley often seemed more indignant about the public's being fooled than about their being poisoned; but in his combination of complete integrity, massive scorn and contempt for his opponents, and remarkable skill at political infighting, he was a lot like such younger men as Hiram Johnson and Harold L. Ickes.

Wiley's crusade for a pure food law, beginning before the turn of the century, was aimed more at fraud than at danger. He thought oleomargarine was excellent stuff, but he wanted it to be against the law to call it "butter." He thought that if you distilled the fermented mash of cereals, ran the raw distillate through a multichambered still to eliminate everything but pure ethyl alcohol, and added water, burnt sugar, and phony flavoring, you might get something that people would want to drink but you shouldn't be allowed to call it "whiskey." And he thought that artificial preservatives were used mainly to fool the public about the freshness of the products that contained them and in some cases allowed manufacturers to use inferior raw materials and inferior processing methods.

Which they did, as in some cases they still do. There was at least some point in what the food business did then, for little was as yet known about refrigeration, and processors relied on preservatives instead—mainly borax, salicylic acid, and formaldehyde (formaldehyde is cleared today as a preservative in defoaming agents—an additive to an additive). But Wiley didn't think it was fair to the public.

In 1902 the Department of Agriculture got the first authority to set standards (the authority was taken away in 1906 and given back in 1938). Over the almost total opposition of the food industry, they began in 1903 with standards for milk, meats, milk and meat products, and vegetable products. More followed during the next two years. As always, the food industry fought any restriction on its being allowed to do exactly as it pleased. They were particularly unhappy at the unheard-of idea that artificial coloring, artificial flavoring, glucose, or preservatives had to be listed on labels.

Glucose, incidentally, is only a kind of sugar, produced by the hydrolysis of cornstarch, but manufacturers didn't want to admit they used it, partly because it was known to be cheap (they didn't want the customers to know how cheaply the food was produced, nor that the sugar in the food wasn't "real"), and partly because they thought that the public thought that glucose was made out of glue.

Also in 1902, Wiley organized in his division what was to become world famous as his "poison squad." For some reason all of America at the time, and every writer on the subject since, has taken the "poison squad" to be one of the greatest things the American government ever did. It attracted worldwide attention and is partly responsible for whatever food laws we have. But it was deliberate experimenting with poisons on human beings all the same, and why no one recoiled in horror at the idea is a historical mystery.

At any rate, that's what Wiley did: he took a bunch of young volunteers and completely controlled their eating. Overall, the experiments lasted for five years. They consisted of feeding the subjects perfectly normal, healthy food, except that particular preservatives were added—first boric acid, then salicylic acid, sulphurous acid (prior sanction today for use in the production of caramel), benzoic acid (GRAS today for several uses), and formaldehyde.

Every report was condemnatory. Put in simple terms, the "poison squad" got sick every time. Several reports have been attacked, but none has ever been discredited on serious scientific grounds; the only genuine objection is that autopsies were not, of course, performed on the subjects. And most of the "scientific" attacks were by scientists in the pay of firms making or using preservatives, another forerunner of common practice today.

The widespread publicity given to the "poison squad" was enhanced by muckraking articles, in *McClure's* and *Collier's*, by Samuel Hopkins Adams and Mark Sullivan (today they would be writing "scare books," according to the Mad Scientists). Finally, Upton Sinclair's novel, *The Jungle*, dealing with meat packing, sparked a government investigation that bore out all of the novel's accusations, and the resulting public indignation was the last straw. Wiley got his Pure Food and Drug Act in 1906.

Not without heavy opposition right down to the end, opposition that sounds very much like the opposition to any tightening of regulations today. No less than three prominent university scientists testified on behalf of benzoic acid, a coal-tar derivative then just coming into use; it was later discovered that the three scientists were there under the aegis of one Elliott Grosvenor, who had been food commissioner of Michigan but who was then associated with a Detroit catsup-manufacturing firm.

From the minute the law was passed, the situation started to deteriorate as the food business moved in on the level of "administrative discretion." Almost the first action under the new law was to limit the amount of sulfur dioxide that could be used in processing dried fruit. California fruit interests promptly persuaded Secretary of Agriculture James Wilson that they wouldn't make as much money if the standard were enforced, and Wilson blandly told them to go on as they were and he wouldn't do anything about it. A Congressional committee (and Wiley, privately) screamed about it, but the decision stood.

Oyster interests, California wine interests, producers who used any kind of additive, all moved in on the department as hard as they could. Wiley tried to fight back, particularly with regard to bleached flour, saccharin, and benzoic acid (sodium benzoate, or benzoate of soda, are pretty much the same thing). Finally, President Theodore Roosevelt himself appointed a "scientific advisory panel," headed by Ira Remsen, to circumvent Wiley. There was no provision for the Remsen Board in the law, but it was appointed anyway.

Remsen, president of Johns Hopkins, claimed to have discovered saccharin, but that was obviously no impediment to his appointment. And exactly 50 years later, when the FDA was making up the GRAS list, it dismissed objection to the inclusion of benzoate of soda by noting that it had been approved in the report of the Remsen Board.

The Remsen Board approved benzoate of soda, by the way, on the ground that it hadn't been clearly demonstrated to be injurious. That was not the intent of the 1906 law, and it was clearly not a sufficient finding to meet the terms of the 1958 Food Additives Amendment. No matter. It's still GRAS.

By what appears to have been a mistake, Wiley's "poison squad" report on the dangers of benzoate of soda was issued, against the orders of the Secretary of Agriculture, in July, 1908. At about the same time, bacteriologist Arvill Bitting demonstrated that catsup could be made, and would keep for normal use, without preservatives, if it was made right and if top-grade ingredients were used. The H.J. Heinz Company, by the way, backed Wiley completely on this issue. They didn't use the preservative and didn't mind causing trouble for their competitors who did.

It was in this fight that Wiley slowly began to change his public position on fraud vs. safety, a change that had begun in his own head as he assessed the results of the poison squad experiments. He was now ready to argue that even proper labeling is insufficient protection for the customer; safety must be a primary consideration.

Wiley not only attacked the Remsen Board's 1909 report on benzoate for embodying too loose a construction of the law; he also attacked its scientific data. Counting in the report itself 196 instances of harmful effects, every one of them ascribed to some other cause, he said that he thought "the conclusions were drawn for the purpose of shielding the criminal and laying his sin upon the innocent."

But Wiley couldn't get around the fact that the Remsen Board's report *did* exist, and it was official. By about 1916, the Department, while it had "misgivings" about benzoate, believed itself unable to act unless experimental methods were refined or the law changed. Wiley began to campaign for a change, an amendment that would simply ban "any poisonous or deleterious substance" without the added phrase about it rendering food "injurious to health." That would have enabled the government to take a substance out of circulation if there was any question about its safety; but no such amendment ever succeeded.

Wiley was thinking not only of benzoates but of Coca-Cola, some of which his department had seized back in 1909. They charged misbranding because the syrup contained no coca and little if any cola, plus a poisonous ingredient, caffeine. In 1911 it was ruled that it wasn't misbranded because Coca-Cola was a "distinctive name," and that caffeine might be poisonous but it wasn't an *added* ingredient, it was an integral part of the syrup.

The Supreme Court reversed the last part of the ruling in 1916, and sent the case back to a lower court for determination as to whether caffeine is in fact poisonous or deleterious. Coca-Cola then changed its formula and lowered the amount of caffeine. Since the old court case would no longer apply to the current product, it consented to a judgment against the syrup that had been seized seven years before. As was the case with benzoates, the Department then wasn't sure whether it had a case against Coca-Cola that would stand up in court. Or that's what they said.

Wiley thought this was nonsense. If caffeine was injurious, he argued, then it was injurious. Nonsense it may very well have been. The political and legal clout of Coca-Cola has been repeatedly demonstrated (in the original seizure case, the U.S. attorney believed that jurors had been bribed).

Today, of course, Coca-Cola admits that it adds caffeine, and doesn't worry about it. It's in the standard. From an attempted prosecution in 1909, the government has come all the way around to *requiring* the presence of caffeine in Coca-Cola, despite continuing evidence that it's a harmful substance, including some studies that link it to heart disease.

Wiley left the Department of Agriculture in 1912 and worked for *Good Housekeeping* until 1926 (he set up that "seal of approval" which, while he was there, actually stood for a program of testing which at least tried to avoid taking advertisements for potentially harmful products). It was in that year that the difference between enforcing the law and exercising "administrative discretion" came before the public as clearly as it ever will.

Paul Dunbar, at that time, was assistant chief of Wiley's old Bureau of Chemistry.[2] He wrote an article explaining Bureau policy on the food laws.

Dunbar said that of course deliberate violations of the law would be speedily prosecuted, but that in most cases offenders were simply warned:

> It is the Bureau's theory that more is to be accomplished by acting in an advisory capacity under such conditions as will insure legal products than by accumulating a record of successful prosecutions with attending fines turned into the Treasury of the United States. . . . Its policy . . . is to pursue educational methods as a preliminary to legal action.

Wiley, furious, wrote a rebuttal with regard to the 1906 law that can be used today, with equal pertinence, with regard to the laws that are now in force:

> A few sentences of a year and a few $1,000 fines would do more than all educational endeavors with violators of the law to bring the offenders into respect and obedience of the law. . . . There is no warrant in the law, nor any suggestion of a warrant, that offenders should be called before the Bureau of Chemistry for the purpose of receiving instructions in ethics . . . Congress laid down the plan which was to be followed and the administrator of the law has no authority to deviate from the plain spirit and word of the law itself.[3]

But they went on doing it anyway, and they still do, while you literally eat the results. "If the Justice Department," writes Turner, "held regular meetings with the Mafia suggesting that it knew of gambling at certain addresses which if not stopped would lead to a raid on the premises, it would be following a procedure not unlike that used by the FDA to convince the food industry to obey the law."

Both Turner and Wiley's biographer, Oscar Anderson, Jr., make one thing quite clear: during the tenure of Harvey Wiley in the Department of Agriculture, an excellent scientific staff functioned with miserably low morale, administrative decisions continually overrode scientific fact, and, as far as enforcement of the law was concerned, the chickens pretty much built the chicken coop to suit themselves. And precisely the same situation, without a Wiley, obtains today.

One more man tried to repair the situation, Rexford G. Tugwell, named Assistant Secretary of Agriculture by Franklin Roosevelt in 1932. Tugwell determined that he would do something about the FDA, and even before he took office wrote in his diary that he would "do the best I can for the consumer regardless of politics."

The idea of doing anything at all with the FDA "regardless of politics" is almost funny. It took five years to get the new Food, Drug, and Cosmetic Act passed (it was drafted in Tugwell's department at his instruction), and if it had passed as it was originally written we'd be a little better off today. But he ran head-on into those same food manufacturing interests who so much enjoy telling us how they have fought for good food all these years.

Tugwell wanted grade labeling on food products so that you could look at any can of peaches, say, whether or not you have ever heard of the brand, and it would be plainly marked Grade A, Grade B, or Grade C, according to government quality standards. The proposal was scuttled by, among others, the Associated Grocery Manufacturers of America and the National Canners Association. What we've got instead is a law that foods which don't meet certain minimum standards have to be labeled as "below standard."[4]

Tugwell also wanted complete label identification of ingredients, and he lost that to the same opponents. To the argument that complete labeling was necessary for the protection of people suffering from allergies, the counsel for the Associated Grocery Manufacturers replied that allergy sufferers would be completely protected if the manufacturers filed a list of their ingredients with the FDA!

Tugwell tried to close the "distinctive name" loophole by means of which such products as Miracle Whip, Velveeta, and Ovaltine avoid product standards. He won only the compromise that "distinctive name" products have to list their ingredients on the label. We've already talked about what that means.

Finally, another burst of industry activity went into the Wheeler-Lea Bill, which, when it passed, transferred control of food, drug, and cosmetic *advertising* away from the FDA and put it under the Federal Trade Commission, thus effectively dividing the enforcement responsibility. That commission is, of course, a completely independent agency, so that there is not even a member of the cabinet to coordinate the work of the two organizations.

Tugwell was assailed as both a fascist and a communist as soon as he had proposed the legislation. One newspaper took to calling him "Rex the Red." Actually, he left the government before the Food, Drug, and Cosmetic Act became law. Turner comments wryly on Tugwell's efforts:

> Among the recommendations of the December, 1969, White House Conference on Food, Nutrition and Health were proposals for various kinds of grade labeling, the elimination of loopholes in food standards leading to uninformative labeling, and a more vigorous control over misleading advertising. These proposals were just what Rexford Tugwell hoped to achieve when he began planning in 1933.

And nobody tried to stop him except the food business.

Today, the Food and Drug Administration has 4,250 employees and a budget (fiscal 1970) of $72 million. Forty percent of its resources are spent on regulating the $5 billion-a-year drug industry. Considerably less is spent on the $125 billion-a-year food industry. One reason, probably the principal one, is easy to figure out: a $125 billion-a-year business has a hell of a lot more political clout than a $5 billion-a-year business.

The FDA is charged also with enforcing The Fair Packaging and Labeling Act of 1965. The entire staff working on that job consists of two part-time employees.

Very little research on foods is done by the FDA, either directly or by contract. Almost all of it is done by the industry, which is like relying on a White Citizens' Council to do research on racism. And of the research that is done, most people in the FDA can say nothing.

When I talked with Larry Cornell, who is assistant to the director in the San Francisco regional office of the FDA, he couldn't answer any questions about studies being conducted by the agency.[5] "We don't do any research at all here—the district offices are strictly for enforcement and for contact with industry. All the research is out of Washington, and they don't tell us what they're doing. Anyway, they farm most of it out to universities."

If you live in Berkeley, you know about universities. Scientists who get government research contracts do not come up very often with controversial results, or with results likely to displease the contracting agency. If they do, they don't get any more government contracts. It's a shame that it should be so, but everyone who has spent any time around a university knows at least one juicy story about what happened to somebody who got out of line on a government contract.

In the agency itself, of course, there are serious and dedicated scientists. And there are others, as Dr. Louis Lasagna noted a few years ago:

> [T]he most dangerous aspect of the FDA setup [is] the well traveled, two-way street between industry and Washington. Men from the drug industry have gone on to FDA jobs and—more important—FDA specialists have gone on to lucrative executive jobs in industry. . . . It does not seem desirable to have in decision-making positions scientists who are consciously or unconsciously always contemplating the possibility that their futures may be determined by their rapport with industry.[6]

We don't know why people have to keep discovering that over and over again; it has been the pattern in regulatory agencies since there have been regulatory agencies, as any reporter knows. No "defense" contractor worth his cost-plus contracts is without a retired general or admiral on the board of directors (General Lucius Clay is on the board of Continental Can along with Dr. Stare). The Federal Communications Commission issues rulings that favor the broadcasting industry with monotonous regularity;[7] the utilities have their way with the Federal Power Commission;[8] the railroads ran the Interstate Commerce Commission for years until the truckers got big enough to take it away.

Turner's study provides a graphic picture of the low morale among scientists in the FDA, caused by bureaucrats interfering with scientific research. Some scientists keep "atrocity logs" to record instances where scientific integrity is routinely violated. Turner also provides a number of examples of connections between the FDA and the industries it is supposed to regulate. One example is the 1954 appointment of George Larrick as Commissioner of Food and Drugs, an appointment made on the suggestion of the general counsel of Pillsbury Mills. The Pillsbury man then became Assistant Secretary of Health, Education, and Welfare, with responsibility for overseeing the FDA.

In July, 1971, Ralph Nader and Rep. Benjamin Rosenthal of New York protested President Nixon's appointment of Peter Hutt as FDA general counsel. Hutt has been counsel for the Toilet Goods Association, the Institute of Shortening and Edible Oils, and the cosmetics industry.

With morale in collapse inside the FDA, with various food laws hanging in shreds, and with the food industry virtually running the operation to suit itself (the food and drug industries are not all that separated; drug companies make additives), it's no wonder that there have been so many commissioners at the head of the agency over the last few years. Dr. Charles Edwards, the current commissioner (at this writing!), is obviously faring no better than his predecessors, some of whom were fine, tough men until they got the top job.

Nor is the new HEW Secretary, Elliott Richardson, likely to be able to do much to shake up the bureaucratic mess below him, even assuming that he wants to. The degree of pressure that can be brought on the Secretary can be deduced from watching, in the next chapter, the squirming that was done by the previous Secretary, Robert Finch, on the cyclamates issue.

And Finch, according to *Science* for December 13, 1969, had "exerted a direct line of authority over [FDA's] business that is virtually unprecedented." You can guess what that means if you're a Californian. Maybe *only* then, since the rest of the country seems to have gotten the idea that Finch is some sort of "liberal" within the Nixon administration. He is a liberal only by comparison with the worst Republican troglodytes. He has been an aide to Nixon since the latter's days in the House of Representatives in the late 1940s, and he wasn't too liberal to stay right alongside while Nixon was rivaling Joe McCarthy and doing the character assassination for the Eisenhower administration.

The idea that under Finch, or his successor, there might appear a commissioner with the zeal of a Wiley or an assistant secretary with the dedication of a Tugwell is simply ridiculous; probably, with the rise of the food industry to its present size and prominence, no such man could gain any headway again, under any administration at all.

For it is quite clear that the FDA is solidly and bureaucratically determined to go all the way with the industry. The agency has published a program outlining its objectives until 1974, and if you think they have any overall concern about the growing numbers of untested or barely "tested" food additives, try this quotation:

> The use of direct food additives in food manufacture will have approximately doubled by 1974 from the level of use prior to [1958]. Over 1,000,000,000 pounds of 2,500 food additives chemicals will be consumed. . . . The primary need at present is for the development of multiadditive detection methods to facilitate surveillance and measurement of actual additive intake levels.

In other words, they don't even blink at the idea that we will eat a billion pounds of this hidden nonfood, almost none of which has anything to do with making our food any better and much of which may actually make it poisonous. Their "primary need" is simply to figure out some way to determine how much of it we eat.

Don't be surprised. The love affair between the FDA and the people it's supposed to regulate is as public as it is illicit, reminiscent of the public figure who defiantly flaunts a mistress (even "mistress" may be too dignified a word). Blandly, the FDA avers that most manufacturers and processors not only obey the law, but go out of their way to do so; that the interests of the industry are really the same as the interests of the public; that everything is really all right.

Standards (which, as we have seen, are usually written to benefit industry if, in fact, the industry does not write them) are worked out in private conferences with trade associations, as the American Bottlers of Carbonated Beverages (a Coke- and Pepsi-dominated trade association) worked out the caffeine standard described earlier. The FDA expresses, in intraagency communications, the opinion that the biggest problem it has with food additive applications is that they aren't being processed fast enough.

The agency talks a great deal about "voluntary compliance," leading to Turner's being able to note that "thirty-four thousand representatives of five thousand regulated firms have attended 325 workshops and forty national conferences between 1965 and 1970," under FDA aegis. In contrast, *Science* notes that "an advisory council on food and drugs, on which consumers are represented, has not met to advise the Commissioner in more than a year."[9] Thirty-four thousand industry representatives; no consumer representatives.

Congress has obligingly gone along with this liaison on occasion, including that notable occasion when it forbade FDA employees to give away food manufacturers' "trade secrets" under penalty of a $1,000 fine or a year in jail. Since nobody knows what a trade secret is, this leads to a cozy game in which manufacturers put all kinds of information into the FDA files, and otherwise conscientious employees of the FDA (even though they may feel that the information demonstrates some danger to you and me) are afraid to tell us about it for fear of doing a year in the stony lonesome or coming up a thousand bucks short. At government salaries that's a healthy chunk.

And the industry, of course, returns the agency's affection. A trade magazine a few years ago made it exquisitely clear:

The most frequently heard comment [from industry sources] today about FDA's handling of the law is that it has been "entirely reasonable." Many of the companies praise FDA for its "able handling of a difficult set of regulations." "The provisions," says one industry man, "could have been hideously misapplied. They have not been. Much of the original anxiety was simply fear of the unknown."[10]

It's as though vampires were saying to each other, "Shucks, fellas, they're not going to use those pointed stakes after all." It is not the job of a regulatory agency to be regarded as "reasonable" or "able" by the people they're supposed to control. If they are so regarded, that ought to be evidence enough that they can't possibly be doing their job.

A perfect example of how the love affair works is provided by a regulation which has been doggedly allowed to stand since 1959, despite screams of outrage from heart specialists all over America. It says simply that

any claim, direct or implied, in the labeling of fats and oils or other fatty substances offered to the general public that they will prevent, mitigate, or cure diseases of the heart or arteries is false and misleading, and constitutes misbranding[.][11]

The regulation justifies itself by saying that "the role of cholesterol in heart and artery diseases" has not been established, which is a pretty poor reason when a great many heart specialists are convinced that there is such a role. But the effect is not only to prevent "false" claims; it prevents any possibility of labeling for fat content.

They say they're protecting you from misbranding, but that's not what they're doing at all. The NRC-NAS, the American Medical Association, the Heart Association, and a number of other organizations don't think it would be misbranding; they're all for fat labeling (and remember it's only labeling they're asking for, not a ban). What the FDA is really doing is protecting the meat industry and the National Dairy Council and Procter and Gamble and a few other people who don't want fat labeling. And inside the sacred walls of 200 C Street, they're quite honest about it.

In 1965, when the matter was up for review, an internal memo referred to "a considerable number of large firms who believe more time is needed," and the agency's legal counsel said that if it went to a hearing, it "would involve us in a controversy between the dairy industry and the corn and vegetable oil interests." In 1967, when it was reviewed again, the commissioner and some other FDA types had a quiet meeting with some Procter and Gamble people. No consumers or even independent cardiologists were present. The regulation was again allowed to stand. Instead, it was decided, an education program would be conducted—by the Grocery Manufacturers Association!

To show you how much it's all intended to protect *you,* the Justice Department refused in 1968 to file charges against some products which insist on telling you that they're high in polyunsaturated fats. The department said that it can't prove that you *shouldn't* be told, regulation or no regulation. Attorneys for margarine and other manufacturers who make the claims have told their clients to go right on making them. An odd posture for attorneys unless they're damned sure they can win.

But in the meantime, heart patients don't have low-fat cheese, pastry, or other foods to choose from, because the FDA regulation, written as a valentine to the meat and dairy industries, makes it pointless for other manufacuturers to develop those products for sale.

What the law intended was that the FDA should set standards and enforce them against the industry; they should educate, but they should educate the public, not pretend to educate businessmen who know as much about the law and how to avoid it as the FDA does. The FDA should require that *we* be "educated" through full and understandable labeling, through extensive consumer education programs, through any device that will help us to understand the hazards we face in all of the food we eat. Their only relationship with the industry should be to make damned sure that the industry does nothing—no matter how profitable, no matter how commonly practiced—that even looks as though it will increase those hazards.

Debate about whether FDA should take a "cop approach" is complete nonsense. They are cops. The law says so. And cops do not, and cannot, proceed on the assumption that those guys standing outside the bank with the tommyguns under their arms should be given a good talking-to. Much less do they hold joint meetings to decide on the best kind of ammunition.

There should certainly be workshops and national conferences, and there is no reason why industry leaders and the Mad Scientists they employ should not participate. But they should be held for you and for us. We, or people who are in some real sense our direct representatives as consumers, should participate equally and with the knowledge that the focus is to be on our benefit.

Instead, they entrust our safety to the Mad Scientists themselves, although even Republican Representative Craig Hosmer of California, hardly the most wild-eyed liberal in Congress, knows that the food business "is the lowest industry on the ladder in R&D [research and development] expenditures." Overall, the food business spent 0.4 percent of its net sales on research and development in 1964, perhaps the lowest percentage of any major industry in America. Twenty-two of the largest food manufacturers spent 18 percent of their sales on advertising, more than six times the percentage spent by automobile companies, who aren't cheapskates in the advertising world. From that you can get a pretty good idea of how much they care about your safety, compared to how much they care about your money.

Crucial questions about our health are entrusted to "scientific" panels of trade associations. The associations represent only the industry point of view, and they are completely dominated by the largest firms. The food business is as tightly controlled as autos or steel, and the proliferation of brand names to which we are accustomed is largely razzle-dazzle to conceal that fact. Turner points out, for example, that four firms produce 85 percent of our breakfast food, two produce more than half of our cheese, two produce almost all our salad dressing, and one company—Campbell—sells us more than 95 percent of all our prepared soups.[12]

In the meantime, the food business continues to treat us with contempt, as demonstrated by the caffeine standard for Coca-Cola, or the industry's use of mineral oil, which it doesn't like to have talked about either.

Until 1941, the FDA officially went around telling everybody that mineral oil is okay. No food value, but okay. Experiments, however, began to show that it *isn't* okay: it interferes with the body's use of certain vitamins, it all too frequently gets into infants' lungs, and in general it's pretty bad to take into your system, other than under doctor's orders for specific reasons and then never in conjunction with meals.

The industry to which the FDA entrusts our safety fought for five years, all the way through a long court battle, for the right to keep using this plainly harmful ingredient. Finally, the courts ruled that mineral oil can't be used as an ingredient in any food product.

But it's still a legal food additive. The Food Protection Committee (see Chapter Four) listed it in 1965 as a defoamer used in processing beet sugar and yeast.[13] It also lists it as a "lubricant and binder for capsules and tablets supplying small quantities of flavor, spice, condiments, *and vitamins*." We put in those italics. Here is this stuff whose use has been restricted after a five-year court battle because it interferes with vitamin intake, and the same industry that lost the battle is now blandly using mineral oil in combination with vitamins!

The FPC lists some other mineral oil uses which can and sometimes do result in residues in food. Among them:

> coatings for fresh fruits and vegetables,
> sealant in food production to prevent access of air and to retard evaporation,
> lubricant in food-processing equipment,
> lubricant in meat packing plants,
> release agent in drying pans,
> release agent and sealant in confectionery.

A "release agent" is simply something that makes the food come loose easily from a pan or baking slab; butter is often used on baking pans in home kitchens as a release agent, although people don't usually use the term. There are standards for how much mineral oil can be left in the food in each of the above cases, ranging up to 0.4 percent in confectionery, which hardly sounds as though the food industry is bending over backward trying to keep potentially harmful substances out of food. Surely there is something they can use for grease which isn't known to be hazardous to health. There just may not be anything quite so cheap.

But the FDA doesn't chase down things like this, nor even investigate them. Instead, when they decide to be cops, they either go around telling people not to talk about polyunsaturated fats, or they play private-eye with hidden microphones and tapped telephones and whatnot, trapping such suspicious hotbeds as health food stores, sellers of vitamin supplements, and other tiny and relatively harmless individuals they can label as "quacks."[14]

Some of them *are* quacks. In fact, some of them are outright frauds, milking the uncertain consumer and possibly damaging his health just so they can line their own pockets. They should be prosecuted.

True, the customer may be confused about nutrition and health because the FDA not only hasn't educated him, but has actually contributed to the confusion. Still, anybody who deliberately misleads customers, who encourages them to believe things that aren't true, who keeps them from learning other things that *are* true, and who does all that for private profit, should absolutely be prosecuted.

Even if that "person" is a key corporation in a $125 billion-a-year industry.

But the FDA only goes after *little* quacks, and while they're at it they go after a lot of little people who are not quacks at all. At worst they are in disagreement with official FDA edicts about what is good to eat, for which no one can conceivably blame them. At best they may be trying to provide the food products that will fill the growing deficiencies in the American diet, or trying to provide us with ordinary food which has not been injected, emulsified, bleached, stabilized, and otherwise faked out of existence.

It should be noted that the FDA actually gets after these insignificant defendants, not necessarily because their products are potentially unhealthy, but simply because the FDA thinks they are unnecessary. But even when they actually come up with a fraud, they remind me, at least, of the FBI.

The FBI regularly issues lists of the "ten most wanted criminals" and of the vast number of arrests it has made, or in which it has assisted. But none of the "ten most wanted criminals" has ever been anything but a minor, if sometimes vicious, individual criminal (and some of course may be innocent). And despite nearly 50 years of existence, the FBI has yet to arrest a single major figure in organized crime, or to make any inroads whatever on that massive group which the Attorney General doesn't want us to call the Mafia because it will offend Italians.[15]

The analogy is very close. The FDA's ten-most-wanted list would probably be ten perfectly healthy-looking young men with beards who run health food stores; and the agency would issue the list while kissing the feet of the Flavor Extract Manufacturers Association as it rushes to make sure that the use of brominated vegetable oils isn't interrupted.

The FDA has spent nobody knows how many hours and dollars trying to push through a requirement for what has come to be known in the trade as the "crepe label," a labeling requirement for vitamin and mineral supplements that would tell the buyer that he already gets enough vitamins and minerals and doesn't need the supplements. Specifically, the "crepe label" would read: "Vitamins and minerals are supplied in abundant amounts by commonly available foods. Except for persons with special medical needs, there is no scientific basis for recommending routine use of dietary supplements." This is pure hogwash.

The FDA and the food industry want desperately for us to believe that we are well-fed, that between them they provide for our welfare. If we find out that they don't, we may ask too many questions, and somebody's profits may suffer. So they insist, over and over again, not only that we don't have to worry about additives, but that the food "normally available to us" supplies all of our nutritional needs and keeps us in the pink of condition. If a doubt arises, they haul out their Mad Scientists, like columnist-board member Stare, to reassure us.

But it's getting more difficult. More and more of us are becoming aware of the stratum of the very poor, and of the much larger stratum of people just above, whose dollars cannot stretch to provide a decent diet as food is now sold in urban markets. And over at the Department of Agriculture, they're crossing up the food business by doing nutritional studies, such as the massive one just completed as this is written, and reported by columnist Sylvia Porter:

> Only half of all American households are eating a good diet today, a drop of 10 percent from 1955, according to a massive and alarming Agriculture Department study. Nearly one in 10 families in the $10,000-and-up bracket have diets rated as poor, and, over-all, one in five families have diets rated as poor, up 15 percent from 1955.[16]

And this, the columnist points out, is despite our spending more than $100 billion a year on food. The Mad Scientist's quick answer, of course, is that the good diet is available, but people just don't buy it. The answer to that is that in the first place it isn't true, and in the second place the massive load of nutritional misinformation and nonsense put out by the FDA and the food industry has a great deal to do with what people do and don't buy.

The FDA's mini-war on vitamin supplements and on health food stores isn't designed to combat fraud at all, although some of its enforcement personnel may be convinced that it is. It's designed to keep us thinking that "normal" food is all right (as if we ever saw any except in unusual cases). It's designed to keep us thinking that we can't know anything about our own health. This ties into the American Medical Association's position against health food stores, against over-the-counter vitamins, against anything that keeps us from going to the doctor if we don't feel well (nutrition is still not taught in most medical schools). It's designed to keep us thinking that the FDA is doing something about enforcing "safety," so we won't notice what it's not doing.

And of course it does occasionally turn up some real nut who thinks French-fried pistachio nuts provide a complete, nutritious diet and cure warts. This keeps attention focused on "quacks," so that anybody who disagrees with the FDA-AMA-food business "line" can be easily dismissed as a "quack" or a "food faddist" or even a "crook" (or, of course, a writer of "scare books").

When some activity like the mini-war on people who are at worst mini-offenders serves so many interlocking purposes, it becomes extremely difficult to look at it as coincidence, or simple self-importance, or an error in conviction. Saying it as we're about to say it is of course an oversimplification, but it is also closer to the truth: the industry *owns* the FDA, and the FDA does what the industry wants it to do.

Another constant theme which is becoming extremely tiresome in FDA propaganda is the one about their budget problem, and about all the things they could do if they could only afford more investigators, or more scientists, or more studies, or more anything.

In no year since 1960 has the FDA spent its complete appropriation. In every year since 1960 the FDA has given some money back to the Treasury. Since 1960 the amount the FDA could have spent, but did not spend, is more than $7 million. In 1963 alone it failed to spend $1 million of its budget. In 1966 it failed to spend $3 million. We really do not want to hear any more about the FDA's budget problems.

And in fact, we really don't want to hear any more about how good everything is. We are a little tired of statements like those in an FDA "fact sheet" issued in 1967, which included such charmers as:

> FACT: Chemical fertilizers are not poisoning our soil.

Or:

> FACT: . . . Today's scientific knowledge, working through good laws to protect consumers, assures the safety and the wholesomeness of every component of our food supply.

Like mineral oil, or brominated vegetable oils, or caffeine, or benzoic acid, or monosodium glutamate, or saturated fats. But as Turner says, "It is difficult for the agency to correct a problem that it claims does not exist."

And we are tired of hearing the FDA say that its positions and its regulations are not controversial. They are obviously controversial. The only point in saying that they aren't is so that anyone who criticizes them can be characterized as a food faddist or a quack or some kind of a nut. Or they can insinuate that the only motivation for criticism is either insanity or crooked profit.

If you start out with the idea, as the FDA publicly does, that the food industry is really devoted to our interests, then any conclusions you draw are going to be pretty far off the mark, as theirs are. It takes, at best, a reasonably decent third-grade education to figure out that the food industry in America, like any other industry, is interested in making as much money as possible as soon as possible by whatever means it can get away with. Historically, it has fought every restriction on its freedom to do whatever it wants to do, to use whatever ingredients it wants to use up to and including poisons, to sell whatever it wants to sell for as much as it can get. Its sole criterion has been its own profit.

Our interest is in good, healthy food, food that is what it appears to be, food the content of which we can easily determine, sold as cheaply as we can buy it. The seller wants to sell the cheapest stuff he can produce for the highest price he can get; the buyer wants the highest quality for the lowest price. That's about as basic as you can get, and an agency that proceeds on any other assumption is either nuttier than any food faddist you'll ever meet, or is in the pocket of the seller.

Dr. Herbert Ley, who was a fine scientist, became an ineffectual Commissioner of Food and Drugs under Finch and was finally replaced in December, 1969. On the last day of the year, he burst forth with the simple truth in a statement to *The New York Times:*

> The thing that bugs me is that the people think the FDA is protecting them—it isn't. What the FDA is doing and what the public thinks it's doing are as different as night and day.

CHAPTER EIGHT

The Hopeful Ostrich and the Hypersusceptible Mouse

Cancer terrifies us, of course, because there is no cure for it, and because we don't understand it. And because we don't understand it and most of us don't understand how animal experiments are done, or what their significance is, or how to evaluate a "scientific study," it has been fairly easy for the food business and the FDA to run games on us about food additives that are possible carcinogens. The various compounds of cyclohexyl sulfamate, better known as the cyclamates, are an excellent example.[1]

Artificial sweeteners have been in use in America since before the turn of the century. One of them, dulcin, was used for 50 years before it was definitely found to cause cancer; but by far the most common was saccharin. There has always been some controversy about its safety, too. Dr. Wiley wanted to ban it, but President Theodore Roosevelt vetoed the idea because his own doctor had prescribed saccharin for him, and the President was convinced that it must therefore be safe. It remains today the only legal artificial, nonnutritive sweetener. All the others are now banned, for a number of medical reasons.

Most Americans didn't and still don't like saccharin. It leaves a bitter, metallic aftertaste, and except for a small number of people who were told that they must use artificial sweeteners for medical reasons, there weren't a lot of customers for a full line of "diet" foods and drinks. In 1950, Abbott Laboratories thought it had an answer to this problem, and applied to the FDA for permission to market a new drug for use "in foods and beverages by diabetics and by others who must restrict their intake of sugar." They called their drug Sucaryl Sodium; it was the first of the four varieties of cyclamate to come into common use in America.

There was no 1958 Food Additives Amendment then, and Abbott did not propose the use of cyclamates as food additives but as drugs. They supplied some data which they said demonstrated the safety of cyclamates, and the FDA's Dr. A.J. Lehman promptly said that the application was "an illustration of how an experiment should not be conducted." There were not enough animals, not enough autopsies, not enough use of control groups, and too much vagueness about the whole thing.

The FDA decided nevertheless to allow Abbott to go ahead anyway, because FDA laboratories had done feeding studies with cyclamates. Dr. Lehman said at the time that "it is on the basis of our own work that recommendations are being made to permit this application to become effective."

This is astonishing because the FDA's own two-year feeding studies showed "a highly suspicious frequency of lung tumors," and a number of rare tumors of the ovaries, the uteri, the skin, and the kidneys that cropped up in the rats about 600 times as often as they should have.[2] These warnings were ignored (and the studies have never been discredited).

In 1954, the Food and Nutrition Board of the NRC-NAS (not the same as the Food Protection Committee, though certainly not antiindustry) examined the known data about cyclamates and said in a policy statement, "The priority of public welfare over all other considerations precludes . . . the uncontrolled distribution of foodstuffs containing cyclamate." They were, they said,

> impressed with the fact that cyclamate has physiologic activity in addition to its sweetening effect, that there is no prolonged experience with its use, and that little is known of the results of its continued ingestion in large amounts in a variety of situations in individuals of all ages and stages of health.

In 1955 the Food Protection Committee itself referred to "possible deleterious effects," and the Food and Nutrition Board followed that up by repeating its earlier warning.

Also in 1954, by the way, scientists at the Harvard School of Public Health reported a study on the use of cyclamates by diabetics and by people suffering from obesity. They said that use of cyclamates instead of sugar in "normal" amounts had no effect whatever on weight control and that there were at least serious questions about their use by diabetics. This study is particularly interesting because the head of that same school's Department of Nutrition, Dr. Frederick Stare, who speaks so loudly and so often for the industry, ignored his own school's study 16 years later when he happily noted in a column that cyclamates would still be available for diabetics or those desiring to lose weight.

Until the passage of the Food Additives Amendment in 1958, however, it was possible to regard the use of cyclamates as a serious but not overwhelming problem. They were presumably limited to special dietary foods, and manufacturers were required to label such foods to indicate that they "should be used only by persons who must restrict their intake of ordinary sweets." There was nothing, of course, to prevent other people from buying them, but the major uses of cyclamates were still to come.

The GRAS list provision of the 1958 law gave Abbott, and others who were to become cyclamate manufacturers, the chance they wanted. Despite all the previous studies, showing a number of potential dangers from cyclamates including their identity as possible carcinogens, cyclamates were included on the list circulated to the 900 scientists (so was saccharin). One responding scientist complained, but the FDA ignored him and said on August 10, 1959, that "safety has adequately been established for" both cyclamates and saccharin compounds. Despite the fact that the old studies were (and are) still valid, and that there was no new evidence for the safety of cyclamates, they became GRAS from the beginning, and thus never had to be tested any further.

Other substances, also potentially dangerous and much more widely used at the time, were also GRAS-listed, and drew most of the criticism when the list was issued. But once cyclamates were GRAS as food additives, the food business was of course ready and eager to find new uses for them.

Most familiar to Americans is the diet-drink business, which mushroomed into existence with the appearance of Diet Pepsi, Tab, Wink, Like, Fresca and dozens of other products. A previously insignificant branch of the "food business," it has now risen to a billion-dollar-a-year industry. But other uses of cyclamates, unsuspected even yet by most Americans, also became common. The Agriculture Department allowed the curing of bacon and ham with cyclamates, with no label declaration. Vitamin tablets for children have long been coated with cyclamates for sweetness, again without notice to the purchasing parent. FDA standards for artificially sweetened foods did not require that cyclamates be mentioned on the label.

We should, perhaps, point out again that the GRAS designation has nothing to do specifically with cancer. The Delaney Amendment is about cancer, but the GRAS designation means that the product in question is generally recognized as safe *in all regards.*

And the evidence about cyclamates, prior to their banning, did not only concern cancer. There is clear evidence that they are probably both mutagenic and teratogenic—that is, that they cause both long-term genetic changes and immediate birth defects. We'll talk more about mutagens and teratogens in the next chapter, and that evidence about cyclamates will be discussed more fully there.

The point here is that the FDA knew perfectly well, unless it chose to allow itself a frightening stupidity, that cyclamates were not "generally recognized as safe" at all, either in 1959 or later as more evidence began to pile up. Extremely important in this regard was a Japanese experiment in 1966 that showed that cyclamates can, in the human body, be transformed into cyclohexylamine (also sometimes called CHA).

Cyclohexylamine is damned dangerous stuff. Prior to the Japanese experiment, it was used in some food processing, with an FDA-imposed limit of ten parts per billion in boiler feed water and a provision that the water must not be allowed to come into contact with milk or milk products. CHA is extremely toxic and is known to speed heart action, raise blood pressure, interfere with the action of other drugs (including those taken by diabetics), produce chromosome damage in animals, produce a high rate of stillborn young, decrease the rate of growth of young animals, and produce birth defects.

Although Turner says that the FDA regulations for cyclohexylamine were set in 1958, "long before the relationship between CHA and cyclamates was known," *some* relationship must have been known, because the manufacture of cyclamates starts with cyclohexylamine.[3] Following the 1966 Japanese experiment, it was found that about 30 percent of humans convert as much as 40 percent of their cyclamate intake into cyclohexylamine.

This is important because the Mad Scientists insist on arguing that cyclamates mostly pass through the body unchanged and thus don't do any harm. The discovery that this isn't necessarily so—certainly not for everybody—is put in everyday product terms by Turner:

> For example, one package of artificially sweetened Kool-Aid contained 28.5 percent cyclamates, which convert to 3,200 parts per million of CHA in a significant portion of the population.

Compare that to the ten parts per million which the FDA formerly allowed, in a usage where the CHA wasn't allowed to touch the food!

By 1969 it had been amply demonstrated that cyclamates cause diarrhea in children, that they block the action of certain antibiotics, that they are at least possibly dangerous to diabetics taking certain oral drugs, that they probably keep blood from coagulating properly in case of injury, that they somehow affect liver function and the intestinal tract, and that they were an almost certain cause of birth defects and genetic changes in the animals usually used to test such effects. Cyclamates are more readily absorbed into the body, instead of "going on through" as the industry scientists claimed, when they are taken at the same time as fats, caffeine, or citric acid (the last two of which are widely used in diet drinks). And once absorbed into the body, they are distributed to babies through breast milk and to fetuses across the placenta.

But by 1969 the FDA wanted no part of another controversy like that over thalidomide, the drug which had been banned after it was found to cause gross birth defects when taken by pregnant women. Thalidomide, at the time of its banning, was a big story in America, but it had not yet become big business. In 1969 Abbott Laboratories alone was devoting about 4 percent of its business to the manufacture of cyclamates, and raking in about $14 million a year in profits. Besides, the FDA had been publicly embarrassed by exposure of its own sloppiness in dealing with thalidomide, and it didn't want to get caught again.

Which it could have. Biochemist Jacqueline Verrett reported teratogenic effects from cyclamates to an FDA seminar in March, 1968, and in her year-end report nine months later told her FDA superiors that "it has been found that calcium cyclamate [and cyclohexylamine] are specific teratogens, having the ability to produce phocomelia and similar defects in the embryos [of chickens]."

Phocomelia is the name for the kinds of gross and horrible defects (flipper-like appendages instead of arms, for example) that were also caused by thalidomide, and the effect from cyclamates was present in more cases than had been the effect from thalidomide. But the FDA did nothing, as it had done nothing in the thalidomide case until its hand was publicly forced. Now, once again, it was sitting on the reports of its own scientists.

Dr. Verrett continued to report these effects, and Dr. Marvin Legator, also of the FDA, began to report in 1968 on chromosome breakage from cyclamate use. Early in 1969, Dr. Legator wrote a memo to Assistant Commissioner Daniel Banes on the effects of cyclohexylamine in humans—a report derived from a study on prisoners which was intended to test something else, but in which it was found that cyclamate intake led (as the Japanese had found previously) to cyclohexylamine production in the body. Dr. Legator found clear evidence of both mutagenic and carcinogenic properties, and concluded:

> The use of cyclamates should be immediately curtailed, pending the outcome of additional studies.[4]

A funny thing happened on the way through the office. Dr. Legator's memo went, of course, through channels, and somewhere along the way, someone didn't like the conclusion. So the sentence quoted above was *taken out*, and the altered memo sent on to Dr. Banes, still over Dr. Legator's signature. Neither Legator nor Banes, however, was told that the memo had been changed!

There things might have stayed, except for the fact that a few members of the press focused on the increasing scientific discussion about cyclamates and wondered whether there wasn't something going on that might be as important as the big thalidomide story of a few years before. Dr. Legator had presented some of his findings as early as October, 1968. After his memo was altered early in 1969 without his knowledge, he began to prepare his material for technical publication, and in the meantime word began to get around.

On September 4, 1969, Maria Torre interviewed an Abbott scientist, Dr. Claire Dick, on KDKA-TV in Pittsburgh, Pennsylvania. Dr. Dick, while generally touting the safety of her employer's product, admitted that "no pregnant woman should consume any kind of chemicals unless she's advised to by her doctor" (as if it were possible to avoid it in America).

Dr. Legator's findings then appeared in *Science* for September 21, 1969, and were alertly spotted by *Newsweek*, which ran a story in its September 29 issue, and by Paul Friedman of NBC News in Washington, who queried the FDA and was put in touch with Dr. Verrett. She in turn went by the rules and consulted her superiors; Deputy Commissioner Winton Rankin authorized that she be interviewed. On the air, Dr. Verrett, too, said that pregnant women should avoid cyclamates, and said that "an effect on reproduction has been demonstrated."

Note that none of this growing press interest had anything to do with cancer. Any reporter with 15 minutes' experience could see the story potential in the possibility that a chemical which all kinds of Americans ingested in their Tab or Fresca, in their "diet canned peaches" or in their bacon, caused birth defects more widespread than those associated with thalidomide in the sensational story of a few years before. You can almost see Commissioner Ley and Secretary Finch squirming, because you can be sure that, by now, they knew that the FDA had been ignoring the evidence about cyclamates for years, and that it would be their administration that would be stuck with the scandal.

Fortunately for them, but not for us, cyclamates also cause bladder cancer in rats.

With what motivation no one knows, Abbott had started its own rat feeding tests with cyclamates, contracting with the Food and Drug Research Laboratories of Maspeth, Long Island, to do the actual studies.[5] This was extremely unusual behavior for an additive manufacturer whose product is on the GRAS list and doesn't require testing—especially years after the product has gone into general use. Maybe it was those Japanese tests; it's tempting to suggest that they might have been looking for something to offset the facts that were sure to break sooner or later—but there is no evidence about Abbott's actual motivations.

Abbott claimed to be looking for mutagenic and teratogenic effects and said they didn't find any. In any case, it happened, possibly by total coincidence, that a few days after the *Newsweek* story on Dr. Legator and Dr. Verrett's appearance on television, Abbott ordered the Long Island rats killed and autopsied. As the Abbott scientists tell it, it was to everyone's complete and total amazement that they found bladder tumors in the rats.

With commendable rapidity, the results were shown first to Abbott scientists, then to National Cancer Institute investigators, then to the FDA. A six-man scientific panel reviewed the data on Thursday and Friday, October 16 and 17, and on Saturday, October 18, 1969, Secretary Finch and Surgeon General Dr. Jesse Steinfeld called a press conference and announced that, because they were forced to do so by what Finch called "the so-called Delaney Amendment," they were taking cyclamates off the GRAS list.

In the course of the press conference and in subsequent statements, Finch, Steinfeld, Abbott's spokesmen, and various Mad Scientists proceeded to confuse the American public so much that nobody ever really grasped the extent of the scandal which was averted by the fortuitously timed discovery of the bladder cancers.

Obviously about to break was the major story, a story that would have dwarfed the thalidomide furor, telling America that a product was being sold to them in billion dollar-a-year quantities which might very well cause more gross birth defects and genetic damage than thalidomide ever could, and that the Food and Drug Administration had sat on the evidence while big drug companies continued to reap millions in profits. The public would have seen Abbott and the other firms as cynical profiteers, and the Nixon administration, Finch, and the FDA as lackeys to their profits.

Instead, Abbott came out as heroic. They discovered the cancers themselves and told the FDA, right? And the FDA and Finch came out as having acted within the shortest possible time to protect America, despite the "fact" (as they continually repeated) that the danger was probably nonexistent and that only a clumsy and restrictive law forced them to deprive you of your Fresca.

The fact remains that evidence that cyclamates are dangerous had been in FDA hands for years, and they had not acted as the law clearly requires them to do; that they used the Delaney Amendment to get out of their clear responsibility for allowing a probable mutagen and teratogen to circulate in vast quantities; and that when they did act, they lied in their teeth.

The line that seems to have reached most Americans was the one put forth by, among others, Dr. Frank R. Blood of Vanderbilt, a member of the six-man panel that recommended the ban:

> [The amount of cyclamate fed to the rats in proportion to their body weight] is 100 times to 120 times greater than even the highest cyclamate users could consume. We recommend the cyclamate ban because of the law, not because there is any reason to believe it causes cancer in man.[6]

Those figures are incorrect, but in haste that can happen. More important, the statement distorts the relationship between animal experiments and their possible application to humans, as we will shortly see. Not all doctors are animal experimenters, of course; but the FDA knows better, and it was their job to prevent such distortions.

Finch's behavior at the October 18 press conference was abominable. If it was because he was misinformed, we can only reply that it was his business *not* to be misinformed. Any information we can find, he could have found.

"Who's to say," he asked ingenuously at one point, "that using Fresca or some other diet drink . . . isn't better for you than the problems of overweight or diabetes?" Cyclamates had clearly been shown for years to have no effect whatever on weight control and to be possibly dangerous, in combination with other necessary drugs, to diabetics. In fact, only three days before Finch's statement, Assistant Surgeon General Arthur S. Wolff had pointed out in an internal memo that diabetes "appears to be etiologically associated with a higher prevalence of congenital defects as well as still births and neonatal mortality"—that is, the kind of teratogenic activity that cyclamates appear to have is specifically dangerous for diabetics!

And Finch even descended to castigating his own scientists, particularly Dr. Verrett for her television appearance and Dr. Legator because his scientific publications were picked up by an alert press:

> I was unhappy and expressed my unhappiness about the . . . doctors in the Food and Drug Administration who chose in the case of the eggs to go directly to the media without having consulted with their superior and with the office. This is not a procedure I approve and certainly they did not act in a very ethical way.

That is an extraordinary thing for a cabinet member to say, especially when his facts are completely wrong. Finch has yet to apologize publicly to the doctors he called unethical, but it is about time he did.

Dr. Steinfeld, as Surgeon General, cannot even be allowed the flimsy excuse of misinformation. At the same press conference, he came up with this authoritative-sounding gloss on the facts:

> We have no indication that human bladder cancer from whatever cause is increasing to any significant degree. Our data to this effect are obtained from studies underway for at least two decades in the state of Connecticut. It is the only good source known to us for such data, and it was brought up to date two days ago by what we have obtained in the way of data up to now.

Very strange. Because 12 days *after* that statement, the National Institute of Cancer reported that, according to the same Connecticut study, bladder cancer had in fact *doubled* in Connecticut between 1945 and 1965!

Aside from that rather grotesque error, however, Dr. Steinfeld must know that the type of bladder cancer in question takes at least ten years, and more often nearer 20 years, to develop in humans. Cyclamates were put on the GRAS list in 1959, and the food business developed its billion-dollar widespread use of cyclamates only after that. In other words, cyclamates had been used widely for only about ten years. Whatever the results of the studies in Connecticut, it is simply too early yet to know whether the use of cyclamates has led to a sharp increase in bladder cancer in America; even though they're now banned, it may be another ten years before we can even begin to measure their carcinogenic effects.

Steinfeld has gone on issuing misleading information about cyclamates. In *Science* for February 20, 1970, he joined Commissioner Ley and some Abbott scientists in assuring readers that the FDA was forced to act against cyclamates by the Delaney Amendment, and that the bladder cancer experiments involved extremely high dosages. As a matter of fact (and as Dr. Steinfeld must certainly have known by then) further studies had already been done, showing that cyclamates cause bladder cancers in much *lower* dosages, and in other strains of rats.

But HEW and the FDA were by no means through squirming. While Finch talked darkly about repealing the Delaney Amendment, almost the only *real* protection Americans have against poisons in their food, the astonishingly powerful fruit interests in Finch's own California, concerned about their summer canning schedules, decided to let their former lieutenant governor know that he had gone a little too far to suit them.

Finch's original order on October 18 was that soft drinks containing cyclamates must be out of the stores by January 1, 1970, and artificially sweetened foods ("diet" fruits and vegetables) by February 1. On a quiet weekend in November (government agencies traditionally issue statements on weekends when they don't want the press to take too much notice of them) it was gently announced that the fruit and vegetable deadline would be moved from February 1 to September 1, 1970.

The announcement that we could be poisoned for seven more months was hailed in California, where R.L. Gibson, president of California Canners and Growers, called it "the best possible news for the canning industry." Gibson helped the confusion along by telling a meeting of the giant organization that "the amount of cyclamate used in tests which showed tumor developments in rats was so massive that it has no relation to normal consumption in humans."

Inside HEW a fight was going on between FDA scientists, who hoped (and still hope) that the Delaney Amendment might be not only retained but strengthened, and Finch, who (cheered on by the industry, if not acting at their urging) wanted to repeal it. Also in November, Finch appointed a "scientific" panel to study the cyclamate situation further—actually and obviously an attempt to find a way around the law.

The panel, as might be expected, found a way around the law. In fact, they found two, although one way was never taken seriously and may not have been intended to be. They recommended that cyclamates could either be made prescription drugs, or could be restored to use as nonprescription drugs with a label requirement, like the one used before 1958, warning buyers that cyclamates should be used only under medical supervision.

Calling cyclamates "drugs" instead of "food additives," while doing exactly the same things with them, would not of course change anything in the real world. Obviously the labels would be about as prominent as the ingredient labels now are on the canned and bottled goods you buy, and the salesmen in the billion-dollar business were going to go on telling you how great all this junk is. Never mentioned was the fact that the Delaney Amendment is there to protect the American people from cancer, and not to be circumvented by semantic games.

Still posing as a stalwart defender of the law, Finch made another announcement in February, 1970. Cyclamates, he decided, *could* be used as nonprescription drugs, added to foods, if they were labeled. Finch picked up obesity and diabetes as excuses for bailing out the industry, despite, again, the evidence that cyclamates are useless and possibly dangerous in those cases. Soft drinks were still to be banned, and unlabeled foods containing cyclamates still had to be off the shelves by September 1.

By this time the industry propaganda machine was working frantically to minimize the damage, and Helen Gurley Brown was gushing to the readers of *Cosmopolitan* that she was so depressed by the whole thing that she'd rushed out and bought just *cases* of cyclamated soft drinks and "diet" foods. Finch's assistant for health and scientific affairs, Dr. Roger Egeberg, told an audience at Harvard that he'd advise his daughter to do the same thing (we wonder whether he'd advise her to take a little thalidomide, too, while she's at it).

Libby sent its cases of diet liquid off to Laos, and Lipton donated thousands of cases of cyclamated soda water to mental hospitals and prisons in Ohio, where they were accepted without a murmur of protest, cancer apparently being okay for the mentally sick and cons. Finch even thought it was so funny that he wrote a poem about it for reporters:

Damage in brains, mutations and cancer,
Who in the world can give us the answer?
Chromosome breaks, chick malformation,
Pot and pollution and desegregation,
So revising Delaney should just be a cinch—
But why in the hell must it always be Finch?[7]

We could die laughing.

Not laughing was a Congressional subcommittee on intergovernmental relations chaired by a feisty North Carolina Dixiecrat named L.H. Fountain. Fountain blasted the "drug" designation as a subterfuge—which of course it was—and announced an investigation. Consumer groups were suspicious of the new rules, and another of Finch's assistants, Elliot Richardson, took the scientists' side on Delaney. The pressure was beginning to come from the other direction.

By June, and not entirely because of cyclamates, Finch was out as secretary, and Richardson replaced him. FDA Commissioner Ley had in the meantime given way to Commissioner Charles Edwards. Suddenly, a different line was leaked to reporters. On June 23, the Associated Press reported blandly, and without giving a source, that the FDA was "reconsidering" the cyclamates question, and that Edwards "will reconvene soon a medical advisory group" to study the whole thing.[8]

So he did—and in a magnificent display of telling the boss what he wants to hear, the same panel, which in November told Finch that cyclamates were really okay if they were labeled, now told Richardson that cyclamates aren't so hot after all. Somehow the panel discovered the evidence that cyclamates don't do anything for weight control, and that there were studies showing cancer even at relatively low dosages.

On August 14, 1970, the Food and Drug Administration firmly banned *all* products containing cyclamates, ordered them off the shelves for sure on September 1, and gave up on its wishy-washy "drug" classification. Said Representative Fountain:

> It is unfortunate . . . that the American people, as well as the affected industries, were confused and misled for the nine months or more that it took FDA and the Department of Health, Education and Welfare to face up to their responsibilities in this matter.

Well, yes. Except that the "affected industries" were about as "misled" as Falstaff was by Prince Hal, and the FDA and HEW never did face up to the responsibility of telling Americans that they had condoned for years a clearly unsafe substance, in defiance of the law, before political and other pressures forced them to take the least damaging way out.

So much for cyclamates, at least at this writing. But the industry propaganda campaign is not finished by any means. They were willing enough to use the Delaney Amendment to get out of a bad spot, but the Amendment is still a restriction on their arrogant determination to feed us anything that brings a profit, and they still want to get rid of it.

Millions of Americans, therefore, are still convinced that a "rigid and unnecessary" provision in the law has somehow deprived them of the Tab and Fresca and Diet Pepsi that they loved.[9] The press has not even begun to make clear that the amendment may have protected us, even though it's concerned solely with cancer, from millions of cases of stillbirth or deformity. We will probably never know how many such cases we weren't protected from, thanks to FDA and HEW permissiveness to industry.

The food business, almost needless to say, didn't want the Delaney Amendment in the first place, but it certainly shouldn't bother anybody except a manufacturer who wants to sell you something that may cause cancer. The amendment became law over the official, and very loud, protests of "your" Food and Drug Administration itself. Congressman Delaney had to threaten that, as a member of the Rules Committee, he would block the entire Food Additives Amendment; only then did the FDA give in and agree to accept an anticancer clause.

Helpful as the Delaney Amendment is, it isn't nearly so good as Mr. Delaney wanted it to be. He, and a number of cancer experts who were backing him, wanted a law that would not only ban known carcinogens, but would require that all food additives be tested to see whether they are carcinogens. They also wanted some sort of specific criteria that additives had to meet, instead of a vague provision that leaves it up to the FDA to pick its own evidence.

A Delaney Amendment "with teeth" would immediately cut down on the number of useless substances thrown into our food for no other reason than to fool us about its content, or to enable a wholesaler to keep it around longer. While nothing can be proved *safe,* it would at least require fairly extensive—and expensive—testing of every additive, and the banning of anything that showed suspicious results.

Congressional testimony by experts (real experts) has indicated, for instance, that the results of tests can be controlled. Carcinogens fed to animals may prove to be inactive if certain other substances are added to, or kept out of, the animals' diets. Tests can be done on several strains of animals, with negative results in some cases and positive results in others; then only the positive results may be submitted. The choice of test animals can have a marked effect;

hamsters, for instance, are much less susceptible to most cancers than other animals, and a manufacturer may submit glowing reports from hamster tests without making any tests on rats or dogs. Not all manufacturers would do those things, of course; but you'd have to be starry-eyed indeed to believe that none of them would.

Furthermore, the Food Additives Amendment requires that new additives either be approved, on the basis of the manufacturer's evidence, or that cause for disapproval be shown by the FDA within 180 days. There is no way in which such technical information can be evaluated, much less verified, in so short a time. In effect, the FDA is barred from checking on whether even the manufacturer's own evidence is good enough.

Then, even if the FDA *does* disapprove, the manufacturer can go to court, where a judge, necessarily ignorant in the field, may well decide the case on the basis of who has the cleverest lawyer. And you know who has the most money to pay clever lawyers.

In 20 years of covering both politics and science, I've had ample opportunity to learn that attorneys and judges, accustomed to the concept of *legal* proof, rarely understand the entirely different nature of *scientific* proof (and remember how many legislators are lawyers). Much of the public has the same difficulty.

Legal proof is a positive demonstration that something is, in fact, true. In criminal cases, it must be true beyond a reasonable doubt; in civil cases, it must be shown to be true by a definite preponderance of the evidence. But in either case, if you can't prove that it *is* true, then legally it isn't.

In science, on the other hand, nothing is "true" in that sense. For something to be "true" means that there is a very high probability of its truth, and nobody can prove that it's false. A lawyer or a Congressman or a judge, confronted with a question of the safety of a food additive, is likely to think in terms of proving that it's safe. That can't be done. The scientist thinks in terms of coming up with the most rigorous possible experiment to prove that it's *not* safe. Then if it passes that experiment, he calls it "probably safe," until somebody comes up with a better test.

That is the best that science can ever do. In the meantime, to arrive at its working hypotheses (the ideas on which it acts about things that *appear* to be true), science uses methods that law generally will not and cannot accept, like mathematical extrapolations and statistical correlations.

Nobody has yet proved that cigarettes have anything to do with lung cancer, as the tobacco companies are telling us over and over again whenever they can get us to listen. But they're trying to confuse us by talking about *legal* proof. It has been shown that heavy cigarette smokers get a lot more lung cancer than other groups in the population.

If that kind of correlation runs fairly high, science calls it a "significant correlation" and adopts a connection between the two facts as a working hypothesis. If the correlation runs extremely high (for instance, if all cigarette smokers got lung cancer and nobody else did), scientists will usually call it "proof," even though the law may regard it technically as only an interesting and possibly indicative coincidence. The higher the correlation, the more convinced scientists will be. In law, you'd have to show exactly that there is a physical connection between the smoking and the cancer; in science, you only have to show that a very high probability exists, even if you don't know anything about the connection at all.

One of the best examples I can remember came during the debate in the late 1950s on nuclear testing. Geneticists generally were arguing that any increase in the amount of radiation to which we are exposed will result in a corresponding increase in genetic damage (that is, both mutagenic and teratogenic damage). A great many Congressmen and other public figures, however, untrained in science and unable to shake off legal concepts of proof, were honestly convinced by the argument that genetic damage from radiation below a certain level had never been flatly demonstrated.

It can't be demonstrated. It's impossible to look at one high-speed proton knocking one electron off the shell of an atom in one gene. We know about it because we know in theory how it works, we can see the effects in large numbers, and we can extrapolate downward to smaller numbers. According to the theory, there is a very, very high probability that the mechanism doesn't change just because the amount of radiation gets smaller. But there is no way to "prove" it in the legal sense, and it is meaningless to talk of such "proof" in science.[10]

There is no such argument in the law, and no such use of statistical or other mathematical methods. Possibly we will never have really meaningful food laws until there are enough congressmen who understand the distinction between law and science when words like "proof" are being used.

No such confusion, of course, underlies the continuing and steady effort of the food business and its Mad Scientists to get the Delaney Amendment repealed. They know perfectly well what they're doing. When Dr. Steinfeld tells you that "there is absolutely no evidence to demonstrate in any way that the use of cyclamates has caused cancer in man," he isn't lying—as far as Perry Mason concepts of "evidence" are concerned. But he knows that a scientific demonstration is something else again, and that the Delaney Amendment doesn't require that kind of proof anyway; it requires only that any scientific question be resolved in favor of your safety instead of the industry's profits.

The Mad Scientists will tell you, nonetheless, that the clause is "unscientific." Any test, they say, no matter how poorly done, can force an additive off the market if it appears to show cancer. That's just not true. The FDA and HEW, with their continuous history of bending over backward for industry, would have to agree to the validity of any such test. And even if the agency and the department were doing the job they are supposed to be doing, they would of course reject a test done poorly.

The fact is that there can be only one reason why the food business wants the Delaney Clause out of there. They don't want a law that requires an additive to be banned if tests show that it causes cancer. That's all there is to it. And that tells you all you really need to know about the food business.

In 1957, when a Congressional committee was holding hearings on the proposed Delaney Amendment, Dr. William E. Smith caused a minor flurry by naming a number of scientists who were fired for not taking the food industry line. Longgood reports:

> Dr. Smith charged that government, university and industrial research in the field of cancer control had been obstructed, while apologists for carcinogens were in great demand; scientists who advocated caution in the use of carcinogens "are apt to end up without a job."

Among people who had been fired he included himself, and Dr. W.C. Hueper, a world-renowned authority who, Smith said, had been fired as assistant medical director of DuPont after confirming that beta-naphthylamine, which was used in DuPont dyes, caused bladder cancers (beta-naphthylamine has since been banned). Both Smith and Hueper have continued to be serious critics of the FDA-industry "line" on possible carcinogens, and both have raised a number of questions about specific additives that have yet to be answered.

What the food business wants to do, as far as the law is concerned, is well illustrated by the brief but chilling history of a chemical awesomely known as beta-chloroethylbeta-(para-tertiary-butylphenoxy)-alpha-ethyl-methyl-sulfite. The U.S. Rubber Company, which developed it as a pesticide, called it Aramite.[11]

Aramite was introduced in 1951, licensed for sale by the Department of Agriculture in 1953, and came under the coverage of the Miller Pesticides Amendment (which has no clause comparable to the Delaney Amendment) in 1954. Under that law, the Food and Drug Administration sets "tolerances" for how much of any given chemical can remain on foods when they are sold for eating. They are *not* required to conduct any tests.

In this case, however, they did. U.S. Rubber was asking for a tolerance of two parts per million on some foods and five ppm on others. But the FDA found that Aramite caused liver tumors in rats, and set a zero tolerance—*no* Aramite on foods (that didn't ban the use of Aramite; it only said that there couldn't be any left on the food). U.S. Rubber then came back with a new application for a tolerance of one part per million and asked for a scientific panel of experts. A lot they cared about liver cancers.

They got their panel, and the committee recommended the one ppm tolerance—pending further studies (meaning that they didn't know enough, and we could go ahead and develop cancer while they found out). They suggested that the doses in the FDA experiments might have been too high (sound familiar?). In an argument that would serve just as well in the dispute about cyclamates, the FDA's own Division of Pharmacology said in response that "an experiment with any lower dosage level will not remove the onus that Aramite is a known carcinogen," and insisted that it should not be used on human food. The FDA, predictably, ignored its own scientists in favor of industry.

For two-and-a-half years, the known carcinogen, Aramite, was used on our food. In the spring of 1958, the additional studies were completed: Aramite still caused liver tumors, at much lower dosages than those originally used, and other kinds of tumors as well. The chemical was removed from use—after all of us then alive had eaten at least some of it.

The committee's original recommendation for a tolerance for Aramite followed by the FDA regulation permitting the use of Aramite, which didn't even mention a possible cancer hazard and was never made known to the public, was partly responsible for Representative Delaney's insistence on his amendment, although it couldn't be made to apply to pesticides. At the time of the recommendation, Delaney blasted the scientists on the panel:

[They] admitted that they felt that the data which they reviewed were insufficient and incomplete, and, in particular, suggested that more information be secured regarding the cancer-inducing propensities of

Aramite. Yet, at the same time, they were perfectly willing that the public be exposed to a certain amount of it.

Beyond that, the principal significance of the Aramite story is that it set a precedent which the industry has been trying to keep in force ever since: the idea that there can be a maximum safe dosage, a "tolerance," for a carcinogen.

To a reporter who remembers covering the argument about the genetic effects of nuclear testing, this is a weary controversy. Is there a "threshold" below which no damage will be caused? With cancer, as with genetic damage from radiation, the answer is almost certainly "no," no matter how hard the Mad Scientists squirm to avoid it.

It is possible, with some known carcinogens, to get down to a small enough dose so that no effect is observed in a given test on a given number of animals. This is often called a "no-effect level"; strictly speaking, it's a "no-effect level" for that particular series of tests only, a distinction which is sometimes blurred because a particular substance is not tested over and over again. Some scientists genuinely believe that a "no-effect level" in animals may indicate the existence of a safe dose for humans, but they admit that even if that's true, there is no way to figure out what the safe dose might be.

Most experts, however, don't believe that there is any such thing as a safe dose of a carcinogen at all. The Food Protection Committee, in a booklet on carcinogen hazards in food additives, looked at the "safe dose" arguments and then said:

> Despite this evidence, the possibility exists that doses at "no-effect" levels do in fact exert carcinogenic effects but that the effects are too weak to detect with the numbers of animals feasible for routine testing. Such a possible effect . . . might become evident in a large population such as . . . the population potentially exposed to food additives in our culture.

Hence, they said, all the facts put together "make meaningful extrapolation from the 'no-effect' level . . . to a 'safe level' of use by man currently impossible." And they recommended that the use of any amount of a carcinogen, however small, as a food additive could be justified

> only if (1) values to the public are such that banning the use would constitute an important loss or hardship, and (2) there is no reasonably good non-carcinogenic alternative.

No such case has ever been demonstrated, or even argued. But they didn't say that.

There are a lot of reasons why the "safe dose" argument, which the industry likes to use, simply won't stand up. One of them is that there is a limit on the number of animals you can use in a test. If a substance, for example, is strong enough to cause one cancer in every 200 rats, and if you use only 50 rats in your experiment, there are three chances out of four that you'll get no effect at all. But in a population of 200 million people, the same percentage will give you a million cancers.

Substances tested on healthy laboratory rats are not used only, in the human population, by linebackers for the Los Angeles Rams; smaller amounts may be deadly for the weak, the sick, the aging, or for babies and children. Each of us, individually, may vary in our susceptibility from time to time depending on any of a number of other factors.

In terms of our inner ecology, one of the most important facts about chemical carcinogens is that their effects accumulate. A small dose now may cause an effect too minor to be noticed, but another dose, even some time later, adds to that effect. Two different substances, both of which cause the same kind of cancer, accumulate their effects in the same way, another argument against the idea of a "safe dose."

Possibly even more important are what scientists in this field refer to as synergistic effects. Almost all of us eat at least hundreds of food additives, in addition to all the "natural" substances we consume. There is no possible way to know, or even to guess, what combinations may trigger that multiplying malignancy we so fear in the ongoing and complex processes that are our bodies. For once again, what is inside us is an ecosystem, to which we can add thousands of strange and active compounds only at our peril.

All of these are reasons why Dr. Hueper insists, for example, that "there is no scientifically valid and practical method available for determining a 'safe dose' of carcinogens for humans."[12] And if some other American scientists are less blunt, we ought at least to note that whenever scientists get together internationally, away from the influence of the American food business, their recommendations on cancer, food, and food additives are always far stronger than those to be found in this country.

The International Union Against Cancer, an organization of cancer specialists from 50 nations, has more than once passed resolutions against the whole idea of "safe doses," and insisted that *any* substance which induces cancer in *any* animal, in *any* dose, by *any* method of administration, should be barred from food supplies for humans. And a committee of the World Health Organization was just as unequivocal:

> The uncertainty of the extrapolation of the safe dose to man, and lack of knowledge of the possible summating or potentiating effects of different carcinogens in the total human environment, preclude the establishment of a safe dose at the present time on the grounds of prudence.

Some scientists believe that cancers in humans come about, sometimes if not always, in two steps: a carcinogen creates a condition that lies dormant, and is then triggered, perhaps a long time later and perhaps by another substance, into malignancy. The second substance may not be what is usually called a carcinogen and may be innocent if the first, dormant change hasn't taken place. We like our causes and effects to be simple, but ecological interactions rarely are.

The word "cocarcinogen" has been coined to describe substances that may act with carcinogens, as promoting agents, to cause cancer. The idea was first applied when it was discovered that weak solutions of benzpyrene, applied by themselves, don't seem to bring about skin cancers, but that they do if a fraction of a creosote oil, itself not carcinogenic, is applied at the same time.[13]

Because food additives are eaten by children and even transmitted or fed to babies, the cumulative effect of carcinogens over time is extremely important. In almost all the animal experiments that have turned up carcinogens, the feeding was begun early in the animals' lives, but the cancers turned up in middle life or after. A given carcinogen may take longer to work than a rat lives, but it may work on a human who starts to ingest it early enough and builds up its effect over a long enough period.

And the effect of any carcinogen is irreversible. If you happen to drink something that is immediately poisonous, and it doesn't kill you, your system eventually gets rid of it, and in many cases whatever damage was done will repair itself. But carcinogens do their damage irreversibly; you can't go back and fix it.

Some people are confused about this with relation to smoking and lung cancer. Smoking cigarettes builds up deposits in the lungs of materials which most experts now agree are carcinogenic. The deposits are there whether you develop cancer or not. If you quit smoking, the body begins to clear away those deposits, and thus your chances of contracting cancer are diminished. So the effects of cigarette smoking on your chances of developing cancer are reversible. But once the cancer itself starts, it has started. It can in some cases be arrested, but the damage is not reversible.[14]

Therefore: *reduce* the dose of a carcinogen, and cancer will develop more slowly; but it will develop. Give it early enough in life; and keep the subject alive long enough, and cancer will develop. The effects of small doses are irreversible for an entire lifetime. We can never go back and repair the damage done to a lot of us by dulcin, Aramite, beta-naphthylamine, FD&C Orange No. 2 (of which until 1956 we ate thousands of pounds a year), or a dozen other now banned coal-tar dyes, or cyclamates.

When the Food Additives Amendment was under consideration in 1957 and 1958, Dr. Hueper tried to argue against the idea that additives should be exempt from testing if they had been around and in common use for a long time. He reasoned partly that long use is in no way an indication that a substance isn't carcinogenic. People do get cancer, it comes from somewhere, and a lot of long-used additives are now banned. He also referred to the long "incubation period" before carcinogens show their results, and to the synergistic effects of two or more substances acting together.

But since the FDA is concerned with protecting industry profits rather than the health of Americans—at least whenever the two come into conflict—what has happened instead has been a series of weaseling interpretations of the law. The 1938 Food and Drug Act, for instance, makes it quite clear that nothing poisonous can be added to a food. From the beginning the FDA has interpreted this to suit the industry, with the charming argument that if the substance is ingested in amounts small enough so that no effect can be immediately seen, it is not a poison. The *law* doesn't say that; the FDA does.

They even have an ingenious law enforcement argument for this interpretation. They say that unless they take that position, they can't control the amount of the substance which is used. They don't bother to explain that if the substance were banned, there'd be no need to control the amount used.

Since the passage of the Delaney Amendment, the FDA has resolutely applied the same sort of thinking to carcinogens (and their decision is a major reason for all the talk about "safe doses"). The deep and difficult problem, the FDA argues—as do the Mad Scientists who provide their rationale—is to define a carcinogen.

In 1957, for example, Dr. Herbert Carter of the University of

Illinois told Congress, "Certainly I would agree wholeheartedly with Dr. Hueper that a proven carcinogen should not be used as a food additive. However, I would underline that both 'proven' and 'carcinogen' must be properly defined." He didn't go into detail, but it should be noted that Dr. Hueper was for banning *possible* carcinogens until they had been tested.

You would think that it might be fairly simple to define a carcinogen. The International Union Against Cancer seems to think so, anyway; they have a formal definition:

> A carcinogen is a chemical, physical or animate agent which is capable of producing cancers in any organ or tissue of any species following exposure to it in any dose and physiochemical state and when given by any route either once or repeatedly.

Not, however, according to the FDA. They argue, for instance, that a substance which causes cancer when injected (or implanted, or administered by some other method) is not necessarily a carcinogen when taken by mouth.[15] This argument is originally based on experiments showing that some dyes, carcinogenic when injected, do not appear to be absorbed into the system in digestion.

But it ignores the fact that they are not eaten alone. Emulsifiers, for example, may have a cocarcinogenic effect by making it possible for these dyes to be absorbed in the stomach and intestine, so that they will cause cancer in ways that would not happen if they were taken by themselves.

The situation isn't helped much by such Mad Scientists as Dr. Paul Cannon[16] who meets criticism of carcinogenic artificial food additives by confusing the issue. He attacks "enthusiasts for natural foods" because they "overlook" the presence of possible carcinogens that may occur naturally. Another Mad Scientist, Dr. Emil Mrak, makes the same irrelevant argument. Mrak is always identified as a professor of food technology at the University of California, and occasionally as chancellor emeritus of that university's Davis campus. It is never pointed out that a chancellor's job at that vast university is virtually a political appointment, and that the Davis campus is and always has been a taxpayer-supported research and development department for California's huge and politically dominant agriculture industry. Nor is it ever mentioned that Mrak, who like Cannon is a member of the Food Protection Committee, is also a member of the board of directors of Universal Foods, a Midwestern firm which is both a food processor and a manufacturer of food additives.

Their argument is, of course, a form of telling you that your two broken legs really aren't so bad because you've also got a scratch on your finger. Of course we may ingest possible carcinogens naturally. That's no excuse for piling *more* carcinogens on top of them, and possibly adding to cumulative effects while we're at it. The argument also overlooks any possible difference between foods to which humans have adapted over hundreds of thousands of years and those invented in a laboratory within the last couple of human lifetimes.

Instead of worrying about the fact that there may be a little apiole in our celery, or arguing about "what is a carcinogen?" we could be worrying about what a possible carcinogen might be. We might, for instance, worry a little about a class of chemicals known as polycyclic aromatic hydrocarbons.

The basic aromatic hydrocarbon is benzene, C_6H_6,[17] a chemical (and incidentally a carcinogen) in which the carbon and hydrogen atoms are arranged in a "ring," something like this:

```
          H           H
           \         /
            C — C
          //       \\
   H — C             C — H
          \         /
            C = C
           /         \
          H           H
```

For convenience, the benzene ring is usually represented like this:

From that construction, the other aromatic hydrocarbons can be seen as variations.[18] Take away one hydrogen atom (H) from the right place and replace it with an atom of bromine, and you have bromobenzene. Replace it instead with a molecule of NO_2, and it's nitrobenzene. One or more of the hydrogen atoms can be replaced with all sorts of simple or complex structures, yielding, among other things, such well-known substances as benzoic acid, salicylic acid, and vanillin—the stuff that's in vanilla extract. Also based on a benzene ring are the structures of toluene, phenol (the original antiseptic of Lister), aniline (on which a number of dyes are based), oil of wintergreen (a food additive that is also a poison), and aspirin.

It is also possible that the chain attached to the benzene ring in place of a hydrogen atom may eventually lead to or include another benzene ring, or even more. One simple example, phenethyl benzoate, looks like this:

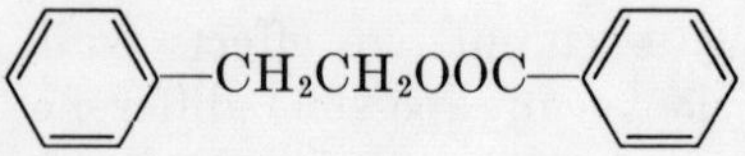

$$-CH_2CH_2OOC-$$

Phenethyl benzoate is an artificial flavor, used to give a fruit or honey flavor to beverages, ice cream, candy, baked goods, or chewing gum. It is on the flavor manufacturers' GRAS list.

Aromatic hydrocarbons are called that for the obvious reason: they are aromatic. One which has two or more benzene rings in its structure, like phenethyl benzoate, is called a polynuclear, or polycyclic, aromatic hydrocarbon.

These compounds are important to us here because it is very, very likely that a polycyclic aromatic hydrocarbon will be carcinogenic. The coal-tar dyes which have been shown in the past to be carcinogens, the "tars" in tobacco smoke, and a number of other carcinogenic agents fit the description. It is the one class of chemicals of which it can be said that it is deeply suspect.

In its examination of the hazards involved in using food additives which might cause cancers or tumors, the Food Protection Committee itself was compelled to note that:

> Of the more than 450 compounds for which acceptable evidence of tumorigenic activity has been presented, almost half (more than 200) are polycyclic aromatic hydrocarbons or their derivatives or analogs.[19]

Coming from a bunch of industry-oriented and industry-financed scientists, that ought to be enough to send an even halfway rational FDA rushing into a vigorous program of testing with extreme thoroughness all the polycyclic aromatic hydrocarbons that are in use as food additives. Instead, the chemicals in question might as well have been the moon rocks Neil Armstrong didn't pick up. We took a look at the chemical structures of flavoring agents alone, and not even all of those by any means, and found 29 polycyclic aromatic hydrocarbons in use today, 26 of them cleared by the FDA and three more in use because the flavor manufacturers say they're GRAS.[20]

Any one of these may be perfectly safe, for all we know. Or none of them. Despite the known strong possibility that any chemical in this class may be a carcinogen (even Holum's elementary chemistry book gives this fact without qualification), the testing of such compounds has been extremely limited, as the Food Protection Committee itself acknowledges.[21]

Of course many polycyclic aromatic hydrocarbons which have been shown to have strong carcinogenic effects *have* been banned. The definitional problem now goes back to the legalistic idea that we ought to be sure it's "proven" to be a "carcinogen," by some clear standard, before we ban it. The FPC puts it this way:

> The problem of minimizing carcinogenic hazard from use of food additives thus is largely one of experimentally detecting low levels of carcinogenic activity.

That isn't the way it seems to us. It seems to us that the hazard would be much better minimized by ruling that no additive can be used unless the manufacturer can show, with a very high degree of scientific probability, that there are no "low levels of carcinogenic activity."

There is no reason why we should have to keep discovering that something is a weak carcinogen after it's been in our food for years. But it happens all the time, partly, we suspect, because so many people are confused about what a "weak carcinogen" is.

A weak carcinogen does not cause weak cancers. It's a name for substances which take longer to work, or cause *fewer* cancers in a given population, or appear to require higher doses. The Mad Scientists of the FPC talk about "weak carcinogens" like this:

> [T]here is often doubt about the activity of some of the agents reported to be weak tumorigens or carcinogens. . . . To establish that an agent exerts this low order of carcinogenicity, special consideration must be given to the many known factors that govern the induction of tumors. . . . In addition, it is necessary to show that the data on the experimental and control animals are statistically adequate for the conclusions drawn.

Notice that there's no mention of what's being done with the substances in the meantime. What is being done is that we're eating them while somebody figures out whether the data are statistically adequate. And we can wait a long time and develop a lot of cancer.

There is an experimental relationship between the weak-carcinogen problem and the "safe dose" argument, as you can see by imagining a simple experiment.[22] Suppose, for demonstration purposes, that we're testing a substance which (although we don't know it yet) will in fact cause cancer in 1.0 percent of the rats who eat it, but only in 0.1 percent of the people.

If we test it on 50 rats, there is an even chance that it will cause no cancer at all. Somebody will immediately start talking about "no-effect levels" and "safe doses." If we test it on 100 rats, we'll get one cancer and a lot of questions about weak carcinogens, statistical adequacy, and possible experimental errors; we may or may not get a further and more elaborate test.

But if we feed it to 200 million people, we will come up with 200,000 cases of cancer.

That illustrates the possible cost of assuming the safety of suspect but untested chemicals, or of quarreling about statistical adequacy when there are indications of carcinogenicity, while we go on eating the stuff and waiting for an adequate test. Clearly, these are not in any sense scientific decisions. They are political and economic decisions.

Somewhere in the complex processes of preparing our food, eating it, digesting it, and making bodily use of its components (in the whole ecological process which our "safety" regulations largely ignore), a lot of things happen to us. A lot of us metabolize cyclamates—regardless of rats and bladder cancers—into cyclohexylamine, a known and banned carcinogen. The simple application of heat in processing or cooking can change apparently inert substances into carcinogens. Detergent residues on our cooking pots or in our water can do the same thing that deliberately added emulsifiers can do: change the way in which our bodies deal with other substances (there are experiments showing that polyoxyethylene compounds, widely used in America today as emulsifiers, cause animals to absorb extremely abnormal amounts of both iron and Vitamin A into their systems, resulting in a number of diseases including cancer of the liver).

The Food Protection Committee could hardly ignore this ecological question, and nodded to it in passing in their study:

> The effects of mixtures of compounds of similar and dissimilar chemical types and carcinogenic activities have not been studied except in a few preliminary experiments[.]

We acknowledge, they said dutifully, that we ought to know more about it. Yet they didn't say anything about taking suspected substances out of our food until they find out.

Our ignorance is so vast, especially in the inner-ecological area of what happens when hundreds or even thousands of artificial chemicals are put together inside us in unpredictable combinations, that we cannot, of course, track down carcinogens by observing what happens to the human population. "Even in the case of thalidomide," as Jean Carper pointed out, "where the deformities were grossly observable and the drug was being taken by a specific population (pregnant women), it was five years before the link between the two could be established."

That's the big reason why we have to use animal studies, and why we have to know enough about them to keep the Mad Scientists from doubletalking us to death, as they tried to do in the case of cyclamates.

The first thing we have to get rid of, and stay rid of, is the idea (which, I'm afraid, a lot of ignorant reporters have done a lot to spread) that they feed so much of these substances to rats and other experimental animals that it has no relation to the amounts usually eaten by men. This is the game that Finch and Steinfeld and the food business tried to run on us with regard to cyclamates, aided by the kind of reporter who figured out (wrongly) that to get the same amount of cyclamate, a man would have to drink several dozen cases of soda water a day.

In the original Long Island cyclamate experiments, the rats got, proportionately to their body weight, about 50 times the amount of cyclamate that a human being might be expected to ingest. The Food Protection Committee, however, explains how substances are normally tested with animals, not only for cancer but for any deleterious effect. There has to be some ratio between the amounts fed to animals and the proportional amounts used by humans:

> In practice the terms used in this ratio are 1) the highest chronic feeding level that has no significant adverse effect in any group of the species tested, i.e., the "no-effect level," and 2) the maximum level of the substance reasonably obtainable in the human diet under the proposed conditions of use. *A ratio of these two levels of 100 or more to 1 has been considered an adequate "margin of safety" for translation to use by man . . .* The arbitrary selection of this "margin of safety" allows for such factors as the assumed greater sensitivity of man to toxicants in general, variation in individual human response, the presence of disease, and the extremes of age [emphasis added].

In other words, 100-to-1 is scientifically "standard," and it has to be so in order to provide some kind of guarantee of relative safety for a human population in which some people are sick and aging and some are babies and children, and in which all of us are a little more sensitive to toxic substances than specially bred healthy laboratory animals.

The 50-to-1 ratio in the cyclamate experiment was thus not too high, but too low by half. Umberto Saffiotti of the National Cancer Institute explains another reason for using what at first glance appear to be relatively high doses (he was responding specifically, by the way, to a question about the cyclamate experiments):

> Since you can't test millions of animals, to detect in small numbers of animals the level of effect comparable to what is expected in man, you have to use maximum tolerated doses. If a chemical produces cancer in 50 percent of the animals at a high dose level and if this is taken by man in amounts fifty times less, we could expect that in man it might produce a one percent incidence of cancer, and in man that's quite a number of cases—two million. So the factor of one in fifty is a *low* safety factor—half the 100 times higher dose level which is usually accepted as the minimum[.][23]

That also explains how high doses help to find weak carcinogens. To be strictly accurate, we should be sure to note that results are not necessarily transferable in any given case from rats to men, and that even when they are, they may not follow such a neat mathematical exactitude as in Dr. Saffiotti's example. What he said is that "we could expect that it might." In other words, he is describing, as scientists always do when they're really being scientists, probabilities rather than absolutes. The point is that those are the probabilities.

And it is extremely important to understand that, in any discussion of animal experiments. We all know, for example, about the correlation between cigarette smoking and lung cancer, and most of us think it's a pretty good guess that lung cancer is also associated in some way with air pollution. Yet tobacco industry apologists constantly remind us that animals exposed to similar respiratory conditions don't show any signs of lung cancer. Are we being conned?

No; we just don't understand the probabilities. It happens to work out that over a period of 25 years ending not too long ago, about 2.5 percent of the entire population seem to have developed lung cancer. But 2.5 percent is too small to find in the laboratory except under extraordinary conditions involving more animals than can really be practical.

Even to test something that has its effect right away (unlike carcinogens which work over a long period of time), you'd have to have 100 animals in the test group, and another 100 animals in the control group, in order to find a 4 percent effect within the limits of what statisticians would call normal significance.[24] It would take many hundreds of animals to spot a 2.5 percent cancer incidence, even if the response of the test animals were exactly the same as in man.

Whatever you're testing for, you need more animals if you want to measure smaller effects. If you have only ten animals in each group, and one group is unaffected, you have to show an effect on five animals in the other group (or 50 percent) to achieve "statistical significance." But if you have 50 animals in each group, and one group is unaffected, you achieve the same statistical meaning by affecting only six animals (or 12 percent) in the other group.[25]

Thus if it's cancer you're looking for, you need more and more animals to find weaker and weaker carcinogens; and after a while this gets beyond the abilities of most laboratories. So while we wait for someone to invent a new test (or breed a strain of half-sized rats), we eat the stuff.

There are compounds which cause cancer in humans but not (as far as we've been able to find out) in other animals; so "safety" in a rat test doesn't mean safety in your food. However, *almost* every substance known to cause cancer in humans also causes it in laboratory animals, which is the big reason for continuing laboratory tests on animals. Even though it doesn't necessarily transfer in the other direction either, there are excellent reasons why any substance that causes cancers in rats, mice, dogs, or monkeys should be regarded with at least the highest suspicion as an additive to human food.

But it doesn't work that way, thanks partly to the FDA's "definitions." Carboxymethylcellulose (or CMC) is on the GRAS list, a stabilizer widely used in ice cream, bakery goods, chocolate milk, pressurized whipped cream, jellies, salad dressings, cheeses, and cheese spreads (to name only some products). Injected into rats, it hangs around the injection site and produces tumors in 80 percent of the rats. By the definition used by Dr. Hueper or the International Union Against Cancer, it's definitely a carcinogen, and a powerful one. According to the FDA, it's no such thing.

One of the most fascinating things about animal studies of food additives is the double standard the FDA seems to apply to them. When a study (however sketchy, clumsy, or inadequate) appears to support the safety of a proposed additive, the agency seems to fall all over itself to accept the results, even if it already has information to the contrary. On the other hand, when studies go in the other direction (even those by its own scientists), the FDA and its "advisory panels" are apt to rumble vague phrases in their throats about further study, while we go on eating the stuff being studied. One observer compared the habitual attitude of the FDA, the industry, and the industry's scientists to that of "a hopeful ostrich."

"I would not necessarily condemn an additive," said one such scientist loftily, "which produced skin cancer in a hypersusceptible mouse." He didn't say what it was that made a mouse hypersusceptible; possibly it's the profitability of the additive being tested.

Finally, while we're still talking about animal experiments, we may as well get rid of another myth that has been making the rounds in newspapers and magazines lately: the idea that if you feed high enough doses of *anything* to animals, you can cause cancer. This just isn't true.

It is true that there have been a couple of experiments in which common substances like salt and sugar, injected repeatedly into mice in high doses, have appeared to cause cancer at the injection sites; but it isn't known for sure whether it was the substance injected that caused the cancer, or just the fact of repeated injection (to keep the record straight, this possibility does not seem to be present in the carboxymethylcellulose injections mentioned earlier).

What is much more important is that of a great number of compounds fed orally to rats and mice in the highest doses they could tolerate without dying, only about 25 percent cause cancers.[26] While this disposes of the myth, it also points up a danger, in a country in which we're eating 3,000 known food additives, most of which are untested.

If 25 percent of the chemicals fed to animals cause cancers—without testing synergistic effects, without waiting for the long slow build-up of "weak" carcinogens, without any cocarcinogenic help from other substances—don't you get uneasily curious about the unnecessary and untested chemicals in your food?

For the important thing, the overriding thing about food additives with regard to cancer is that they are untested and they are unnecessary. Carcinogenic effects, when they are found, are usually found by accident.

Cancer is up, drastically, in the United States and in other industrial-technological countries. Leukemia is up drastically among children, who don't make choices about what they eat, and is continuing to rise fast enough to alarm medical authorities. Yet despite these facts, and despite the fact that we know almost nothing about dozens of ways in which what we eat may be contributing to the rise in cancer, we continue to allow known carcinogens to circulate in our food, as well as thousands of substances which have never been tested.

The cheap and easy way to deal with the problem is to ban, right now, any substance known to be or suspected of being a carcinogen. Let the company that wants to use it demonstrate, to the satisfaction of truly independent experts (a panel from the World Health Organization, for example), that no known testing procedure shows a carcinogenic effect.

Most such additives would then disappear completely, because they are unnecessary to foods. They are colors to fool you, phony flavors to con you about what you're eating, substances that allow wholesalers to maintain inventories. They are not in our food for our good, and there is no reason why we should have to pay to test them. If it costs a corporation too much to test an additive, that is just too bad. Think of what it costs an average family when one of its members develops cancer.

The food business will survive the banning of cyclamates, and it would survive the banning of saccharin, the only remaining non-nutritive sweetener and also, very likely, a carcinogen. In 1969, Dr. George Bryan of the University of Wisconsin induced cancer in the bladders of 64 out of 130 mice with a combination of saccharin and cholesterol; cholesterol by itself (it is a weak carcinogen) induced cancers in only 13 of 106 mice in a control group. In this case, the saccharin was neither fed nor injected, but surgically implanted.

Because of the cyclamate fuss, the press picked up Bryan's experiment, and in March, 1970, the FDA announced that the good old Food Protection Committee would take a look at saccharin. It did, and it came up with a good old FPC report. The available information is incomplete, they said; saccharin may interact with medicines and with insulin, and further study "would be wise"; it may or may not be safe. "It is essential," says the report, "that a definite study be conducted in at least two species, with adequate numbers of animals."[27]

The report went on to dismiss, by implication, Dr. Bryan's study, saying that feeding studies are the only ones that should be used. In the meantime, the rest of us will go on consuming 4 million pounds of saccharin a year.

What nobody mentioned in all this brief public flurry is that saccharin has been decidedly suspect since the days of Dr. Wiley's "poison squad." In 1951 three FDA scientists conducted a *feeding* study of saccharin, and while they admittedly used too few animals, they reported that "the fact that in four of the seven [cancerous] rats abdominal as well as thoracic lymphosarcomas were present is unusual, since ordinarily the ratio is about 1 to 15-20."[28]

This report was completely ignored for 17 years (during which saccharin went on the GRAS list), after which another FPC report, in 1968, mentioned it in passing and said that "this incidence cannot be considered significant without additional experiments." There have been no additional experiments. There is nothing in the FPC report in July, 1970, or in the FDA statement that accompanied it, to indicate that there will be any additional experiments. The FDA has not required them of the manufacturers who should be responsible for conducting them (DuPont and Monsanto have most of the American patents on saccharin). If the FDA has itself begun or farmed out such experiments, we can't find it out.

Obviously to everybody who isn't in the FDA or the food business, saccharin ought to be banned, at least pending the completion of those "further studies" we've been waiting for since Wiley. That would mean, of course, that there wouldn't be any permissible nonnutritive artificial sweeteners left, and the billion-a-year "diet"-food-and-drink business would be forced to fall back (as some soft drinks already have) on smaller amounts of sugar.[29]

We already know that the use of saccharin instead of sugar has no effect on weight control, and there is at least some indication that it may be harmful to diabetics. But filling ourselves with the nonnutritive calories of sugar is not the answer to the problem of potential carcinogens in our "diet" drinks. It's just one more easy way out that leaves profits intact.

We must recognize, as do genuine scientists, how primitive is the level of our understanding about our inner ecology. The importance of the story of cyclamates is precisely that it demonstrates how primitive that level is. The editor of *Science*, Philip H. Abelson, made that point in an editorial in 1969, and noted sadly, "Often, for what seem rather trivial benefits, risks of undetermined magnitude have been incurred by substantial fractions of the total population."[30]

I want to be there when Paul Cannon or Emil Mrak calls *him* a "food freak."

INTERLUDE

Holden

One thing remains to be said about animal experiments—a point that might have eluded us except that as the previous chapter was being written, we learned that our cat is dying of leukemia.

Holden (he is named for an Angolan rebel leader) is small and delicate and black and affectionate and beautiful. He is just over two years old. He will not live to be three.

His death has nothing to do with food additives and is not traceable to the Mad Scientists or to their profiteering bosses. It has simply led us to think, because of its timing with relation to the chapter just before, about animal experiments in general.

We love animals. We have a dog as well, and at one time or another have been friends with a group of mice, and exasperated but loving guardians of a guinea pig. Yet we are not in any way opposed to the use of animals in medical research.

Far too much good has been done to make such a position tenable, and the evidence overwhelms any Shavian criticism. For us, although the thought of individual animals is genuinely poignant, there could be no question of giving up experiments which have done so much for human health.

But when we consider that virtually none of the thousands of food additives in use in America do anybody any good—that their sole purpose is to magnify the profits of manufacturers and in most cases to do so by fooling you or pleasing your eye—and when we think back over what we have just written, things seem to take on a different tinge.

We look with love and sadness at our dying cat, and go back to work on the book, and think about cake mixes. And suddenly we wonder.

Would you kill a dog just to make your cake look a little more chocolatey?

It is no argument, we know. It is just a thought, about the strange values that have developed in the society in which we seem to live.

CHAPTER NINE

Dirty Pool

Some time during the year 1818, in one of the testicles of Edward, Duke of Kent, a few molecules, a few dozen atoms, shifted their arrangement. Possibly it was related to some chemical of which he ingested a small amount. Possibly it was due to the normal background radiation of the earth. Nobody knows; and of course, nobody knew at the time that it had happened at all.

In 1853, Leopold, son of Queen Victoria and grandson of Edward, was born. He was a hemophiliac.

In 1870, Frederick William, son of Queen Victoria's daughter Alice, was born. He was a hemophiliac. In 1884, Victoria's daughter Beatrice bore a son, Leopold, and in 1891 she bore another son, Maurice. Both were hemophiliacs.

Frederick William's sister, Irene, bore three sons. Two of them, Waldemar (b. 1889) and Heinrich (b. 1900) were hemophiliacs. Frederick and Irene's sister, Alexandra, had one son, who became Alexis, Tsarevitch—also a hemophiliac, born in 1904.

Leopold's daughter, Anastasia, bore two sons. One died in infancy. The other, Rupert, born in 1907, was a hemophiliac.

Beatrice, mother of Leopold and Maurice, also had a daughter, Victoria Eugenia, who in turn had four sons. Alfonso, born in 1907, and Gonzalo, born in 1914, were both hemophiliacs.

Gonzalo was born nearly 100 years after those few dozen atoms shifted their positions in the testicle of his great-great-grandfather. All in all, Edward, Duke of Kent, was ancestor to ten known hemophiliac males, and seven known hemophilia-carrier females (the line also includes four males who died in infancy and about whose blood nothing is known).[1]

There were not 3,000 food additives in use in England in 1818. They were certainly not the cause of the hemophilia that spread through the ruling families of Britain, Russia, Spain, and central Europe. But hemophilia is only one of the more spectacular mutation effects—and it is just as certain, statistically, that food additives *are* causing mutations and birth defects in the United States right now, very possibly in your family.

We have been talking about cancer, including leukemia. We mentioned the startling rise of leukemia in recent years. This means more than larger cancer figures, or the tragedy and grief behind the growing number of childhood and adolescent deaths. In *The Journal of the American Medical Association* more than ten years ago, on April 11, 1959, Dr. M. Burnet wrote this:

> [A]nything that is increasing the incidence of leukemia may also be breaking through the other barriers that in the past have protected the germ cells from mutagens. I think one would be wise to pay even more attention to leukemia than its intrinsic interest and lethality demand, just because it is possibly the best indicator one can have of the last and greatest danger to civilization, active and avoidable genetic deterioration. . . . The exposure to physical and chemical mutagens should be reduced as much as possible.

"The last and greatest danger to civilization" is a very impressive phrase; dangers of that order often seem so far out of the reach of understanding as to make us a little uncomfortable. Possibly, if we stay a little closer to details, the size and shape of the danger may emerge. At the moment, the important connection we want to make is that the rise in leukemia may be connected to a rise in genetic damage—a rise we would not necessarily be able to detect, just as no one could have made even the first guess about Edward's hemophilia gene in the 35 years before his first hemophiliac grandson was born.[2]

For instance, Paul Amos Moody, author of a recent basic text on genetics, describes the basis for mongoloidism, which is known to be due to genetic factors, and then notes that "the incidence of childhood leukemia is about twenty times as great in mongoloids as it is in normal children." The implication is clear that the leukemia may involve genetic factors, and Moody believes that soon "we shall doubtless know much more than we do now about the chromosomal basis of leukemia in children."

There is not—and with knowledge in its present state, there cannot be—any evidence that a particular food additive may cause a particular genetic effect. As we will see, in most cases it wouldn't work that way. But there is enough evidence that some food additives may cause some genetic defects so that a number of authorities, including Nobel Laureate (in genetics) Dr. Joshua Lederberg, believe that mutagenic effects are far more of a danger than carcinogenic effects. Writes Jean Carper of Lederberg:

> He says the present reckless policy on additives could create undesirable mutations that might not become noticeable for 100 years or more, if they were ever spotted. Such effects could be subtle: a lowering of intelligence, a loss of physical vigor, sterility—in short "a general reduction in the viability of succeeding generations."

Lederberg and others believe that keeping the Delaney Amendment is not enough; it should be extended to cover *all* food additives and pesticides, and to include birth defects and mutations within its scope as well as cancer.

The cyclamate ban, said Senator George McGovern in October, 1969, "opened a Pandora's box in the food-additive field. Now that the box is open it should not be closed until every food additive in it is proven safe beyond doubt for human consumption." Of course that is what the law is supposed to require now—but we've been through that.

It's interesting that McGovern made his statement on October 24, just before Gerber Products Company, Beech-Nut, Inc., and the H.J. Heinz Company announced that they were taking monosodium glutamate out of their baby foods (this struck us as being another one of those remarkable decisions that three competing American corporations just happen to reach on the same day).[3] Not, the companies insisted, that there is anything wrong with monosodium glutamate; it is fine stuff. It's just that the public is confused.

The public is confused, all right, and the food industry is doing its best to keep you that way, while the FDA starts its now familiar beating about the bush all over again. The fact is that there are tests strongly indicating that MSG, which has no nutritional value whatever, is not only harmful to the individual (and especially to babies) but also teratogenic.

The Great MSG Brouhaha of 1969 actually began as long ago as 1957, when a researcher published a paper showing damage caused by MSG to the retinas of newborn animals. Eventually, the paper came to the attention of Dr. John W. Olney of the Washington University School of Medicine in St. Louis. Dr. Olney is usually described in the press as a psychiatrist, which he is, but not of the type who spends all his time in earnest conversations with patients; he is extremely interested in the effects of chemical substances on human beings, and particularly on their brains and nervous systems.[4] He wondered whether MSG has any nervous-system effects and set out to find out.

Olney reported his first findings (for an injection test) in *Science* for May 9, 1969, saying in part that "when MSG is injected into infant mice, it causes brain damage. . . . Nerve cells, particularly in the hypothalamus, swell dramatically and within several hours these cells die." More fully, he described "skeletal stunting, marked adiposity [a picture of one of the mice resembles a grotesque furry sphere with a tail], sterility of females, coupled with histopathological findings [abnormalities in the tissues] in several organs associated with endocrine function."

Those results were found in the mice after they had grown to adults, having been given the injections during the first few days of life. Olney said that the symptoms "suggest a complex endocrine disturbance," and thought that an explanation might be that the MSG affected the regulatory centers in the hypothalamus that control endocrine function.

But there was more to it than that. MSG is very like an amino acid that the body makes naturally, called glutamic acid. You've probably had enough chemical diagrams, but glutamic acid is a more or less complex molecule made up of 19 atoms (when it's not attached to anything). If one hydrogen atom is replaced with a sodium atom, you have MSG, which the body sometimes—but not always—treats in the same way.[5]

MSG does not itself occur naturally in the body, however, and is metabolized, in human adults, in the liver. Olney took careful note of the fact that the enzymes which help to metabolize MSG in adults *are not functioning* in infants. He also noted that in the case of a pregnant woman, there is twice as much glutamic acid in the placenta, normally, as in the maternal circulation. And he knew that a lot of the body's glutamic acid seems to be used somehow in the brain.

It was logical enough to think that MSG might cause brain damage in newborn mice *because* there are no enzymes to metabolize it and *because* it goes to the brain, causing a dangerous excess.[6] But this result from baby mice led Dr. Olney to raise a question about pregnant women:

> The finding that neuronal necrosis can be induced in the immature mouse . . . raises the more specific question whether there is any risk to the developing human nervous system by maternal use of MSG during pregnancy. . . . The possibility that brain lesions could occur in the

developing primate embryo in response to increased glutamic acid concentrations in the maternal circulation, therefore, warrants investigation.

Taking MSG out of baby food does not, of course, affect the diets of pregnant women.

Olney and a colleague, Lawrence Sharpe, went on to other experiments, feeding (not injecting) MSG to 40 mice and finding even greater toxicity, and trying it out on rats, rabbits, and one rhesus monkey (at this writing Olney has a grant, not from the FDA, to do further monkey studies). There were similar effects on the rats and rabbits. "In the monkey," Olney reported, "there was no clinical evidence of damage, but histological examination [examination of tissue] . . . revealed gross abnormalities."[7]

In July, 1969, the Senate Select Committee on Food, Nutrition, and Health (you may remember it as the "Hunger Committee," a press appellation) held hearings on MSG, with Senator McGovern as chairman. After hearing from Olney, McGovern asked the FDA to comment on MSG, and Commissioner Ley responded. There were, he said, four "refined toxicological" tests which supported the FDA's position that MSG is perfectly safe.

The hastily prepared report from Ley to the committee ignored the committee's questions about use of MSG by infants. One of the four studies he mentioned was by Dr. Jacqueline Verrett and involved chick embryos. MSG, reported Ley, "caused no adverse effect on the embryo." Dr. Verrett was reported as furious; the experiment had included only 180 eggs and was considered to be preliminary at best.

Ley's report on another study said:

> When fed to pregnant rats at a level of 10 percent of the diet . . . no adverse effect on the pregnancy occurred. The young born from those mothers were normal and developed normally.

The trouble was that the test in question, being conducted by Dr. J.S. Adkins, was still in process. Again, the data were only preliminary, and in fact there was another parallel test that *was* showing damage from MSG.

The third of the four tests mentioned by Ley had never occurred at all. It had to do with chromosome breakage, and Dr. Legator, the FDA's chromosome-breakage expert, says that there weren't any such tests in the FDA until after Ley's report. Finally, there was another test that nobody could find either, involving the feeding to rats of MSG "at levels of 30 percent of the diet."

This was an astonishing performance by the FDA Commissioner, and he compounded it by attacking the Olney studies—but getting the amount of MSG in those studies wrong. At the time, James Turner and his Ralph Nader study group were working on the FDA, and they hurried to check out Ley's statement. As the errors were gradually found, Ley apologized to the McGovern Committee for getting the amount of MSG wrong in speaking of the Olney tests, but the FDA did nothing about the other errors when the Nader group pointed them out. Turner and the others then went to the McGovern Committee, who asked the FDA for an explanation.

Ley did then apologize for reporting on a test that hadn't been made on chromosome breakage, and another staff report that went with his letter acknowledged that Dr. Verrett and her coworkers regarded *their* test as preliminary only. The report stood by Adkins' feeding test, but acknowledged that "L-glutamic acid [a substance similar to MSG] fed to young rats at 25 percent of the total diet resulted in a reduction of growth rate of about 50 percent as compared with controls."

Ley's response on the fourth and, at the time, unfindable test was interesting. Nader had entered the fight by saying that the figure should be 30 percent of the protein in the diet, not 30 percent of the diet; Ley properly pointed out that if Nader could correct the figures, there must be figures to correct, and therefore there must be a test. But, it turned out, it wasn't an FDA test. Said Ley:

> The reference for this "non-existent study" is: Hepburn, F.N., and Bradley, W.B., "The Glutamic Acid and Arginine Requirements for High Growth Rate of Rats Fed Amino Acid Diets," *Journal of Nutrition*, 84:305–312 (1964). FDA scientists were the source of my information about the study which is why I cited it in testimony.

He admitted that Nader was right about the amount.

But his citation is even more fascinating, especially since the Mad Scientists keep yelling that MSG must be okay because glutamic acid is a natural amino acid that is synthesized by the human body. It turns out that Dr. Adkins and Dr. Edwin Hove had written a memo about the Hepburn-Bradley study back in February, 1969, advising their superiors that:

> When monosodium glutamate was fed to growing rats at six percent of diet . . . it reduced growth rate by 16 percent; glutamic acid . . . had no effect.

So they're not the same thing, no matter what anybody tells you.

We have, then, the FDA Commissioner, who is supposed to be protecting us, reporting to a concerned Senate committee and quite cheerfully citing four tests—two of which were incomplete, one of which doesn't prove what he said it proved, and one of which had never taken place at all.

According to *Science News* (October 4, 1969), Legator explained it all as "simply a matter of human fallibility compounded many times over." The publication itself said that the FDA "is going to have to explain itself and answer speculation that what happened in this case is not unique."

Not unique, indeed. Dr. Olney decided that *Science News* had been a little easy on the FDA, and he wrote them a letter, which they published on November 22. It's too long to reprint, but a little of it may give you the idea:

> Perhaps the most important observation is not that they made a few errors or even that they told a few bold-faced lies, but rather that every one of the numerous distortions and misrepresentations in their paper was slanted in the same direction, which suggests that those who wrote the paper were not only unconscionably careless with facts, but that they must also have had an ax to grind.
>
> [How could] a consumer protection agency . . . possibly become so turned around as to end up, in effect, defending industry at the expense of infant health?
>
> It would be entirely possible for them to do the appropriate studies to confirm my findings in less than one month, but they would then have to admit that the practice of adding MSG to baby foods is in violation of all sound principles of toxicology. They would also have to remove MSG from the GRAS list . . .

Please note that MSG was not banned from baby food. It was voluntarily removed by the manufacturers as a public relations gesture. Note also that they didn't have to make an announcement about it. And they wouldn't have to make an announcement *to put it back.* It's still perfectly legal to use it.

Note, finally, that MSG is still being eaten, all over the land, by pregnant women—and that they can't necessarily avoid it by reading labels.

"We know from the long-established research," said Dan Gerber of Gerber Products Company, "that monosodium glutamate is a safe and wholesome ingredient in baby food" (they had been using it for 18 years before they took it out). A spokesman for Heinz said that the use of the additive will be "vindicated."

Probably the only reason it was taken out at all was that on the day before, October 23, the Women's National Press Club of Washington heard an address from Dr. Jean Mayer, who is the President's nutrition advisor. Dr. Mayer sharply denounced the use of MSG in baby foods and got a lot of press attention.

Obviously Gerber was irritated; he let it be known in a letter to his stockholders a month later. He described Olney's first injection experiment with newborn mice, and then said that the publicity "was further augmented by the publication of a report . . . by Olney to the effect that MSG injected into a single monkey caused similar brain damage." At the end of his letter he said that "qualified scientists" who have reviewed Olney's work "are not convinced that his work is relevant to human feeding."

Gerber didn't mention to his stockholders that Olney had done feeding experiments as well as injection experiments, although by then the FDA had acknowledged as much to the McGovern Committee. Gerber did say that "an attempt was made" to present "the accumulated data demonstrating the safety of MSG," but the press gave it no coverage. Apparently he's referring to that now infamous report by the FDA.

Gerber's letter ends, "It is the generally accepted consensus [among baby food manufacturers?] that further work must be done with different species of animals in order to validate the conclusions drawn by Dr. Olney."

First of all, Gerber is arguing that "accumulated data" can prove MSG safe—which is impossible. You can't prove a substance safe. Olney's were new experiments, providing new data.

In the second place, Gerber is of course arguing that because some people are "not convinced" by Olney, the additive should stay in our food until further studies are done. That is not, again, the way the law is supposed to be enforced. MSG should come out of everything right now and stay out at least until there is a valid test showing that Olney's results don't relate to human feeding.

Gerber did not tell his stockholders either, by the way, that Olney's were the first tests in 20 years done on animals less than 20 days old. Since the whole controversy is about the effect on newborn infants, the rest of Gerber's "data" are irrelevant anyway.

The MSG, of course, is put into baby food only to make the baby food taste better to the baby's mother. It does nothing whatever for the baby. But this is okay with the FDA. According to Turner:

> [T]he Director of the Bureau of Science of the FDA responded to the assertion that "baby foods are made largely to be pleasant for mothers" by saying that it was important for the baby that his mother enjoy baby food. If she did not, the assessment went, the baby might not eat it.

She might like it even better with a shot of bourbon in it, but most mothers probably wouldn't feed that to their kids.

But why shouldn't, say, a jar of strained peaches or a can of pureed beef taste good by itself? Peaches are peaches, aren't they? Not if you replace a lot of the substance of the baby food with starch "fillers" and sugars in order to cut your manufacturing costs. A little monosodium glutamate can be a lot cheaper in the manufacturing process than full-quality food, and it fools Mom.

Some of the fuss about MSG has died down—to the relief, no doubt, of FDA officials—but for a while the industry was pretty worried, lest MSG in the rest of our food receive the scrutiny it should. One manufacturer, International Minerals and Chemical Corporation, ran a full-page ad in the December, 1969, issue of *Food Technology*, claiming in large type to give "the facts about monosodium glutamate" in seven sober, numbered paragraphs.

Paragraph three was typical of the whole approach:

> Its use [as a "food flavor enhancer"] has been researched for safety and effectiveness for more than a quarter of a century; it has been used, in some form, by millions of people for centuries without ill effects.

It has *not* been researched, for a quarter of a century or for any other length of time, for its effects on babies, except for Olney's tests (with their implications for pregnant women). It has not been used "in some form" by millions of people for centuries; glutamic acid has, and the Hepburn-Bradley study makes clear that the effects may not be the same as those of monosodium glutamate.

And finally, there is no way to say whether there have been "ill effects" or not. The kinds of results that Olney got in mice and other animals may show up in humans as minor imbalances or smaller disabilities in brain function that are not "strange" enough to result in extensive study.

It is only recently, to take just one of many possibilities, that we discovered the importance of a compound called gamma-aminobutyric acid (GABA) as a "transmitter substance" in the brain.

MSG converts to GABA in part; at this point in our knowledge, we couldn't even begin to predict what an imbalance might mean.

International Minerals, not at all incidentally, is the manufacturer of more than 40 million pounds of monosodium glutamate a year—70 percent of all the MSG used in America. That should make it funny when they say in their ad that cyclamates are "man-made" chemicals manufactured in laboratories, but MSG is a "natural" food.

International Minerals' president, Nelson C. White, sent a telegram to the White House Conference on Food, Nutrition, and Health during the same month that that ad was running. Turner quotes it at length, but you can get an idea from a few of its phrases, like the one in which he expressed the hope that the conference

> not be used for rubber stamping the unrealistic thoughts and proposals of completely uninformed and biased so-called protectors of the consumer.

For the president of a firm that makes 70 percent of the MSG in the country and sells it all at a profit, he has a strange idea of what makes people "biased." He goes on to refer to

> the utterly distorted statements of not only pseudo-scientists, but those in government circles who sought political favor with the public through association with so-called consumer protectionism.

Protecting the consumer is being "uninformed and biased"; scientists who don't like MSG are "pseudo"; and democratic processes, when they go against your side, are "seeking political favor." White ends up complaining about the spotlight being turned on monosodium glutamate

> on the basis of limited and inconclusive experiments being taken as proven findings against a quarter of a century of true scientific research.

As we've already seen, the "quarter of a century of research" is irrelevant to Olney's work, but that's not the point of the distortion. They are inconclusive experiments (and incidentally, can you imagine an "unlimited experiment"?). Almost all scientific experiments are inconclusive. But Olney's tests are also as "true" as any other "true scientific research." And the law says that when there's doubt, we're not supposed to be fed the stuff.

International Minerals even went so far as to demand free television time to rebut critics of MSG; but in spite of the position of IMC and the FDA, monosodium glutamate is certainly not, as of right now, "generally recognized as safe," and it should come off the GRAS list immediately. The FDA, of course, will do nothing until it's politically forced, and will do as little as possible then. And the manufacturers' great gesture of taking the MSG out of baby food bears some resemblance to the first partial banning of cyclamates: it confused the public without correcting the problem or informing the public of the real danger.

For one thing is clear: the principal danger from cyclamates is not that they are carcinogenic but that they are teratogens and mutagens. And if MSG is potentially dangerous in baby foods, then it is equally dangerous, possibly more so, for pregnant women. It, too, appears to be a teratogen.

The distinction between teratogens and mutagens is not easy to draw, especially when some writers persist in mentioning "congenital birth defects" in the same stories or articles. To call a birth defect "congenital" simply means that it is present and evident at birth, no matter what its cause.[8] The distinction between the other two words is more important.

Generally speaking, a mutagen causes a change in the germ cells—the egg cells in women, the sperm cells in men—which is then passed on to future generations if it does not destroy life entirely, or the power to reproduce. A teratogen causes a change in the fertilized egg after it begins to divide and specialize—in the embryo or, more rarely, in the fetus—and damages only the individual in the womb, not its descendants.

Difficulties arise because both mutagens and teratogens can cause birth defects, and sometimes the defects are quite similar. Phocomelia (flippers instead of arms) in babies is caused by the teratogen, thalidomide, taken by pregnant women during early pregnancy. Similarly, malformations of the heart, eyes, and other organs are caused in the embryo when the pregnant woman contracts German measles (rubella) during the first 12 weeks of the pregnancy.

Acheiropody, on the other hand—the name for a condition in which babies are born without hands or feet—is congenital too, but is hereditary. At some time in the family history of anyone suffering from the condition, a mutation took place in an ancestor's genes.[9] To the anguished parent, it matters little whether the condition of the baby results from a teratogen or a mutagen; but it matters a great deal to future generations.

In the interest of being a little more accurate—and perhaps to help in avoiding some of the deliberate confusion of propagandists and the inadvertent confusion of reporters—we should note this comment from geneticist Victor McKusick:

> The majority of congenital malformations, e.g., cleft palate, harelip, clubfoot . . . and congenital heart malformations are probably the result of a collaboration of genetic and environmental factors, none of which is understood in any great detail.

We do know that some animals are genetically more apt to be affected by a given teratogen than other animals. For instance, cleft palate can be caused as a birth defect in baby mice by feeding cortisone to mothers during pregnancy; but some strains of mice get a lot of cases and some get only a few. Apparently the susceptibility is hereditary.

Having said all that, we can now make a sharper distinction for our own use. We can say that mutagens can affect anyone, and will usually result in genetic changes that will be passed on to future generations (*if* the affected person reproduces, and *if* the effect is not to cause sterility or some other bar to reproduction in the next generation). Teratogens affect pregnant women, and will result in spontaneous abortion, stillbirth, or the birth of defective babies, but not in genetic changes.[10]

It is difficult to test for teratogens for a number of reasons, not the least of which are that a lot of animals have to be involved, and that to be complete, tests must involve at least two generations. No such tests, of course, are required for food additives, and there is no protection in the law against teratogens except for the broad phrases about "safety."

This is the importance of Olney's studies, particularly his feeding studies. If monosodium glutamate affects animals when they eat it the day after they're born, it's almost certain that it would affect them the day *before* they're born, and probably for some time before that.

Since Olney's are the only studies of their kind in more than 20 years, and since there is a mechanism, related to the high concentration of glutamic acid in the placenta, by which monosodium glutamate might reach the hypothalamic area in the fetus (or whatever area it affects to cause its damage), MSG clearly needs much more testing before it is allowed in the diets either of babies or of pregnant women.

Further, it may have teratogenic effects throughout a pregnancy—unlike thalidomide, which is dangerous only in early months.

But it is another case of the FDA letting us eat the additive in question while we wait around for somebody, someday, to run more tests. We hardly have to ask why the FDA once again resorts to being a collection of hopeful ostriches; the production figures for monosodium glutamate—reportedly 46 million pounds in the United States in 1966, and about 60 million today—tell us all we need to know.[11] The average American eats more than a gram of it a day (which means that some of us eat a lot more than that), and it has been widely used in the United States since the 1940s.

One reason there may seem to be a lot of detail in this book is that it always takes longer to tell the truth than to do, for instance, what *Time* did—we presume hastily—in its issue of August 31, 1970:

> Although MSG causes brain damage in mice when injected in large doses, researchers have found no evidence of harmful effects when the chemical is used as a food additive. Taking no chances, the FDA banned the addition of MSG to baby foods . . . .

The damage is caused whether injected or eaten, in large or in medium doses. A 100-to-1 safety factor is common in food safety tests. The "no evidence" phrase misses the entire point of the law and of the science of food testing. And the FDA did not ban MSG from baby foods or from anything else; on the contrary, they're still trying to defend it. The feeding tests on MSG, by the way, were reported in *Science* more than five months before the *Time* story appeared.[12]

Once more, the absurdity of it all arises from the fact that monosodium glutamate does nothing whatever for consumers. Found in everything from canned food to bouillon cubes, mayonnaise to frozen fish sticks its sole purpose is to fool you into thinking that the food is better than, in fact, it is. The description in the *CRC Handbook* tells you why it's there:

> It acts on one, and possibly two, types of nerve endings (those in the taste buds and the tactile receptors of the mouth). It contributes no flavor of its own when used in normal small quantities, and it seems to affect the impressions of basic taste as well as produce a feeling of satisfaction throughout the entire oral cavity.

We of course have no objection to Americans' feeling satisfaction throughout their entire oral cavities. The question is whether Americans want to achieve that satisfaction with a product that has so many attendant risks—and at the moment Americans, including pregnant American women, can't even make the choice. As noted previously, you can't always tell from the labels what foods contain monosodium glutamate; and remember, people eat in restaurants, too.

Of course, monosodium glutamate is not in baby food to produce a feeling of satisfaction in the baby's entire oral cavity, but in the mother's, and the reason is simple: the three-firm baby food monopoly puts less food in its baby food. Instead, it replaces much of the food with starches and sugar—sugar which it also says is there "for taste." The baby, at least at the beginning of the time when it eats baby foods, can't taste either the sugar or the MSG, although it might become, in a sense, "addicted" to the sugar.

Mothers, who often don't realize that the baby doesn't taste what *they* taste, used sometimes to add sugar to babies' milk; some still do, although pediatricians have been trying to talk them out of it for years. Feeding sugar at so young an age, they say, makes the baby used to consuming it even as his taste is just developing. Salt in baby foods, too, can develop "a salt appetite which then must be satisfied the rest of his life," according to Dr. Lewis Dahl—who thinks salt in baby food may have something to do with the fact that Americans now develop hypertension ten or 20 years earlier than they used to.[13]

The starches added to baby foods are even more interesting. When the manufacturers first began to add them to stretch the baby food and increase profits, it was found that the starches would break down into a watery mess after the baby food had been opened and partly used. On examination it turned out that the baby's saliva—transferred from spoon to jar (or can or dish) so that some of it remained in the leftover baby food—was "digesting" the starches in the jar.

What the manufacturers did, of course, was not to take the starches out, but to use starches that wouldn't be digested by the baby's saliva. The question now is whether the rest of the baby's system will break down the starch at all. But the manufacturers couldn't care less.

The complete uselessness of monosodium glutamate except to satisfy our entire oral cavities and to mislead us about our foods is enough to make us wonder why men like Dr. Stare rise so hastily and vehemently to its defense. He was another who (like International Minerals in their ad) came up with the argument that MSG *must* be okay, because glutamic acid is a "natural amino acid" present in our "normal" food.

The argument is pure hogwash. Glutamic acid *is* in some natural food (it is about one-fifth of the composition of milk protein). Presumably our inner ecologies are accustomed to a certain amount of it—although, according to present knowledge, the body seems to be able to manufacture all that it needs out of other substances. That's not what's wrong with Stare's proindustry argument.

The point is that they are not the same thing. Monosodium glutamate fed to growing rats reduced their growth rate by 16 percent; glutamic acid fed in the same proportions did not.

In any case, one can get very tired of hearing industry apologists, with wide-eyed innocence, saying that "monosodium glutamate is only the sodium salt of glutamic acid, a normal food substance." We can say with equal accuracy that sodium chloride (ordinary table salt) is "only" the sodium salt of hydrochloric acid; but we don't recommend sprinkling hydrochloric acid on your breakfast eggs.[14]

It is easy to say, as was said on television shows about cyclamates, that "pregnant women shouldn't take chemicals into their systems except on the advice of their doctors." It is easy—but it is macabre. For in fact, pregnant women have no way to avoid dosing themselves with untested chemicals, almost none of which has ever been tested for teratogenic effects. Dr. Verrett's study of chick embryos—a study so extensive that 13,000 embryos were involved, in a test directly analogous to that which first exposed the teratogenicity of thalidomide—showed severe teratogenic effects from cyclamates and cyclohexylamine, but only the barest whisper ever reached the public.

Even in the face of such evidence, there is no specific law, and not even a specific government regulation, which requires the banning of such a substance; there is only the general law about "safety"—although, of course, that should certainly have been enough in that case, as it should be enough with MSG.

Instead of using Dr. Verrett's exhaustive study as a basis for banning cyclamates as teratogenic, however, the FDA used the incomplete, small-scale, preliminary test that Dr. Verrett did with MSG on only 180 eggs to convince the McGovern Committee that MSG was safe. Once more: when a test, no matter how useless, appears to come out on the side of industry, the FDA is willing to use it; when a test, no matter how extensive and obviously important, demands restrictions on industry, the FDA ignores it. In these cases, the pattern held even though both tests were done by the same person, and an FDA employee at that!

Obviously both the law concerning and the public use of scientific experiments are, at this point, absurd.[15] As Dr. Lederberg and others have suggested, there must be, at the very least, a stiff "Delaney Clause" for teratogens, requiring testing in advance and specifying some standard (written without the overeager help of the food industry) which every food additive must meet.

Carcinogenic effects from food additives are unforgivable, and teratogenic effects almost unbearably tragic when the infant lives; but neither is, by thousands or perhaps millions of individuals, the most serious of the possible effects of food additives. That distinction must go to the mutagenic effects: those that change, literally, the bodily heritage of every descendant we will ever have, forward infinitely into time.

We don't really know what a gene is. It's our handy name for a small unit—pretty certainly, now, a chemical unit something like a small molecule—that determines a single hereditary characteristic. We don't know how many genes a human being has. J.N. Spuhler calculated in 1948 that there should be between 20,000 and 42,000 of them inherited from each parent, and nobody's made a better guess since; but even this highly educated guess rests on some fairly broad assumptions.[16]

We do know something about where genes are, and how they work. What we know about their location is that they occur somehow in sequence along long, thin formations called chromosomes; and relatively speaking, we know quite a bit about chromosomes.[17]

The normal human being inherits 23 chromosomes from each parent. They duplicate themselves every time a cell divides (we each start out, of course, as a single cell), so that each of us has a set of 46 chromosomes in each of the cells in his body. Each of these million billion or more sets of 46 is theoretically identical (we don't know the number of cells in our bodies either, but it's somewhere between 10^{14} and 10^{15}—the number 1 followed by 14 or 15 zeros).

There are exceptions. Some liver cells and a few others have twice as many chromosomes—two sets from each parent, in effect, or a total of 92 chromosomes. And our *germ* cells—the egg cells of women, the sperm cells of men—have only 23, though they may be mixed: some from one of our parents, some from the other.

Because chromosomes are long and thin, and because genes are located somehow along them, a lot of us were taught to think of genes and chromosomes as being something like beads on a string. A row of knots on a string might be a little better; just simply locations on a string might be better yet; no analogy will be exactly accurate. Genes are not really separate things at all, as it now appears. But beads on a string (if we remember that it's an inaccurate model) might help us to think about the different kinds of occurrences inside our bodies that can be called mutations.

Before we can talk about how mutations take place, and how untested food additives may be causing them, we have to talk about what they are. It's the fuzziness of our knowledge in these areas that makes it possible for the Mad Scientists to doubletalk us into believing that nothing serious is happening to us. And even those of us who remember a little about genetics are likely to be carrying around some outdated ideas.

One kind of distinction exists between somatic mutations and genetic mutations. Another kind exists between chromosomal mutations and gene mutations. They are separate kinds of distinctions—just as in roulette you can divide the numbers into red and black, or into odd and even. Each distinction is independent of the other.[18]

We are not so concerned in this section with somatic mutations, but they are important—and sometimes the distinction is deliberately blurred to fool you. A somatic mutation is a change in either a gene or a chromosome in one of your body cells *other than a germ cell.* It is not passed on to your descendants, but it *can* change the way that a particular cell does its specialized work.[19] It is from the genes and chromosomes that each specialized cell gets its information on how to do its particular job; that's why each cell has its own set. A somatic mutation can interfere with, or change, that information.

Many specialists believe, for instance, that cancer can result, in some cases, anyway, from somatic mutation. The theory, in very simplified form, is that some enzyme in the cell, getting its "signal" from a gene or group of genes, doesn't do its job right, or perhaps doesn't stop doing it when it should. The result, it's thought, is that abnormal cells thus develop and multiply.

In one kind of cancer—chronic myeloid leukemia—it's been found that a part of one chromosome is damaged in blood cells only. This is probably, though it isn't certain, a somatic mutation caused by radiation.

Radiation can sometimes bring about somatic mutations that stop cells from multiplying. If this happens to healthy cells in your bone marrow, the ones that manufacture white blood cells, then your white blood cell count drops, one of the first signs of radiation illness. The same mechanism, however, can be used on purpose on cancer cells; radiation is sometimes used as a treatment to stop the unhealthy, cancerous cells from multiplying.

The process of mutation in somatic cells and in germ cells is precisely the same. If a chemical, for instance, can cause a somatic mutation, it can also cause a genetic mutation—if it can reach the germ cells in testicles or ovaries. In either case, the result may be either a chromosomal mutation or a gene mutation.

A chromosomal mutation is one in which, to use our imperfect model, the beads stay in the same relative place but the string breaks. The partial strings may then recombine in the wrong way, or they may not. That is grossly oversimplified, but it will give you the idea.

The 23 different chromosomes are more or less individually identifiable under a microscope, and are given numbers. A *somatic* cell thus contains two No. 1 chromosomes (one from each parent), two No. 2 chromosomes, and so on up to No. 23. A germ cell has only one chromosome of each number. A well-known example of a chromosomal mutation in a germ cell involves a breakage and recombination such that part of a No. 15 chromosome ties up with part of a No. 21. The result is a mongoloid child. The 15-21 recombination perpetuates itself, but mongoloids rarely mate.[20]

We know that certain chemicals result in "broken chromosomes." What we don't know is what significance that has in human beings. The important thing to remember when you read about something causing broken chromosomes is that broken chromosomes in a *somatic* cell—in a blood cell, say, or a skin cell—are one thing; broken chromosomes in a *germ* cell are something else.[21]

It is not even clear whether broken chromosomes in a somatic cell have any effect at all. Remember that if a No. 10 chromosome in a somatic cell is broken, there is another, unbroken No. 10 chromosome in the same cell—the one from the other parent—and it may be that it takes over all or part of the function. Chromosomes (and genes) seem to have remarkable self-repairing powers. The chromosome may even be able, in some cases, to function in parts. Only logic tells us that in so delicate a mechanism as the human cell, something out of the ordinary must be happening.

A broken chromosome in a *germ* cell is something else. Again, knowledge is shaky, but it may mean that the sperm, or the egg, will be sterile.

That wouldn't make the *person* sterile. Half of his sperm cells, or her egg cells, would contain the other, unbroken chromosome of the same number. But it might make him or her "half sterile," with only half the sperm (or egg) cells capable of reproduction.

The abnormal breakage and possible recombination of chromosomes, as in the 15-21 mongoloidism mutation, should not be confused with a perfectly normal and healthy process known as "crossing over," which takes place when germ cells are formed. This is a process of exchanging genes, which takes place only between homologous chromosomes—for example, the No. 10 you got from your father and the No. 10 you got from your mother—and helps to mix up characteristics in the offspring, one of nature's ways of trying out new combinations.[22]

We aren't really sure what a *gene* mutation might mean in a somatic cell either, partly because we might not even know whether one had taken place. Since genes seem to act by producing enzymes which in turn catalyze other reactions, it would seem that even an effect on one gene in one cell could in theory produce some fairly drastic effects—some of the individual processes that go on in individual cells are crucial ones. But again, this would affect only the individual; the hereditary effect of changing a gene in a *germ* cell has consequences that reach much further.

All of us know, of course, that some genes are "dominant" and some are "recessive"—meaning that if one of each arrives in the same fertilized egg (or "zygote"), the dominant one will win out. Thus we say, for example, that brown eyes are dominant over blue eyes.

It is unfortunate that we all know that, because it isn't true.

The eye color thing is too complex to go into here, and irrelevant to our purpose. Let's just say that according to classical Mendelian theory, two blue-eyed parents *could not* have a brown-eyed child—but in fact, they can. Besides, every beginning student of genetics has noticed that the rules don't account for deep green eyes with gold flecks in them.[23]

What is important is that dominance and recession in genes (or in combinations of genes) are relative rather than absolute. Geneticists now speak of some genes as having "partial penetrance," meaning that they're dominant to a certain degree. The terms are still used, but they don't mean separate and opposed things as we were once taught.

Even though a "normal" gene is "dominant," a "recessive" gene may make itself felt in the inner ecology of the individual. That means, in turn, that any change in the genes of the germ cells is just that much more dangerous—since changes are very, very rarely improvements.

Beginning with the work of the Scottish scientist Charlotte Auerbach, who worked with mustard gas, a number of scientists have shown the ability of a number of chemicals to cause mutations if they reach the genes. Caffeine is a known mutagen among food additives; so is formaldehyde. A number of phenols (chemicals whose structure includes a benzene ring) are known to be mutagenic. Hydrogen peroxide (GRAS as a bleaching agent, and cleared also for use in modifying food starches) is a mutagen.

All of these mutagens are widely used in food, without any clear knowledge as to whether they can reach the germ cells. And of course, there are thousands of additives whose possible mutagenic effects are simply unstudied.[24]

Some hint of the natural ability of the body to fight off the effects of chemical intrusion comes from the work of Novick and Szilard, who showed that one chemical—adenosine—functions as an anti-mutagen. It reduces the mutagenic effect of caffeine in particular, and of some other chemicals. Adenosine, it turns out, is one of the natural constituents of the genetic material, DNA, of which more in a minute; the point here is that in addition to its primary genetic function it also seems to serve to protect against mutations.

Damage to the entire chromosome can be caused, we know, by chemicals. It can be caused by radiation "hitting" the chromosome directly or by its "hitting" some other chemical in the cell and changing that chemical to a mutagen. It may be that damage can be caused by any kind of agitation that changes the temperature of the cell in a significant way.

The result, as noted above, may be mongoloidism, or sterility, or any one of a number of other things if it happens in a germ cell. It can be myeloid leukemia if it's a blood cell. Or it can bring a "recessive" characteristic into "dominance."

If, for instance, a mutation broke off a bit of the end of one of your father's chromosomes, one of the ones that went into the set you inherited, that lost bit might have contained the "dominant" gene that allows your blood to clot normally. Your mother might have passed on the "recessive" gene which prevents normal clotting. The mutation of your father's chromosome would thus "uncover" the recessive gene—and leave your blood unable to clot properly. The example may turn out not to be a tenable one, but the theory is correct.

All of this was extremely mysterious until a few short years ago.

We knew a great deal about what happened, but very little about how it happened. Now, at last, we are beginning to gain the knowledge we need in order to understand how the junk they're putting into your food may change the heredity of every descendant who follows you.

For we can now come pretty close to drawing a chemical picture of a chromosome.

It is not accurate (though some popular books give the impression) that a chromosome is a single molecule of deoxyribonucleic acid, more often called DNA. It is true, however, that we can quite conveniently think of it that way; it is the DNA that we're concerned about here, and (remembering that that isn't all there is to it) we can use that model.[25]

DNA is the stuff that's coiled in that double helix about which one of its theorizers (one can hardly say "discoverers") wrote such a fascinating and gossipy book. Unlike a lot of molecules, you can't really "draw a picture" (or build a model) of it that's accurate in every detail, because small details of DNA change from one molecule to another. Otherwise we would all look alike and have the same characteristics. DNA has units that are all alike (four different kinds), but they are joined together in a variety of different sequences.

In basic structure, DNA is quite simple. Just imagine an old-fashioned ladder, the kind you lean against the side of the house, that is made out of something flexible. Twist the ladder into a helix, like a spiral staircase (*both* uprights around *one* common center). You've got it: the double helix.

A DNA molecule goes on like that for some time; it is very long and very thin. In fact, its thinness is almost impossible to describe if you're not used to thinking in atomic "sizes." It's about 0.000002 millimeter thick, or approximately 0.00000008 inch. That is so small that, despite the tiny size of a chromosome, it is nevertheless true that a DNA molecule is more than a million times as long as it is thick.

The structure of the two uprights of the ladder is the same in each case, and (again for our purposes) quite simple.[26] A handful of carbon, oxygen, and hydrogen atoms forms a sugar (deoxyribose), which is connected to a smaller handful of phosphorus and oxygen atoms forming a phosphate group (closely related to phosphoric acid, except that when it joins up in the chain, it isn't an acid any more). The phosphate group connects to another unit of the same sugar on the other side, which connects to another phosphate group, and so on. And on and on and on.

The other upright is an identical sugar-phosphate-sugar-phosphate chain. Each of these two chains has, sticking out "sideways" from each sugar group, another chemical, which forms half of a "rung" of the ladder. When the two chains come together in the double helix, these chemical groups (called amines) join to form the "rungs of the ladder."

The combination of one sugar group, one phosphate group, and one amine, the basic unit of DNA, is called a nucleotide. There are four amines that form nucleotides of DNA: adenine, thymine, cytosine, and guanine (the nucleotide that has adenine in it is the chemical adenosine, mentioned earlier as an antimutagen). The four amines are so important that after Watson and Crick arrived at the double-helix model for DNA, Watson suggested that if anyone on the project had had twins, they would certainly have been named Adenine and Thymine.

The important thing about the amines is that they will link together (to form "rungs" and thus bind the whole thing together) only in certain ways. Adenine will link only to thymine, and *vice versa.* Cytosine will link only to guanine, and *vice versa.*

The sugar and phosphate groups, being identical, can all link together; this means that the *pairs* of amines that form the rungs (they're called "base pairs") can come in any sequence. An A-T pair may be followed by a G-C pair, or by a C-G pair (they can go either way, of course), or by a T-A pair, or by another A-T pair.

Using S for sugar, P for phosphate, and the initials of each of the four amines, *one strand* of the DNA chain might look in part like this:

```
-S-P-S-P-S-P-S-P-S-P-S-P-S-P-S-P-S-P-S-P-S-P-
 |   |   |   |   |   |   |   |   |   |   |
 G   T   C   A   G   A   C   T   T   C   A
```

In order to join that chain and form a double helix of DNA, the other chain at that point would *have* to look like this, because they won't join any other way:

```
 C   A   G   T   C   T   G   A   A   G   T
 |   |   |   |   |   |   |   |   |   |   |
-S-P-S-P-S-P-S-P-S-P-S-P-S-P-S-P-S-P-S-P-S-P-
```

Then, with cytosine joined only to guanine, and adenine only to thymine, that brief section of DNA would look like this:

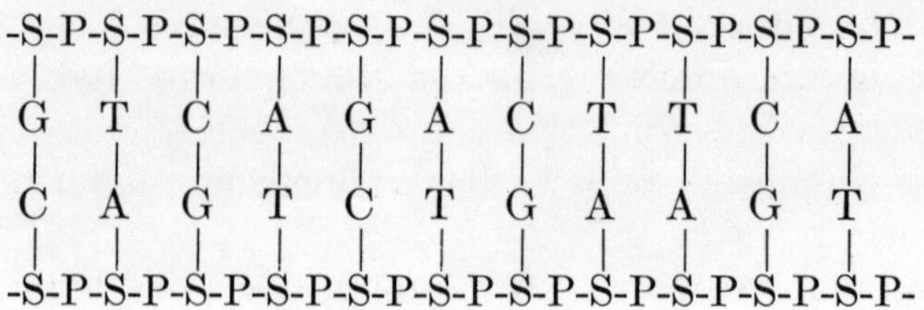

The trick is that while the base pairs can come in any order, the order is not random. It is a code: *the* code by which our chromosomes build our bodies from the information passed on from sperm and egg. The base-pair sequence in each chromosome *is* that genetic information. The right combination in the right sequence gives us our hair color, the ability of our blood to clot (or the absence of that ability), the ability to break down complex materials in our foods, the ability to fire the complex nerve impulses of our brains—literally all the individual characteristics of life.

Portions of those sequences are our genes. There are perhaps 30 billion base pairs in all our 46 chromosomes. It is believed that a sequence of 800 or 1,000 base pairs may be one gene. Altering *one* of those base pairs may be enough to alter a gene.

How does it work? We know a little about that too. The two strands of DNA can separate, and then each strand becomes a pattern for a new opposite, manufactured from chemical material in the surrounding cell. As you can see by looking at the diagram above, the new strand *has to* come out right, because each amine can join with only one other. This is how chromosomes duplicate themselves when cells divide.

By a similar mechanism (using a similar but not identical substance called RNA), the genes form shorter "complementary" structures which in turn go out into the cell and "direct" the formation of other chemicals: the enzymes which make all the rest of the processes of our bodies possible. Understanding this, it is easier to see how genes work, not singly, but in concert, so that some of our characteristics depend on relationships among several genes.

A process, like the ultimate coloring of our skins, may depend on several enzymes which work one after the other; several genes are necessary to control their production. A variant combination (some genes from one parent, some from the other, with possibly a "recessive" or two coming to the fore) makes for a result that differs somewhat from either parent.[27] This is how "crossing over" in the germ cells helps to make for variety.

We can say, now, that a mutation (somatic or genetic) is a change in that DNA molecule. If it somehow breaks in the middle—recombining with a different chromosome, or reconnecting in the wrong way, or simply staying broken—we have a chromosomal mutation. If a gene is somehow changed or deleted, it's a gene mutation.

You will have noticed that nowadays when people use X-rays, elaborate provisions are made to protect the gonads—the testicles or ovaries. This is because radiation is, as noted above, one way to bring about mutations in the germ cells, the kind of mutations we call genetic. During the days when we were all arguing about fallout and nuclear testing, the genetic effects of radiation were one of our principal concerns.

But while the Atomic Energy Commission was trying to convince us that fallout was good for us (an attempt they have, happily, since abandoned), they did come up with some interesting facts—and they *are* facts. For example, geneticist Richard Caldecott, then with the AEC, wrote in *Science News-Letter* in November, 1961:

> By far the most mutagenic agents known to man are chemicals, not radiation. . . . And, in this regard, food additives rather than fallout at present levels may present a greater danger.

Now, ten years later, there are far more food additives and there is far less fallout.

Of course radiation may, as we have said, cause chemical changes in the cell, turning innocuous chemicals into mutagenic ones. "Radiation" and "chemicals" are not necessarily separate "causes." But what does it mean—now that we have some idea of the processes involved—to say that a chemical is mutagenic?

There are three *known* ways to bring about mutations, and a fourth way is suspected (there may of course be others). Most of our information in this area comes from experiments on the fruit fly, *Drosophila*, or on various kinds of bacteria. But that doesn't involve us in the kinds of questions we discussed earlier with regard to animal studies, because DNA is the same chemical in *Drosophila* as it is in Richard Nixon or Mao Tse-tung. Or, for that matter, in elephants, ringworms, or three-toed sloths. Only the code—the sequence of base-pairs—is different.[28]

As we said, one way to bring about mutations that does not involve chemicals is by "radiation targeting." That's when the free particles involved in radiation actually strike the gene itself and knock something out of whack, changing the structure within one of the base pairs so that when the DNA replicates, the result comes out wrong.

A gene can also mutate by "base analog," because of the presence of a particular chemical. One such chemical, which is easy to explain because it works in a relatively simple way, is called 5-bromouracil, or 5-BrU for short. For purposes of explanation, we'll just use the letter "B" for it.

The peculiarity of this chemical is that it can replace thymine in the DNA chain, but it will attach to either adenine or guanine. If the DNA molecule, at a particular point, is supposed to have a T-A pair, and if the 5-BrU somehow gets into the T spot, you have, instead, a B-A pair. In the *next* duplication, the intruder may hook up with a G instead of an A. Now you have a B-G pair. And then, on the duplication after that, the G will tie up with a C as it usually does. The overall result is that where you started with a T-A pair, you now have a C-G pair—and a mutation.

Because the 5-BrU doesn't "fit as well," it is likely to be edged out of the process relatively easily; but the C-G pair is just as stable as the T-A pair was to begin with, so that the mutation is likely to remain in all future replications—for generations to come, if it's a mutation that does not impair life or reproduction.

Other chemicals (acridine dyes, for example, which belong to a family of chemicals used as antiseptics) simply get into the "spaces" inside the DNA helix and in some way, not yet fully understood, cause changes to take place. We know that they can cause one base pair in a series to disappear entirely (thus changing a whole code sequence). It may be that they can enable an additional nucleotide to work its way into a chain, thus forming an extra base pair at the next duplication.

There is at least some suggestion that they may be able to reverse a base pair, so that a T-A pair, for instance, becomes an A-T pair (when the chains are separated, of course, that would change the code on each chain). And they may also be able to cause two base pairs to change places.

Besides radiation targeting, base analogs, and direct chemical deletion (or other change), there is evidence that mutations can be brought about by simple temperature change. "Thermal agitation" is the technical phrase used, because molecules move more as temperature goes up.[29] There is even a theory that mutation rates are higher in human societies in which men wear trousers, and that they might go down if men wore skirts (we would naturally call them kilts).

Our concern here is with chemical mutagens, whether they act directly as do the acridine dyes or by forming base analogs. Caffeine and theobromine, closely related to each other, are also closely related respectively to adenine and guanine; they are both known mutagens in fruit flies and in bacteria, and probably act by a form of the base-analog mechanism. They are among the substances McKusick refers to when he says, "Many . . . chemicals in our food [or medicine] have mutagenic effects in other species." While it's true that mutagenic effects have not been directly proven in man, as he is careful to note, it is also true that our DNA is the same as that in the fruit fly.

Does something bad happen every time one of these changes takes place? Probably not. The genetic "machinery," as we've said before, has a considerable ability to repair itself. In addition, every language has a certain amount of redundancy built into it to minimize the possibility of misunderstanding, which is why you can still read a lot of words even though they're misspelled (it is because there is no redundancy in numbers that Western Union always adds it, by sending numbers twice when they transmit a telegram). The extremely complex language of the genetic code is unlikely to be an exception to that rule.

But the effects, if they do survive as mutations, are far, far more likely to be bad than good. It is as though you hauled off and hit your television set with a sledge hammer; you *might* improve the reception, but you wouldn't want to count on it.

As hinted earlier, genes actually work by directing the synthesis, inside the cell, of enzymes. Enzymes are important, vital, in fact, but they are not magical, as some detergent ads would have you believe. They are simply chemicals, usually proteins, which make it possible for reactions to take place involving other chemicals.[30]

Early in the history of the individual, enzymes produced by genes determine, in effect, that the fertilized egg cell of a mouse will develop into a mouse, and that of a woman into a human being. There are hundreds of enzymes known that work in the human body, every one necessary to some important life process.

Two California scientists, George Beadle and Edward Tatum, first described the relationship between genes and enzymes (and got the Nobel Prize for it). Some of their work, on a mold called *Neurospora,* is still standard as a demonstration of the intricacy with which some enzymes work.

For example, *Neurospora,* in its "normal" or "wild" form, needs the amino acid arginine for life—but it makes its own. By inducing various kinds of mutations, Beadle and Tatum were able to show that even in that simple life form the process is quite complex.

The cells first synthesize a substance call ornithine, with an enzyme there to make the synthesis possible. In the presence of another enzyme, the ornithine becomes citrulline. And in the presence of a third enzyme, the citrulline becomes arginine, which is the point of the whole thing. Three enzymes are necessary to produce one vital amino acid; and a different gene is responsible for each enzyme.

Things like that are going on in our bodies all the time, all of them directed by the delicate chemical structures of our genes, which we may be endangering every time we take in an untested food additive.

How many mutations are there in our genes? How often do they happen? It seems as though you wouldn't be able to tell, and indeed you can't, really. But scientists have an idea. Moody quotes a study which calls it "reasonable to assume that the average mutation rate in man is about 1/100,000 per locus per generation." A "locus" is simply a gene, defined as a location on a chromosome. So the quotation means that for any given gene, one out of 100,000 in the human race mutates in each generation.

It's probably not true, but suppose there is a single gene for hair color. We each carry two, one from each parent. If there are 50,000 of us in a football stadium, then somewhere in that football stadium is one hair color gene that has mutated in the body of the person carrying it.

That mutated gene will then go into *half* of his sperm cells or her egg cells, so that mutation will be passed along to one out of every 100,000 offspring of the people in that stadium.

Each of us, however, carries not merely two hair color genes, but two genes each for, say, 30,000 different characteristics. The odds mean that three out of every ten germ cells (sperm or egg) will carry one gene that mutated during the present generation. Not three out of ten *in each person;* but an *average* of three out of ten of the germ cells in that crowd. "Actual estimates," says Moody, "are as high as this, or higher."

That's a lot. There is certainly no human need to increase the number artificially with a bunch of untried chemicals. There is, in fact, a genuine human need not to do so.

The result of those mutations may be negligible; it may be (and certainly very often is) miscarriage or stillbirth; it may be some deformity or deficiency we can clearly see (mongoloidism, mental deficiency, muscular dystrophy); or it may be something quite harmful to the individual or the species, but not identifiable as a hereditary fault.

It might be a lower intelligence level. It might be an increased susceptibility to cancer or to some other disease. It might be a somewhat slower set of reflexes. It might be a slight impairment of vision. It might, in fact, be almost any kind of human manifestation you can think of in which some humans are different from others.

Because systems are different, but DNA is DNA wherever it is found, we should keep in mind that major effects on the heredity of mice or bacteria may turn up as relatively minor effects in man. A chemical that causes a mutation in mice resulting in severe brain impairment may, in man, cause a similar mutation resulting only in every offspring being a little less quick at mathematics. It's mutagenic damage all the same.

Every one of us carries some "bad" genes—genes which, if they are dominant or have a high enough degree of dominance, will have detrimental effects. Possibly every one of us carries some genes which are lethal; that is, they will make it impossible for an offspring to live if we pass on that gene and if it is not blocked by a more dominant gene from the other parent. Perhaps the greatest geneticist of our century is H.J. Muller; Moody has summed up some of Muller's conclusions in a brief but chilling paragraph:

> Muller calculated that in *Drosophila* one gamete in every twenty [a gamete is a sperm cell or egg cell] "contains a new spontaneously arisen lethal or detectable detrimental gene that arose within the span of the very last (parental) generation." He concluded that the rate was probably higher in man, with his greater number of genes and higher body temperature. Furthermore, Muller concluded that a large proportion of human mutations gives evidence of some degree of dominance . . . Calculations . . . led him to the conclusion that "each individual, on the average, carries eight slightly dominant, detrimental mutant genes in heterozygous condition."

The last three words mean only that the mutant gene is matched by a nonmutant gene from the other parent (if you got the same kind of gene from each parent, you are homozygous for that gene; if they're different, you're heterozygous for that gene).

There are a couple of things about Muller's conclusions that might be overlooked. First, he is talking about *detectable* detrimental genes. A mutant gene which makes your child a little nearsighted, or lowers his intelligence level a notch or two, would not be detected.

Second, a mutant gene which is *recessive* (relatively) might not be detectable either; it will show up in an offspring only if he is homozygous for the mutation—that is, if both parents have the same mutant gene. The odds are against this, of course, but it's not impossible; and the gene can go on and on, in a recessive state, from generation to generation, cropping up in visible form only 100 or 200 years later.[31] How long it can "hide" depends on how common it is and thus how likely it is that carriers will marry each other.

To you and to us the obvious importance of the genetic effects of chemicals is in our children. Since most mutations *are* harmful, we think of the immediate application of that fact. Most miscarriages, of course, take place without the woman's ever knowing that an egg was fertilized at all; others, which come later, are saddening; a stillbirth is far more tragic to most people. The visible effects that are known to be hereditary, from muscular dystrophy or mongoloidism to diabetes, cleft palate, cataract, and many others, impress us most when we think of them in terms of our own families.

But we must think, too, of what is happening to the gene pool—the total number of human genes in the world that are available for reproduction. The gene pool contains all the genetic information developed by our species over the entire span of evolutionary time: in the gene pool are every adaptation our species has made, every painfully evolved defense against disease, every mechanism for warding off danger, every inherent reflex, every natural body function.

We must, in other words, think of the extent to which our inner ecologies affect the outer ecology, the reproduction and the characteristics of the entire species. For like the quiet country pools of yesterday, the gene pool, too, can become polluted. The evolutionary processes which once, ruthlessly but efficiently, eliminated "bad" genes are no longer allowed to work.

For most of human history, the mutation which resulted in phenylketonuria was passed on only rarely—because the phenylketonuriac, an idiot or imbecile, rarely mated. Today, with quick attention to the newborn, the phenylketonuriac can grow into a normal person—and mate, and bear offspring, and pass along the unhealthy gene.

For most of human history, the individual whose eyes were not so good, whose reflexes were a little slow, whose intelligence was less than normal in any one of myriad ways, would be the individual more likely killed by accident or by predator, the individual less likely to get his or her share of food, the individual less likely to succeed in the struggle for mates. Today these weakening characteristics are passed on; the individual is protected, medically and socially, from the failings of his genes.

And since almost all mutated genes *are* weakening, we are steadily increasing the percentage, in the human gene pool, of genes that are changed for the worse. It is not an exaggeration to say that the future of the human species may be as much involved with the ecology of our genes as with the ecology of our oxygen or our potassium.

This is why, at the risk of boredom or puzzlement, we have spent so much time on the mechanisms of genetics, why we have allowed ourselves to appear as though we have wandered so far from our subject. The importance of not causing cancer doesn't take much time to understand, and you have seen how the Mad Scientists are willing to lie about that. They are much more ready to confuse and mislead if we suggest the importance of protecting ourselves against mutation.

When one of them says loftily of some chemical that no mutagenic effect has ever been demonstrated in man, we want to be sure that it is understood that DNA is DNA, in man or mold. What will change DNA in one will change it in the other, or in any living creature. There may be differences in effect, because of other differences in the system, or because of differences in temperature or the chemical environment (the other chemicals in the cell with which the changed gene must work). But the DNA is the same, and the other differences are minor enough that they cannot justify gambling with our children, and their children, and all our descendants to the end of man.

In fact, *Neurospora* and *Drosophila*, and the bacterium *Escherichia coli*, are closer to man, when we're talking about genetics, than are rats or mice when we're talking about cancer. They are used in genetic research because the DNA is the same, and because they breed quickly and can be observed in large enough numbers so that results can be measured statistically. To show the same effects in man we would have to study enormous populations over hundreds of years. But the mechanics of what's happening are exactly the same.

We have been able to test caffeine and nitrogen mustard, formaldehyde and hydrogen peroxide, and a number of other chemicals, and to determine quite clearly that they have a mutagenic effect on the forms of life most frequently used for such research. Three of the four chemicals named are legal food additives today.

The Food and Drug Administration, which *has* to pay attention to carcinogenicity when it is hit over the head with the facts, and which occasionally pays attention to teratogenicity if a scandal appears to be in the offing, has never shown any sign of concern about mutagenicity at all. Almost none of the thousands of additives in our food have ever been tested even on molds or bacteria.

And yet this "general reduction in the viability of succeeding generations," as Dr. Joshua Lederberg calls it, may be, of all the dangers we face from the indiscriminate use of food additives, the greatest of all.

CHAPTER TEN

No Other Gods

I sat in a kitchen once and watched a woman I knew make a purple cake with green frosting. Or maybe it was a green cake with purple frosting. It was a long time ago.

Anyway, she was having fun with a couple of bottles of food coloring. I sat and watched her make the whole cake. I knew that it was a perfectly good cake except for the three or four whimsical drops out of those bottles. I knew that the color didn't change the quality. I knew that the cake would taste exactly the same to me.

And I had a hell of a time forcing myself to eat it.

She had two young sons, who had no such problem. They thought that a purple cake with green frosting was fun. It may even be that in her way, she was trying to teach them something: that they ought to pay attention to facts (the quality and the taste) instead of to appearances.

But we're sure it didn't work. They're both grown men now, and we're sure that they're just like the rest of us. The prettier the food is, the more it conforms to our expectations with regard to color and flavor and texture, the more "attractive" we're likely to find it.

Of course there's an ecological basis for that. During most of our evolutionary history, subtle differences in the appearance or the feel or the taste of a food made the difference between death and survival. It's only natural that we should feel a profound disturbance at the idea of eating pink peas—or purple cake.

If all that the food industry's Mad Scientists did were to use a dye to make a green orange look like an orange orange (a common practice today) then we could say that they're taking advantage of a normal human tendency: the tendency to want our food to look as it is naturally supposed to look. But they do much more than that.

Over more than a generation, they have subtly worked to change that normal tendency, that almost built-in conditioning, to suit their own ends. They have succeeded in conditioning us, as a nation, to the point at which food that looks, or feels, or tastes natural seems to us to look or feel or taste *wrong*.

Today (and they gleefully discuss this in their trade publications), any orange juice or orange soda or orange drink must be an almost precise shade of orange, or most of us will reject it. Margarine was so successfully colored to look like butter (that is, like our idea of what butter ought to look like) that now they have to color butter to make it look like that too. There is an exactly right color for a chocolate cake, for frozen peas, for mayonnaise, for butter brickle ice cream.

Flavors, too: we have seen children, accustomed to canned or frozen fruits or vegetables, reject the fresh product because "it doesn't taste right." And if your children are under ten—probably even if they are under 20—they have never encountered even a minor variation in the texture of ice cream.[1] To them, that *is* the texture of ice cream, and any other texture is disturbing, "funny," wrong.

Food processors are not supposed to use color, or flavors, or flavor enhancers, or texturizers, or emulsifiers, "in a manner that may promote deception or conceal damage or inferiority," or may make "the article appear to be of better or greater value than it is." The language is from the *Code of Federal Regulations*. But in fact, there is no other reason to use any such additive at all.

Think about it. Why should any processor put into a food an additive whose only function is to change the color? Especially, why an additive whose only function is to change the color in such a way as to make the product look like something else, or look like a better grade of the same thing, or remind you of some other product or ingredient?

It may not always be to "conceal damage or inferiority" although we don't know what else you call it when almost every Florida orange grower runs his oranges through a chemical dye called Citrus Red No. 2 so that the green ones and the orange ones come out looking alike. It is certainly always to "promote deception." There is nothing else it can be.

It is not an accident that the most dangerous food additives known to have been used over the last 50 years have been colors and flavors;

dozens of them have been banned, most of them chemically related to substances still in use.[2] And it is no accident that the Mad Scientists and their bosses have fought the banning of every one, and fight every investigation into the ones still in use.

The explanation is in these 1966 figures. The cost to the processors of phony flavorings in that year was $100 million, which means that we paid considerably more than that. Or perhaps you'd like to know how much phony coloring we ate that year. One-and-a-half million pounds.

When the industry talks about using flavors or colors, by the way, they like to talk (in public) about the "natural" ones—caramel or fruit juice or turmeric or saffron. By far the bulk of what is used, however, is synthetic. Read the labels on your next supermarket tour; most references to colors and flavors read "artificial" or "synthetic" or "imitation" or, in the case of colors, "certified," which means the same thing.[3]

To the Food and Drug Administration, needless to say, the use of these additives is not deception at all. Their reasoning, after the usual obfuscations of governmental prose are deciphered, is clearly derived from Alice's Wonderland: deceptive use of food additives is illegal; food additives are widely used by major food processors; therefore, their use of food additives must not be deceptive.

Back in 1956, a Committee on Food Additives organized by the Food and Agriculture Organization of the United Nations and by the World Health Organization met in Rome. They came up with a list of reasons for the use of food additives, and a list of reasons why they shouldn't be used. One reason for using additives, they said, was "making food attractive to the consumer in a manner which does not lead to deception."

What they did not say was how that is possible within the limits of logic and the English language. Of course the food industry itself will tell you as James Noonan does in the colors section of the *CRC Handbook* that because foods with "odd" colors are unappetizing, there is a *need* for food colorings, to meet consumer demand. We have heard that argument before.

And speaking of the English language, another contributor to that volume says, of flavor, that it is "the integration of the sensations of odor, basic tastes, feeling factors [and] texture," and adds that "texture must be considered as contributing to flavor." That may not be what you thought "flavor" meant, but it gives you a clue to the psychological-manipulative orientation of the Mad Scientists.

One investigator (Bicknell) has tracked down the fact that the use of thickeners, emulsifiers, and phony colors goes up in baked goods proportionately to the extent to which the use of cream and eggs goes down. The Delaney Committee, checking on commercial cake formulas, found that as the amount of egg or egg yolk went down, the amount of yellow dye went up.

It may be that "odd" colors are unappetizing, but it's because we know that those "odd" colors reflect the odd things the Mad Scientists are doing to our food. And it doesn't imply a demand for color additives; it implies that a consumer expects eggs in his cake.

But while they certainly try to con you, they don't waste any time trying to con each other, either about the real purpose of food additives or about the profit motive that lies behind it. We went through six issues (July-December, 1969) of *Food Technology* and read the ads they write for each other; somehow, they don't read like the food ads in *Family Circle*. They read, instead, like this:

> Get in league with Green & Green and you'll have a grape flavor for your product so natural with its fresh-from-the-vine taste that your customer will think he's been transported to a vineyard in the south of France.

Or try this:

> Stange's imitation cinnamon has a taste, aroma and texture so similar to the natural cinnamons that we invite you to test and compare . . . All yours for a fraction of the cost of natural cinnamons.

The motivation in the next is even clearer:

> Ac'cent Yeast Extract makes a cheesier cheese product. A brothier, meatier meat product. Better batter mixes, salad dressings, sauces, tomato products, soups and stuffings. And often, you can use less HVP, less prime ingredients (or economy grades of both) yet get fuller, more satisfying flavor!

Or, perhaps most honest of all:

> Tomato juice without tomatoes? There's no limit to the possible uses of specialty synthetic flavors in substitute and natural foods.
>
> Kohnstamm's sophisticated second generation of flavors doesn't stop at a convincing forgery of vine-ripened tomato. It includes . . . hordes of other fool-the-tongue tastes.

In its appearances before congressional committees or in the propaganda materials it churns out for schoolroom use, the industry is never quite so blatant about "forgery" or "synthetic flavors in substitute . . . foods." The clearest of many indications we found of the industry's honesty with itself about its motives are in these three separate ads for a product of the National Starch and Chemical Corporation—and if they aren't talking about deception, it's hard to decide what else to call it:

> That's what Textaid is all about: giving food those important extras of texture, body, and mouth-feel. Textaid works particularly well where there's need to restore natural food appearance lost through heat or complex processing.

> Textaid not only gives spaghetti sauce a home-style pulpiness, but adds body to shrimp sauces, tomato soups, and barbeque sauces.

> With [Textaid], you get that full-bodied, delicious mouth-feel which is what texture is all about . . .
> Textaid makes a natural crushed-fruit look and feel in juices, nectars, jams, pastry fillings and ice cream toppings.

Of course when "natural food appearance" is "lost through heat or complex processing," it probably means that something more than appearance is lost. But even these don't exhaust our supply of examples; for a time we considered making a complete chapter of food industry advertising copy, without comment.[4]

Chas. Pfizer and Company advertises a series of additives and adds that "all are sales enhancers when they're from Pfizer." Hoffman-La Roche says that "you can almost taste the colors." Givaudan Corporation, with a lofty and precious reference to "the flavor scientists' art," says that that art "starts with ceaseless searching," and "ends in repeat sales."

Notice that nobody's talking about "odd colors" or "making food more attractive," whatever that might mean. They're talking about covering for inferior ingredients, disguising the effects of processing, conspiring to fool your eye and your taste (and your "mouth-feel"), and increasing their profits.

What does this mean in your shopping cart? More than you think, especially if you believe, as so many people do, that the phony flavors and colors exist mostly in the more complex kinds of processed food. Do you think of butter? Breads? Meat? Even eggs?

Butter is both colored (especially winter butter, which is paler in its natural form) and flavored. When it's going to be stored for a while, or shipped for some distance, butter is "washed"—a process which both destroys its aroma and, probably, removes its Vitamin A. To make up for this, the Mad Scientists add a dash of a reactive chemical called diacetyl, which makes it smell like butter again, but doesn't do anything about any missing Vitamin A.[5] Diacetyl also lends its aromatic fakery to margarine, cottage cheese, buttermilk, ice cream, candy, chewing gum, shortening, gelatin desserts, and baked goods—at least.

Yellow dye to make you think that there are more eggs or butter in baked goods is so common a practice that even a Congressional committee could find it. Sodium nitrate and sodium nitrite, both used as preservatives in prepared meats, also function to keep them looking nice and red even when they've been on the shelf a lot longer than they should be (we mentioned their use in hot dogs back at the beginning of the book).

That same yellow color is usually associated with Vitamin A in things like egg yolk, but even that may be a fake. At least three color additives—dried algae meal, tagetes meal, and tagetes extract—are added to chicken feed in order "to enhance the yellow color of chicken skin and eggs" (that's CFR language).[6] There isn't any Vitamin A in algae of the genus *Spongiococcus*, or in the petals of the Aztec marigold, the raw materials for those additives.

Red and yellow color is associated with Vitamin A, usually, because of the natural presence of carotene, out of which Vitamin A is manufactured by the body. Actually there are several "carotenoids" ("things like carotene"), some of which are completely natural in the foods they're in, some of which are extracted and put into other foods, and some of which are manufactured in laboratories.

That's one reason for the fact, mentioned in a previous chapter, that "certified" and "uncertified" colors are not the same as "natural" and "synthetic," although a lot of people think they are. Beta-carotene is "natural," but it can also be made synthetically. In fact, making it synthetically is a pretty big business.

Most carotenoids have Vitamin A activity, although at least one, called canthaxanthin, does not. As colors, the carotenoids function in nature to give the color to carrots, tomatoes, apricots, orange juice, and lobsters, which will give you an idea of their color range (they actually reach much further down on the yellow side as well).

Originally used as a phony coloring in margarine and oils, beta-carotene is now produced in various forms to color orange beverages, puddings, cheese, ice cream, popcorn oil, and salad dressings. It's useful not only because of its Vitamin A activity but because it doesn't break down in the presence of Vitamin C, as some colors do. Carotenoids are also used in other dairy products like skim milk, buttermilk, and cottage cheese, in order to make them look less "blue-white" and just a tiny bit more "cream-colored." This of course has nothing to do with nutrition.

But the industry, which cares not at all about your Vitamin A and is interested in only one authentic color—green—doesn't like beta-carotene, because it's expensive. In recent years the industry has turned more and more to another carotenoid, annatto extract or bixin, which is a lot cheaper. Of course, wherever possible they use the certified colors, which are cheaper yet; but thanks to the fact that two yellows were finally banned after a lot of us ate a lot of them and a few of us probably got cancer, there are no certified colors available for some uses and they have to use bixin.

As nearly as we can find out, however, bixin doesn't have the Vitamin A activity that beta-carotene does. The FPC's own list of food additives lists both products under colors, but lists only beta-carotene under "nutritive supplements."

Chocolate cakes and cake mixes certainly contain at least two phony colorants, and in a good, rich-looking chocolate cake mix you may find caramel, carbon black, an FD&C Red, and a fourth dye in some shade of brown. If that doesn't destroy any illusions for you, how about phony coloring in rye and pumpernickel bread? It's usually there.

FD&C Red No. 3 (erythrosine) is in cherry-pie mixes and canned fruit cocktail. There's ferrous gluconate in canned black olives, and Orange B (one of the two certified colors without the letters "FD&C" in front of it) in sausage casings. FD&C Yellow No. 5 (tartrazine) is in prepared breakfast cereals. Cured or canned meat products may have carmine or carminic acid as well as saffron and turmeric.

Carmine, by the way, is a derived form of carminic acid. Carminic acid is the active coloring matter in cochineal, which is itself an approved uncertified color additive. Cochineal is the dried bodies of the females of a cactus insect found in the Canary Islands.[7] What is "natural" is not necessarily natural food.

Almost all beverages are artificially colored. So are almost all bakery products, including dough products, sandwich fillings, icings, and ice cream cones. Chocolate is the only flavor of ice cream which is sometimes not artificially colored. Added colorants, certified or not, do *not* have to appear on ice cream labels, by the way;

there's a special exemption in the 1938 Food, Drug, and Cosmetic Act.

If you'd like to conduct a little home experiment for the kids, buy a clear glass bottle of grape soda (realizing that it may not have any grape in it, by the way[8]) and, without opening it, leave it in the sun. It will take a while, but the nice purple soda will turn green. FD&C Red No. 2, the certified color used, "fades preferentially" in sunlight, to use the industry phrase. Doesn't change the content any; you can still drink it, if you like green grape soda. Or conduct another experiment in conditioning: give it to the kids next door, tell them it's lime soda, and see whether they can tell the difference.

Turner has a long description of the way in which FDA standards for orange juice and related drinks have been suspended while the Florida and California orange growers work out differences (the California juice has a naturally deeper color; the Floridians want to add coloring to make up for it; the Californians don't want it that way). The standards, proposed in 1968, have still not been put in force because the FDA is afraid to get into the fight. In the meantime, the whole industry is fighting the idea that the percentage of actual orange juice ought to be listed on labels.

It's reminiscent of the fight that took place some years ago when the FDA finally banned a certified dye, FD&C Red No. 32. Florida and Texas orange growers screamed because they were accustomed to running their green oranges through a bath of dye to make them look orange when they got to the store. "An industry spokesman said," according to Longgood, "that more than half the Florida orange crop was run through a dye bath to give green oranges an orange color."

It still is; they just use a different dye, called FD&C Citrus Red No. 2. This is an "azo dye," a classification about which we'll say more later; here we'll note only that a lot of people have the same questions about it that they had about the now banned and possibly carcinogenic dye that is replaced. The regulation on Citrus Red No. 2 reads in part:

> Citrus Red No. 2 shall be used only for coloring and skins of oranges that are not intended or used for processing . . . and that meet minimum maturity standards established by or under the laws of the States in which the oranges are grown.[9]

In other words, it's up to Florida, not to the Food and Drug Administration, to decide which Florida oranges can be dyed. The same state control exists in California and Texas. If you know anything at all about state governments, you can guess who controls the making of standards in each state. You can also guess how much chance you have of buying an orange without any dye on the outside. Remember, it's cheaper to run the whole crop through the dye bath than it is to go through and pick out the green ones.

Incidentally, the oranges used to make frozen or canned juice aren't supposed to be dyed—and the dye isn't supposed to go through the peel anyway. But investigators have often found Citrus Red No. 2 in the juice. It's pretty hard to keep track of every orange.[10]

As we noted in Chapter Four, almost all the food additives in use are flavorings: 1,077 flavors out of a total of 1,622 additives classified by the Food Protection Committee. Other sources such as the *CRC Handbook* give larger numbers, but the proportion is about the same; and in all lists, about 70 percent of the flavoring agents used are completely synthetic. That's not even including synthetic duplications of "natural" substances like beta-carotene.

Perhaps once again we ought to note that the whole classification of colors or flavors as natural and synthetic is more confusing than helpful. A substance can occur in nature, but be added to a food in which it does not occur naturally—as when the coloring matter that occurs naturally in carrots is used to fake the color of strawberry soda. Other substances that occur in nature can also be manufactured in a laboratory.

When we say, then, that about 70 percent of flavorings are completely synthetic, what we mean is that they are not laboratory duplications of real substances; they're substances that either don't occur in nature at all or, if they do, occur only in small quantities and not at all in the foods to which they're being added.[11]

Numbers of additives, however, don't tell the whole story about the enormous, and enormously profitable, food additive business. There are only 35 colors on the FPC list (34 if you count beet juice and beet powder as the same thing); but there is probably more artificial coloring, by volume, in your supermarket right now than there is any other additive type. Historically, phony colors provide the best examples of how little the food business cares for your nutrition or your health, and how determinedly they will fight for every penny of profit.

The Food and Drug Act of 1906, passed before the industry gained its present control over the people who are supposed to govern it, put a severe crimp in the activities of dye manufacturers and the people who used their products. In 1900 there were at least 80 dyes used in foods in the United States (although not in nearly so many foods as now); the Food and Drug Act banned all of them that were artificially manufactured, except for seven.

All seven were coal-tar dyes (the "certified" colors in use today are also all coal-tar dyes, although the industry would prefer that you forget that fact, and has succeeded in getting the phrase cut out of the law). Of the seven, three are still in use today. Known then (and still in other countries) as amaranth, erythrosine, and indigotine, they are now respectively FD&C Red No. 2, FD&C Red No. 3, and FD&C Blue No. 2.[12] The other four were called Orange 1, Ponceau 3R (a bright, yellowish red), naphthol yellow, and light green; the last was FD&C Green No. 2.

Of course the industry couldn't stand still for only seven coal-tar dyes, and over the next quarter-century pressed continually for additions, which they got piecemeal. Tartrazine (now FD&C Yellow No. 5) was added in 1916, and is still in use. In 1918 they got Yellow AB and Yellow OB, which became FD&C Yellow No. 3 and No. 4 respectively and were ultimately banned, decades later, as carcinogenic.

During the next 11 years were added Guinea Green, Fast Green, Ponceau SX (a red), Sunset Yellow, and Brilliant Blue. Their FD&C names ultimately became, in order, Green No. 1, Green No. 3, Red No. 4, Yellow No. 6 and Blue No. 1; the first is now banned, and the third, you'll recall, is allowed on maraschino cherries despite its poisonous effects.

When the new Food, Drug, and Cosmetic Act was passed in 1938, the certification became mandatory. This means that each batch of the dyes had to be certified as being what it said it was. The new naming system went into effect. Colors labeled "FD&C" are supposed to be safe for use in food, drugs, or cosmetics. Other colors are labeled "D&C" or "Ext. D&C," the latter meaning that they're safe for external use only.

The important fact about the 1938 law and colors, particularly what are now called "certified" colors, is that the safety regulations did not say anything about amounts. Any dye, to be used in food, was supposed to be safe to take into the stomach—period. Since there had been questions raised about the safety of coal-tar dyes since the time of Harvey Wiley, the food industry was anxious that no restrictions be placed on their use, and they got their wish.

But after everyone alive in America had eaten pounds upon pounds of the stuff, the law backfired on the industry in the 1950s. The food industry's public-relations-oriented version of what happened can be found in the *CRC Handbook,* in the colors article by James Noonan:

> [In] the early 1950s . . . there occurred three unfavorable incidents as a result of the overuse of color in candy and on popcorn. This overuse resulted in a number of cases of diarrhea in children. The colors involved were FD&C Orange No. 1 and FD&C Red No. 32. These incidents coupled with chronic toxicity animal feeding studies led to the delisting of FD&C Orange No. 1, Orange No. 2, and FD&C Red No. 32.

First of all, note the double reference to "overuse." There was no overuse; there could be no overuse. The dyes were legally supposed to be safe in any amount, and the law put no limits on the amount that could be used. The choice of the word "overuse" is at least disingenuous.

Second, notice the stress on the "popcorn incident," and the inconspicuous reference in passing to "chronic toxicity animal feeding studies." In fact, the first notice of a hearing for the delisting of the three dyes was published in *The Federal Register* on December 19, 1953. It had nothing to do with that Christmas party during the 1954 holiday season at which 200 children were poisoned by the dyed popcorn. The importance of that incident was only that it attracted wide public attention, and forced the hand of the FDA, which had been stalling for over a year. Even then, the ban didn't take effect until February 15, 1956—and the FDA thoughtfully told the industry that they could go ahead and use up whatever dye they had in stock. They merely cautioned the Mad Scientists not to use too much at a time.

The report on which the ban was really based can be judged from this passage on FD&C Red No. 32 which, before it was banned, was used for 18 years, not only on dyed popcorn but in cheese, fats, oils, candy, bakery goods, and (remember?) on the skins of oranges:

> When FD&C Red No. 32 was fed to rats at a level of 2.0 percent of the diet, all the rats died within a week. At a 1.0 percent level, death occurred within 12 days. At 0.5 percent, most of the rats died within 26 days. At 0.25 percent approximately half the rats died within 3 months. All the rats showed marked growth retardation and anemia. Autopsy revealed moderate to marked liver damage. Similar but less severe results were obtained with rats on a diet containing 0.1 percent of FD&C Red No. 32 [remember that a ratio of 100 to 1 is considered a "normal" safety factor]. In addition to liver damage, however, autopsy also revealed enlargement of the right side of the heart in

> this latter group. . . . A level of 0.2 percent of FD&C Red No. 32 in the diet of dogs caused rapid deterioration and weight loss and sporadic diarrhea, moderate atrophy of vital organs, and muscular dystrophy; 0.01 percent in the diet [note that that's only one part in ten thousand] caused weight loss and the death of one out of four dogs.[13]

But the industry spokesman tells you that they banned the stuff because some kids got a little sick at a party.

We put in the reminder about the 100-to-1 safety factor in the middle of that quotation because some of the smaller doses fed to the animals in that test were close to the actual proportions of the dyes in some of our foods, not 100 times higher. Orange B, for instance, is approved in sausage casings at a tolerance of 150 parts per million; the dog tests of Red 32 actually used a lower dose—100 parts per million. Of course we eat other things besides sausages, but you can see that the safety factor was well exceeded.

Mr. Noonan also—if the phrase may be permitted—whitewashes a couple of yellows. After the popcorn incident attracted attention and the three dyes were banned (it was in that fight, incidentally, that the widespread orange-dyeing practices of the big Florida growers came to light), the food industry decided that maybe there should be maximum amounts set for dyes after all. The law said that the dyes had to be harmless in any amount; but now, the Mad Scientists wanted to be able to go on using poisonous dyes even if they were shown to be poisonous.

They fell back on the argument, which we've encountered before, that a poison isn't a poison if you take it in small enough amounts. As we've seen, this is a half-truth; some substances have cumulative or synergistic effects. But the food industry now blamed the whole thing on "overuse," and started pushing for the FDA to set "maximum tolerances."

There was, however, no legal authority for such a practice, and the whole thing got into court. This is Mr. Noonan's version:

> After several conflicting opinions in the lower courts, the Supreme Court ruled that the FDA, under the 1938 law, did not have the authority to set quantity limitations. Following soon thereafter were delistings of Yellow No. 1, Yellow No. 2 and FD&C Yellows 3 and 4, the remaining oil soluble colors.
>
> The Supreme Court had rendered the old law obsolete and unworkable. . . .

It was all the fault of those nine old men. Except that a few hundred thousand cases of cancer are overlooked in this version.

The International Union Against Cancer had already listed all four dyes under a classification headed "unsuitable and potentially dangerous," and had said that they definitely shouldn't be used in food. FD&C Yellows No. 3 and No. 4, in use since 1918 under the old names Yellow AB and Yellow OB, are both made from beta-naphthylamine, which Dr. W.C. Hueper described as "one of the most carcinogenic substances known" (you'll recall from Chapter Eight that Dr. Hueper said he was fired by DuPont for insisting that beta-naphthylamine was a carcinogen). Some large chemical manufacturers had already stopped making the stuff because between 50 and 100 percent of their workers, exposed to it, came up with bladder cancers.

Not incidentally, the first tests which showed carcinogenic potential in animals from Yellow AB and OB were completed 20 years before the dyes were banned. In the meantime, Americans ate at least a half-million pounds of the stuff. Noonan suggests that the dyes were banned because the Supreme Court was unreasonable in its interpretation of the law. In fact, they were banned because they're damned dangerous, and experiments showed it so dramatically that they couldn't be ignored.

But a big industry is a big industry; if the laws restricting your profits are too tough, get the laws changed.

It was at about this time that the U.S. Rubber Company was pushing the idea of a tolerance level for the known carcinogen, Aramite—and getting it, thus setting a precedent for the tolerance idea. The rest of the industry clearly understood that this might be a way to keep manufacturing, and selling, dangerous and potentially poisonous food additives, in a business getting bigger all the time. Since the 1938 law didn't allow tolerances, the need was obviously to change the law. In public, this involves talking about "reasonable testing" and "realistic regulations," and calling your opponents "emotional" and "unscientific."

When the Food Additives Amendment came before Congress, the dyemakers took one look at that pesky Delaney Amendment and decided that they wanted no part of *that*. Subsequently, the Certified Color Industry (and of course the humbly cooperative FDA) sponsored a separate amendment, which became law in July, 1960.

The new amendment not only allowed the setting of tolerances; it also carried its own version of the Delaney Clause, rigged to permit the HEW Secretary to ban or not to ban as he chooses, regardless of the evidence of carcinogenicity. As an added prize, the amendment finally cut out of the law the hated words "coal-tar dyes," which the industry doesn't like because they exactly describe the products.

The strange part of the Color Additives Amendment is that it contains a "temporary" part, to govern the first two-and-a-half years while legal changeovers were made, and a "permanent" part, to govern the use of color additives afterward and presumably forever. What makes that strange is that now, 11 years later, it is still the "temporary" part of the law that governs most use of color additives in America.

What happened was that the law allowed for the *provisional* listing of a number of dyes, with the idea that after two-and-a-half years of studies, the dyes would either be banned or permanently listed (you will have noticed that once again we were to be allowed to go on eating the stuff while it was being tested). The HEW Secretary, however, was given the right to extend the provisional period and that's just what succeeding secretaries have done. Of the 11 certified dyes now in use, only two are permanently listed, and they have only limited use.

The studies were supposedly completed in 1964. In the meantime, the results knocked two dyes—FD&C Red No. 1 and FD&C Red No. 4—off the list. No. 4, you'll recall, was put back on for use in coloring maraschino cherries despite its clear toxicity. In 1966, two more FD&C dyes—Green No. 1 and Green No. 2—were dropped; Noonan says it was because of "the lack of demand," but the IUAC lists them both as "unsuitable or harmful" and says that they're at least possible carcinogens. A 1953 FDA study shows Green No. 2 as carcinogenic in an injection study.

The two dyes now permanently listed are Orange B (used, as noted, only for sausage and frankfurter casings and with a tolerance of 150 ppm) and Citrus Red No. 2, the orange dye-bath, with a tolerance of two parts per million calculated against the weight of the orange. It would take a pretty good scale to test an orange.

The reason that those are the only two is simple: now that the industry has won the battle for tolerances, it wants to avoid them. The first application for permanent listing of one of the more widely used dyes was filed in February, 1965, on behalf of FD&C Yellow No. 5. As tartrazine, it has been with us since the late nineteenth century—but that does not by any means make it safe, as should be clear by now.

The industry ignored all the arguments it had made about tolerances, and asked for no quantity limitations other than "good manufacturing practice." The FDA decided that that was a little much, and ordered it listed at a tolerance of 300 parts per million. The industry protested, the order was stayed in June, 1966, and there the matter has sat ever since with the dye still being used in uncontrolled amounts. Another petition, on behalf of FD&C Red No. 2 (also an old one formerly called amaranth), has been filed, but it also sits, awaiting the judgment on tartrazine.

Obviously the industry couldn't care less whether the dye is listed provisionally or permanently; what they want is for it to be listed with no controls on its use. They wanted tolerances when it appeared that without tolerances they would have to stop selling some dyes at all. They were happy to accept a limitation on Red No. 4 for use on cherries rather than have it completely banned. They're willing to accept tolerances for the relatively minor uses of Orange B and Citrus Red No. 2, especially since those tolerances are more than ample for the way in which the sausage and orange people want to use the dyes.

But of course they will fight to block a permanent listing with a tolerance (no matter what the animal studies may say about the dyes and no matter what the effect on American health) when they can have a provisional listing without a tolerance, and sell that much more dye. That's why we're still governed by a "temporary" part of the law.

The law, incidentally, would *seem* to say that a permanent listing without any tolerance at all is impossible, although that is just what the industry is asking for. The regulations governing permanent listing *seem* to demand a tolerance, because they provide that

> a color additive for use by man will not be granted a tolerance that will exceed 1/100th of the maximum no-effect level for the most susceptible experimental animals tested.[14]

Since presumably every additive has some effect on some animal at some dosage, it appears that there must be some tolerance. But not according to the Certified Color Industry. And of course, that doesn't affect the provisional listing of dyes that have been in steady use under the present law since 1960.

All this date-juggling is probably confusing. We should perhaps recall that before the 1960 Color Additives Amendment, all the dyes now in use (and some now banned) had been certified under the 1938 law. In 1956, four years before the new law, Wallace Janssen, then assistant to the FDA commissioner, said in a speech that

> under the law, the colors are supposed to be certified as completely safe—so that no matter how much color is used it would not be harm-

ful. No legal action is possible against anyone for using an excessive amount of a certified color.

Under the new law, that is still true, except for the two permanently listed dyes and the use of FD&C Red No. 4 on cherries.

We haven't talked much about the uncertified colors, because they're much less widely used and much less worrisome. Their function, like that of the certified colors, is to deceive you or to cover up for inferior raw foods; but they are probably, for the most part, less dangerous. None of them is provisionally listed any more; they're all permanently listed now—and interestingly enough, despite the regulation quoted above, few of them have tolerances.

Some have limited uses. Dried algae meal, tagetes meal or extract (we mentioned those before), and corn endosperm oil are for chicken feed only, all used to make skin and eggs yellower. Ultramarine blue (an example of a coloring that is synthetic but uncertified) is used only for coloring salt used in animal feed. Synthetic iron oxide goes only into pet food, and ferrous gluconate is used only to color black olives (did you think they were really that nice even black color?).

Only two uncertified additives have tolerances. A synthetic carotenoid, beta-apo-8′-carotenal, is restricted to 15 milligrams per pound or pint of the substance colored. Titanium dioxide, used mostly as a white pigment for candy but not limited to candy, is restricted to 1 percent of the final product.

The other uncertified additives include other carotenoids, fruit and vegetable juices, grape skin extract, paprika, saffron, turmeric, cochineal (those dried cactus insect bodies) or carmine, riboflavin (also a vitamin), carrot oil, and beet powder.

While a few of these are extensively used in particular products (some "flavors" of soft drinks, for instance, use a lot of beta-carotene), they make up a very small proportion of the total coloring matter used in American foods. We are fooled much more often by the certified colors—and endangered by them.

Including Red No. 4, the one used only on cherries, there are now nine FD&C dyes provisionally listed,[15] and the two previously mentioned that are permanently listed: Orange B for sausage and frankfurter casings and Citrus Red No. 2 for dyeing green oranges orange. The other eight "provisionals" are Reds 2 and 3 (amaranth and erythrosine), Yellows 5 and 6 (tartrazine and sunset yellow), Blues 1 and 2 (brilliant blue and indigotine), Green 3 (fast green), and Violet 1, which doesn't have a more common name.

Every one of them is a polycyclic aromatic hydrocarbon, a member of that class of chemicals against which international bodies continually warn because of potential cancer danger. Considered particularly dangerous by some cancer experts are the azo dyes, Reds 2 and 4, and Yellows 5 and 6 (Red 2 has four benzene rings, for example).

Three others, Blue 1, Green 3, and Violet 1, are of a type called triphenylmethane dyes; Blue 1 has four benzene rings. Red 3 is a fluorescein dye with two benzene rings, and Blue 2 is a kind of dye all its own, also with two benzene rings.[16]

Before all the detail piles up to hide the meaning, let's restate it.

Almost all of our food is artificially colored. Almost all of the artificially colored food is colored with certified dyes. All of those certified dyes belong to a class of chemicals generally recognized as dangerous because of carcinogenic properties.

Or to make it even plainer: it is unlikely that any one of us is not ingesting *some* of a carcinogen with our food every day. And there is nothing whatever that we can do about it except to stop eating.

As we have already indicated, the imperfect but important protection we get from the Delaney Amendment with respect to other food additives has been much watered down in the Color Additives Amendment. Mr. Noonan's article describes it almost correctly, and at the same time indicates by his choice of words and the space he devotes to it just how much concern the industry has about your contracting cancer:

> This clause is softened somewhat by permitting, in cases invoking the cancer clause, the appointment of an advisory committee to serve as a fact finding body. The advisory committee reports its findings to the Secretary who makes his decision. He is not bound to follow the recommendation of the advisory committee.

You might indeed say that it is "softened somewhat"!

It is the FDA Commissioner, not the Secretary, to whom the committee reports; and the regulations further provide that the committee must be selected, not by the secretary or the commissioner, but by the National Academy of Sciences, about which you already know. But the important fact is that nobody has to do what the committee says; in other words, the flat statement that a carcinogenic substance can't be used does *not* apply to color additives.[17]

That takes us back to "What's a carcinogen?" And to the answer, which is that the question is more political (or economic) than scientific. The "expert" answers, based on what seem like reasonable scientific criteria, are—scientifically—no better than the "expert" challenges from the scientists of other countries or of international bodies.

Differences in standards from one nation to another are obviously based as much on the political clout of the dyemakers as they are on science. The International Union Against Cancer, the Food and Agriculture Organization and the World Health Organization think that serious attention ought to be paid to the results of injection experiments on animals. The Food and Drug Administration does not. And while they disagree, we continue to eat the dyes, as we have already eaten hundreds of thousands of pounds of other dyes which have since been proved carcinogenic.

The IUAC rates our Green No. 3 and Blue No. 1 as "unsuitable or potentially dangerous . . . should not be added to food or drink for man or animals." In 1953 the FDA itself reported that both cause cancer when injected. The green dye is used in mint jelly, candy, desserts, and bakery goods, among other things; the blue is in such things as icings, cordials, jellies, ice cream and ice cream toppings, syrups, candy, cake decorations, puddings and frozen desserts, bakery goods, and food color solutions for home use.[18]

Dr. Hueper, using the IUAC definition of a carcinogen given in Chapter Eight, flatly and correctly lists both Green 3 and Blue 1 as carcinogens. A joint FAO-WHO committee said some years ago that there was at best insufficient data on either to rate them acceptable; they have not changed their rating since. A great many other countries ban them.

We, on the other hand, have been eating Green No. 3 since 1927 and Blue No. 1 since 1929—with what results, no one can tell.

Both of them are triphenylmethane dyes; Violet No. 1 also falls into this category and is closely related chemically, but it's relatively new and not nearly so much work has been done on it.[19] Aside from the confusion about the safety of these three dyes, the leading British authority, Bicknell, thinks that the azo dyes are particularly dangerous: Yellows 5 and 6 and Reds 2 and 4 are the ones currently permitted in America (Red 4 is the maraschino cherry one).

Calling them "triphenylmethane dyes" and "azo dyes" is simply a reference to their chemical structure; the ones in each group tend to behave alike chemically because of the way the molecules are "built." Bicknell describes for a more technically oriented audience the precise mechanism by which he believes that azo dyes are likely to break down, in the body, into cancer-producing chemicals of a class called *ortho*-hydroxamines.[20] This is what happened in the case of the beta-naphthylamine that caused all the bladder cancers and resulted in the banning of two other Yellows—also azo dyes.

Bicknell goes on to point out, as we suggested in less detail in an earlier chapter, that "simple" animal-feeding tests are not all that simple:

> To test a dye on only two species needs 1,440 animals with another 80 if there is doubt about whether the animals may be unusually resistant to cancer. Even so these experiments will throw no light on whether changes in diet, other chemicals added to food, physical accidents, leanness or obesity, illness, pregnancy, etc.—in fact the inevitable changing background of human life—will cause an apparently harmless dye to become cancer-causing. Such possibilities are not theoretical academic niceties, since fat men are more prone than thin to cancer, while in animals and presumably in man both excesses and deficiencies of essential ingredients of food may enhance the susceptibility to cancer—presumably by altering the amount of dye converted in the body to a cancer-causing substance.

Of course, in a rational society, all these arguments would be enough to make us stop using the dyes on the ground of simple prudence. But in a rational society we wouldn't be using them in the first place. Much less would a few corporations be making millions of dollars a year by putting junk into our food that serves no purpose except to make us think the food is something it isn't.

In the case of the food dyes (the certified colors that by volume make up most of the phony coloring used in American food), the FDA's argument for insisting that they're safe is that they don't show carcinogenic results when eaten—only when injected, or implanted, or otherwise taken into the body.

As usual, the FDA is taking the narrowest possible view, in order to justify supporting the industry that makes millions on this unnecessary stuff. They continually overlook, as they do with other additives, the point that the human body is a delicate instrument that has evolved over hundreds of thousands of years, in a precise ecological relationship to its foods—so that to change the chemical meaning of those foods within a generation or two is almost certainly to wreak an inner-ecological havoc.

More specifically, they overlook the fact that humans, as distinct from rats or hamsters in test laboratories, do not eat a single food additive in isolation. In this case, one of the most important combinations involves foods (or meals) that include both phony colors and members of the class of additives known variously as surface-active agents, surfactants, or most commonly, emulsifiers.[21]

An emulsifier, operationally, is a substance which acts on two things that normally don't mix and makes it possible for them to mix (the result is called an emulsion). An easy and cheap emulsifier is a detergent; although your supermarket distinguishes between "soap" and "detergent," soap is actually just one kind of detergent. Water alone won't wash the grease off the bottom of a pot, because the water and the grease won't mix. The chemical you add to the water mixes the grease with the water so that you can wash it down the drain.[22]

The mixture is, however, an emulsion, and detergents are emulsifiers.[23] Before the food industry got so public-relations-conscious, emulsifiers in foods were sometimes called "soaps."

A detergent emulsifier works something like a towline. A diagram of one molecule of a detergent looks like an old-fashioned hatpin. The fat round end is a polarized combination of atoms (that is, it has an electrical charge) of such a structure that it combines easily and strongly with water. The long thin part is a string of carbon-hydrogen combinations; since most oils and greases and fats are hydrocarbons or are very like hydrocarbons, this part of the detergent molecule combines very well with greases.

When the water and the detergent meet the grease, it is as though the pin part of the hatpin sticks in the grease (hence "surface-active"), the handle part is grabbed by the water, and by rubbing it or shaking it you make it possible for the grease to be pulled away from whatever surface it's clinging to. What you wind up with, instead of a film of grease, is a bunch of little globs of grease that look, in diagrammatic form, like pincushions, with molecules of detergent surrounding them and holding them in suspension in the water.

There are natural emulsifiers produced by the human body itself as part of the digestive process. When you eat fats along with other foods, it all gets mixed up in your stomach, and the fats tend to coat particles of the other food—which makes it resistant to the normal digestive processes. So that your body can make use of the nutrients in those other foods, the emulsifiers (part of the liver bile) act exactly like "soaps" to "peel" the layers of fat off the food particles and thus expose them to the other digestive juices in the system. They also act to keep fats and greases from coating the linings of the digestive system and blocking the absorption of nutrients, and they act directly on the linings themselves to *promote* the absorption of nutrients.

Added amounts of emulsifiers, then, may act in your system to promote the absorption of substances that would not otherwise be absorbed—including some carcinogenic coal-tar dyes. We weren't kidding when we said you might emulsify your stomach, though it is more likely to be your intestine.

The labels on your food may not, if the food is standardized, mention emulsifiers at all. Take it for granted that they're in bread, where their whole purpose is to make the bread softer because the industry thinks it helps sales, or to make the bread hold more water per pound. They're in ice cream; they may be in almost anything else that can be "smoothed" or blended or have its natural texture removed. The label may just say "emulsifier." Otherwise, you have to figure it out from the name, and there are a whole slew of possible names.

Actually, though, there are only two major kinds of artificial emulsifier in use. One kind is a sort of artificial fat, which you can identify by finding the word "monoglyceride" or the word "diglyceride" in its name. The other (known in the trade as "the poly compounds") will usually include some variation on the word "stearate" (like "stearyl") or something with the letters "sorb" in it, like "polysorbate." Sometimes you can get both, as in polyoxyethylene sorbitan tristearate, an emulsifier used in cakes, doughnuts, cake mixes, whipped vegetable-oil toppings, cake icings and fillings, ice cream, custard, ice milk, sherbet, other frozen desserts, milk or cream substitutes for coffee, and a number of dry or powdered foods (it's also called Polysorbate 65).[24]

The glycerides are among the kinds of emulsifier that the body produces naturally in the course of its metabolism; the poly compounds are not. While there is some indication that the poly compounds may be more dangerous, the natural occurrence of the glycerides does not, of course, mean that their artificial use in things like doughnuts or peanut butter is safe. As with other substances, their effects in combination with other additives remain to be tested—and we eat more and more of them all the time.

The glyceride compounds have been in use since about 1930, when they began to appear in margarine and shortening. Their use multiplied in the mid-1930s, when it was found that their presence in shortening resulted in "softer" bread (which the bakers then taught you—or your parents—to want). By 1952 every major brand of shortening was about 2 percent added monoglycerides and

2 percent added diglycerides. Shortly after World War II, polyoxyethylene monostearate (first of the "poly compounds" and, chemically, the basic one) appeared in competition. The glyceride manufacturers were meat packers like Swift; the principal manufacturer of poly compounds was the Atlas Powder Company, a maker of explosives.

Longgood writes of the poly compounds:

> After people had consumed millions of pounds of these new softeners, evidence built up that they were not safe for use in food. Gradually they were outlawed for use in most breads, salad dressings, mayonnaise and, more recently, ice cream . . . .

Alas for the effects of time's passage on reporters: the poly compounds, or at least two of them, are back in the ice cream standards.[25] And it should be noted that they are still used in a number of products, including those in which they're sometimes described as "banned" or "outlawed." They are banned in bread, for instance, only if the bread moves in interstate commerce; most bakery goods do not, and that's also true of a number of other products. And of course, the "ban" applies in particular products only if those products have standards set in the *Code of Federal Regulations*. Unstandardized products can use anything that in general is not illegal—and the poly compounds are legal.

The Delaney Committee, in the early 1950s, uncovered a fantastic pattern of "research"—or, at least, a pattern that seems fantastic until you learn that it isn't all that unusual in the food business—in which the poly compounds were overwhelmingly "proved safe" by scientists working for the explosives company, and just as overwhelmingly "proved" damaging by other scientists in the pay of glyceride manufacturers.

Ten years of studies on dogs, rats, rabbits, mice, and monkeys, said Dr. John C. Krantz, Jr., showed that "prolonged feeding of these compounds at relatively high levels is not harmful." Studies on hamsters, said Dr. Edward Eagle, showed premature death, retarded growth, stones in the urinary bladders, kidney stones, blood in the feces, hemorrhage from the genitourinary tract, liver damage, atrophied testes and a few other "minor" deleterious results.

Dr. Krantz, then with the Maryland School of Medicine, did his tests on the poly compounds as an employee of Atlas Powder. Dr. Eagle was a staff physiologist-toxicologist for Swift.

The committee looked around for other scientists who had worked with emulsifiers, and particularly with poly compounds, and found a couple. B.S. Schweigert and Anton Carlson had worked together; Schweigert testified that he had done the hamster tests at one-fifth the amounts used by Eagle, and that while the results weren't quite so horrible, they did show up extensive damage to several organs, and "definite deleterious effects" to those organs after only ten weeks.

Schweigert also said of the poly-compound emulsifiers that "they would reduce the caloric value of the food . . . so that . . . aside from the question of safety, then, the problem of the nutritive value of the product also must be considered." And he noted that some people tend to eat, relatively, a lot more of a single food than a hypothetical "average American" would eat. There are a lot of ways in which this is true. When I think of Orange B, I think of kids who eat a lot of hot dogs; while I'm sure that I ate at least 20 times as much ice cream 30 years ago as I do now.

Dr. Carlson, who said flatly of the poly compounds that "we should not yet introduce them in human food" (he was a little late), also made, as so many real scientists have, the overriding argument against the use of these compounds that can be made against *any* unnecessary artificial additives. We must consider, he said,

> the subclinical injury, serious injury, we have no certain methods to detect. Small amounts of injury in certain percentages of the people may go undiscovered for generations. This is a serious problem involved in the changes of such fundamental things as that type of food for man.

The use of glyceride emulsifiers in bread led Dr. William Bradley of the American Institute of Baking to tell the Delaney Committee that their only purpose was "to increase the absorption of the dough," which is to say, to get more water into the weight you're paying for. He also said that the glyceride emulsifiers react with the starches in the wheat to make a separate product inside the bread. Somebody asked him whether that separate product is digestible at all.

"I do not know," he answered. "I know that in the test tubes . . . that product remains undigested, but whether it is digested in the alimentary tract or not, I do not know."

Have a sandwich.

Because of their "smoothing" effect, emulsifiers can also be used, particularly in baked goods, to replace fats, oils, eggs, and milk, which have nutritional value where the emulsifiers don't. But by far their most serious effect is in the digestive system, where they act precisely like the emulsifiers that occur naturally in the liver bile or, if you prefer, like detergents. That is, they tear away, with their "pincushion" action, fat or grease coatings from bits of food, exposing them to digestive processes; they affect the stomach and intestinal linings in a similar way to make it easier for other substances to pass into the system; and they break up coherent big globs of some substances into littler, more easily digested globs.

Because emulsifiers do these things, the FDA is wrong about what a carcinogen is. It doesn't matter, they say, if a coal-tar dye causes cancer when it's injected or implanted. It has to cause cancer when eaten, or we won't count it. It might cause cancer in the system, but the feeding tests show that it doesn't get into the system by eating it.

The rats, however (or mice or hamsters or dogs), don't eat a bunch of artificial emulsifiers along with their coal-tar dyes (or their carboxymethylcellulose, or whatever other possible carcinogen is being tested). We do. And it may very well make the difference as to whether we absorb the carcinogens or not.

It is known from test animals, for instance, that emulsifiers increase the intake of iron from ordinary foods—so much so that concentration of the excess iron in the animals' spleens and livers caused both cirrhosis and cancer. In man it is known that at least some emulsifiers (including some now in use in America) can cause an almost fantastic increase in the absorption of Vitamin A. As noted earlier, hypervitaminosis A is not at all healthy, and in pregnant women it's particularly dangerous.

There are even substances (not carcinogens according to the FDA, because they cause cancer when injected but not when eaten) which in fact are cancerous when eaten, if they're eaten with an emulsifier. Still, we go on eating known carcinogens, and emulsifiers, every day, while the food industry gets rich on both and the FDA stays determinedly in its hopeful-ostrich pose.

If there is a test of any size or duration worth noting in which any now permitted U.S. Certified Color has been extensively fed to animals *in combination with* any relatively widely used emulsifier, we can't find it. And yet we eat such combinations every day—not always in the same *foods*, but often in the same meals, and the stomach doesn't know the difference. Almost all ice cream eaten in America—to take just one example in which one food is enough—is both artificially colored and emulsified.

In the meantime, the FDA's hopelessly inadequate definition of a carcinogen is probably responsible for a literally immeasurable, but certainly sizable, number of cancers every year. For these things are scientifically indisputable:

1. The class of chemicals known as polycyclic aromatic hydrocarbons is universally suspect as carcinogenic.
2. *All* certified colors now in use in America are polycyclic aromatic hydrocarbons.
3. Almost all of them are *known* carcinogens in tests involving other forms of administration besides feeding.
4. The feeding tests on which they are judged to be safe involve feeding them apart from other additives.
5. There are many animal tests showing that carcinogens, not normally absorbed into the body by digestion, *are so* absorbed when eaten in combination with emulsifiers.
6. The use of artificial colors and of emulsifiers is so widespread in America that it is difficult to imagine a day's passing without most Americans eating some of both.

Add to those facts a seventh: colors and emulsifiers are used *together* most frequently in margarine and in ice cream, cake, some kinds of bread, and candy. Now think of children. Now think of the cumulative effects of carcinogens, over time.

It cannot be too strongly stressed that the Food and Drug Administration knows all this as well as we do. There is no question of convincing them to change their view of the scientific facts; they *know* the scientific facts. So do the executives of, and the apologists for, the industry. Again, they are in effect (and sometimes, over a career's length, in fact) the same people—and they do not care in the least about the health of you or your children.

It cannot be too strongly stressed that you are asked (almost literally forced) to eat these additives while scientists around the world profoundly disagree. When an international body of cancer experts says they're dangerous, when the Food and Agriculture Organization and the World Health Organization say that at least some of them are dangerous—why should we go on eating them because a few Americans say they're safe? Surely we're being asked to carry a blind patriotism just a little too far.

The safety factors in the dye tests are quite clear. With one of the dyes now banned, there were adverse responses in dogs fed at a rate of 100 milligrams per kilogram of body weight per day. If, like most of us, you're not at ease with metric weights, pretend it's our dog, who is not one of the large breeds but who is big enough to weigh a middle-sized 40 pounds. That dose for her would be less than 3/100 of an *ounce* per day. So don't think to yourself that you (or your children) "don't really eat all that much of these additives." Just remember that in fact you eat a lot more of this junk than you think you eat.

Just as the industry likes to talk about caramel or beet powder when they talk about colors (sliding quickly over the fact that the coal-tar dyes are much more widely used), when they talk about flavors they like to talk about "herbs and spices." But the industry's actual position is better represented in its own publications, as in Robert Swaine's article on flavors in the *CRC Handbook*.

He begins his discussion of spices and herbs by saying, "Materials in this group are, from the flavorist's point of view, simple and crude." He adds, "As a class, they represent our first flavoring and are becoming a smaller percentage of the overall market as time progresses."

After noting that price and availability are among the things that cause the "flavorist" to reject natural materials, he discusses the problem of faking them:

> In most naturals it is not one, but many chemicals that blend to produce the final flavor effect. When the flavor chemist tries to reproduce a natural with a chemical mixture, he usually finds that he must use a relatively large number of single chemicals, or his final compound will be crude.

Please don't take that to mean that he's trying to use the same chemicals that are in the natural substance; it doesn't mean that at all. This is a discussion of laboratory-manufactured, not natural, chemicals. For one thing, many "natural" flavors disappear under the kind of mass processing that the industry uses to cut costs; so having taken the real flavor out, they use the phony chemical substances to put a counterfeit of it back in. They say they don't do this to fool you, but we're still trying to figure out another reason.

Flavors (and "flavor enhancers" and "texturizers") don't have a special law. They *are* covered by the Delaney Amendment. But hundreds of flavoring agents have arbitrarily been deemed GRAS—and remember that a great many of them are GRAS not even because the FDA said so, but because the Flavor Extract Manufacturers Association (FEMA) said so, and the FDA just blinked and went along. Extremely few of these substances have ever really been tested for anything at all.

Why the enormous public relations effort to make you accept phony flavorings, and to train young people and children so well that many people can't, today, recognize an authentic flavor if they taste one? There are a lot of reasons, and one of the big ones has to do with world politics.

In the good old days of colonialism, the "parent" countries (including the United States, which has always benefited from the economics of colonialism even if the colonies "belonged" to Great Britain or France) got their spices cheap, with almost no cost for the African or Asian labor involved. Today, things are not so easy; Swaine says it in extremely delicate words—but he says it:

> [T]he cost of producing many naturals has reached the point where they are so highly priced that they cannot be used. This price rise is a result of competition for both land and labor in the producing area . . . . Patchouly imports were almost non-existent during the Indonesian crisis. Cinnamon and menthol cannot be imported from Communist China.

Thus the public relations effort to convince us, at one and the same time, that the flavor additives we use are somehow related to natural flavors and that the phony chemicals are good for us. Swaine begins, for instance, by talking about "the early cave man" with his "mixture of wild honey with wild berries." But the "flavorist's art" (their term, not ours) ends up with things like 2-(3-phenylpropyl)-tetrahydrofuran.

The "flavorists" are perhaps more sickening to read than any of the other Mad Scientists. They talk about their "art" with all the bubbling enthusiasm of the Three Witches and never, never mention teratogens, or mutagens, or long-term toxicity, or synergistic effects, or (least of all) cancer.

Yet we noted in an earlier chapter that in only a brief glance through the FPC's list of flavoring agents we spotted 29 polycyclic aromatic hydrocarbons (18 of them called GRAS by FEMA). None, of course, has ever been tested in combination with an emulsifier; most have not been tested by themselves, or at best have been tested only minimally.[26]

We talk about cancer and mutations; they talk about "top notes" and "mouth-feel." Using additives to deceive, as we've noted, is against the law; but listen to Monsanto's Abner Salant in the *CRC Handbook:*

> In the area of carbonated beverages, there has been a great deal of work done to impart improved mouth-feel by the addition of CMC [carboxymethylcellulose, a carcinogen], pectin and other materials.

And on the same subject:

> The use of bodying agents can reproduce the viscosity fairly readily, but achieving the same type of mouth-feel at the same time tends to be a more difficult matter. Experience has indicated that perhaps the best approach . . . is by creating the illusion with a suitable flavor that tends to be "heavy in body."

In the ordinary English language that the rest of us speak, an "illusion" is something that is created in order to deceive. But not, apparently, in the world of the flavorist's art.[27]

But that art, secure in its assurance of what a dope you are, cares little for language. The lists are filled with chemicals which are used, for example, for more than one flavor. Sometimes this involves combinations. But sometimes, it is simply a matter of creating a vaguely fruity flavor, adding the appropriate color, and giving it one fruit name or the other. "Peach" or "berry" or "apple" may be the same chemical flavoring, differently colored and named. The eight-year-old eating a jelly sandwich in a hurry won't know the difference. Neither will an adult having ice cream for dessert after drinking a cheap American wine with dinner.

All of them—the colors, the flavors, the mouth-feel additives, the emulsifiers—are there, the industry insists, because you want them. Indeed, you demand them.

We don't think so. We think it probably true that you demand that your food look as it ought to look; but we don't think you want artificial color to cover up for discolored food. We think it probably true that you demand flavor in your food; but we don't think you want artificial flavor to cover up the fact that the real flavor (and attendant nutrients) have been cooked or processed or pounded out of the food. We think it probably true that when you see the word "peach" you demand peach flavoring; but we don't think that what you're demanding is alpha-amyl-beta-phenylacrolein Buxine.

Somehow, a choice for the buyer will have to be made possible. Even beyond all our writing about carcinogens, teratogens, mutagens, and other toxic substances, we need to stop and consider the ecological facts. We have taken hundreds of thousand of years to evolve, to adapt to eating the beef and the carrots and the apples and the potatoes we think of as our basic foods. It may yet be years, perhaps generations, before we discover with horror that we are not, after all, adapted to FD&C Blue No. 1, to carboxymethylcellulose, to polyoxyethylene monostearate, or to 2-methyl-3-(*p*-isopropylphenyl)-propionaldehyde.

The last one is used to impart fruit flavor, particularly citrus flavor, to ice cream, baked goods, beverages, and candy.

But why worry about evolution? The crude methodology of God has given way to the careful practices of laboratories; His infinite and unfathomable ways have been supplanted, at last, by the flavorist's art. The Givaudan Corporation advertises "flavors that nature envies," and Hoffman-La Roche, Inc., is even clearer about the importance of its Mad Scientists in relation to the cosmos that existed before them:

> Nature has ways to make foods good.
> Roche has ways to make foods better.

CHAPTER ELEVEN

Black Is Beautiful

There is a massive lie which you have been told, probably, for as long as you can remember, unless you are very old, or were reared in some other part of the world. It is a lie so massive, and so massively successful, that to challenge it puts the challenger immediately into the classification of "crank" or "crackpot." It is as though one were to suggest that the earth does not in fact revolve around the sun.

For a clear formulation of the lie, an excellent source is the President's Science Advisory Committee's Panel on Food Additives. The President was Eisenhower, and the year was 1960. On the panel were members of the National Academy of Sciences, university professors, and representatives of the Rockefeller Foundation and of cancer research institutes. This was a part of their statement:

> Americans today are better fed and in better health than at any time in history . . . . The integrated contributions of the engineering, agricultural and chemical sciences have resulted in increasing quantities of uniformly high-quality and pure foods which have contributed demonstrably to the physical well-being of the nation.

The food industry says that we are the healthiest and best fed people in the world, and that "healthiest" and "best fed" go hand in hand. The government says so too. The medical profession (which fiercely resists, as a profession, any suggestion that diet and nutrition may be principal factors in most disease) says so louder than anyone.

In *Today's Health* (published by the American Medical Association) for May, 1970, Philip L. White writes, "Scientists agree that the usual American diet is adequate for health. And federal experts and agencies concur." It is another example of The Great Lie.

Americans are neither the healthiest people in the world nor the best fed. In fairness it must be noted that despite the best efforts of the food industry, and the organized propaganda machine of organized medicine, despite the virtual ownership by the industry of the government agencies that are supposed to regulate it, there are people in government who are determined to make the truth available.

Some of them, for instance, are in the Department of Agriculture. The Department issues a yearbook having to do with food, of which we have the 1969 edition. If read quickly, it, too, paints a glowing picture.

For instance: by the spring of 1965, urban families in the lowest third of the income distribution (that's governmentese for the poorest third among city dwellers) were eating more meat, poultry, and fish than the *highest* third were eating 13 years before. Use of milk and milk products had also gone up.

But the overall use of milk and milk products, despite that low-income rise, has gone *down* since 1955. Where once we took in only about a third of our calories in the form of sugar (and other sweeteners), the figure is now 52 percent. The use of grain products, of potatoes, and of fresh fruits and vegetables have all gone down steadily (though in some cases slightly) since World War II.

Compared with the rest of the world, the consumption of animal protein in the United States is startlingly high—as you might guess. A survey by the Food and Agriculture Organization of the United Nations, quoted by the Department of Agriculture, gives the comparative figures as eight grams per person per day in the Far East, 14 grams in the Near East, 62 grams in Oceania (almost all fish), and 66 grams in North America.

As we have suggested before, however, it helps to look very, very carefully at figures. The Department lists, for various food ingredients, a Recommended Dietary Allowance (this is very like the Minimum Daily Requirement with which you may be familiar from labels dealing with vitamins and minerals).

This concept is itself an extremely misleading one, and we have all been given some bad nutritional notions through the use of Minimum Daily Requirements; but that will come later. For the moment let's just note that the Agriculture Department measures "good" diets by whether they include the Recommended Dietary Allowance with regard to seven items: protein, Vitamin A (or precursors thereof), Vitamin B_1 (thiamine), Vitamin B_2 (riboflavin), Vitamin C, calcium, and iron.

In 1968 only one-half of American families had diets which met the Department's standards in those seven items.

One out of every five American families had less than two-thirds of what the Department said they should have of at least one of the seven.

Not listed in the Department's "good" diet were any of the other dozen or so B-complex vitamins, nor Vitamins D, E, F, K, or P, nor essential fatty acids, nor any of a dozen other essential minerals. Thus you can have pellagra (from a niacin deficiency), pernicious anemia (from a B_{12} deficiency), *and* rickets (from a Vitamin D deficiency), and still be rated as having a "good" diet by the Department of Agriculture.[1] And that description applies to only half of America; the other half falls below the Department's standards.

A somewhat more recent study by the Department, apparently using the same criteria since many of the figures are very nearly the same, was reported by columnist Sylvia Porter and provides a few more details.[2] Not only does one family in five have a "poor" diet according to the Department's inadequate standards; there is a "poor" diet in one out of every ten families with an income of more than $10,000 a year. We are not simply measuring poverty; we are measuring an affluent nation badly fed. Remember that a "poor" diet does not, even in the Department's terms, mean an "inadequate" diet. It means, in effect, very inadequate.

The fact is that the nutritional Emperor has no clothes. We eat a lot of sugar, a lot of fat, a lot of carbohydrates—and we do not eat the foods that contain the elements we need properly to metabolize the sugar, fat, and carbohydrates. And most of it is not our fault. We have very little choice.

We are also conditioned to the idea that "nutrition" means that if we eat a little of this and a little of that—so much Vitamin Q, so much hyperflatulate, so much magicium, as if we were putting ingredients into a bowl to make a cake—we will be something called "healthy," and something called "well-fed."

Eating does not work like that. It is an ecological process, with each part bearing on each other part in ways that are understood today in only the dimmest of fashions. It has evolved as a part of our own incredibly complex inner-ecological evolution. We tamper with it at our peril.

But what we eat today is "food" almost completely devitalized, with its important and necessary ingredients largely battered, heated, frozen, or pulverized out of it, and with a very few of those ingredients—barely understood and not at all coordinated—then needled back in along with dozens of other substances which ought to play no role in human feeding at all. And if the Mad Scientists do manage to include in their needles a substance that is actually related to food, they gleefully assure us that the food from which they have taken all but a drop of life is now "enriched."

The outstanding example of this in almost every American's diet (and of course you knew we were going to get to it sooner or later) is bread. Or, more accurately, what we *call* bread.

More nonsense has probably been written about bread than about any other food in the history of man. A lot of sense has been written about it too, some of which must confuse our children.

It must be terribly perplexing to read medieval references to bread as "the staff of life" and then come home to a sandwich made with the featureless, textureless, tasteless, and virtually nutritionless mass of air, water, and glue that inhabits most of our breadboxes. The ordinary adult, in the meantime, is buffeted between perorations like ours on the one hand, and the industry's insistent and repetitive assurance on the other that this same stuff will somehow make children's bodies grow 12 ways.

Perhaps we can clear up some of the confusion, which begins, as most such confusions do, in a question of semantics. Put simply, in any reasonable historical usage that can be imagined, the stuff we eat is not bread at all.

Albert von Haller tells an instructive story about British prison life that dramatizes the point. In 1887 a member of Parliament, John Burns, did six weeks in a prison, and when he came out he complained about the food and, particularly, the bread, which was most of the diet. Public indignation was aroused: the villain of the piece was Sir James Graham, who had set up the prison system, and its bread-based diet, in 1864.

It remained for a British physician, Lionel Picton, to figure out what had happened, and to exonerate poor Sir James. In 1864, Graham had set a ration of 18 ounces of bread per day (along with other food, but with the bread as the major nutritional base). Using present-day terminology, that meant that each prisoner got 1,200 international units of Vitamin A (in the form of its precursor, carotene), 540 international units of Vitamin B, 200 milligrams of calcium, 1,000 milligrams of phosphorus, and 15 milligrams of iron each day.

In the 1870s, however, steel rolling mills replaced stone-grinding in the flourmaking process in England, and the forefathers of our Mad Scientists were able to produce fine white flour and to present it as "progress." English administrators, determined to be humane, saw to it that prisoners were able to share in this progress as well as everyone else. The result was, through no fault of Sir James Graham, that John Burns got a completely different commodity which by historical accident was also called "bread," and which was still assumed to be a valid basis for a diet.

But there was only half as much calcium, less than half as much phosphorus, and less than a third as much iron. There was only one-eighth as much Vitamin B. And there was *no* Vitamin A. Only the word, "bread," was the same.

We still hear of "bread and water" in prisons, and we are struck with wonder when we realize the extent to which it was formerly used as punishment. But the explanation is the same: what was then called "bread" was a different food entirely, quite nutritious and able to sustain life for some time—completely unlike the spongy mass we eat today. The punishment was not the bread, but the deprivation of other, more varied foods.

If a prisoner today were condemned to bread and water for 60 days, and given even the "enriched" version of standard American white bread, there is some question whether he would be alive at the end of his sentence. They've tried it with rats, and few of them ever make the 60-day "deadline."

American prisoners, in fact, are usually fed (if on such a sentence) a special "biscuit" much more nutritious than what we now call bread, much more like the food that has been called "bread" for most of human history. The prisoner *not* under such a sentence, however, eats the same junk that most of the rest of us eat, at least in most prisons. Some armed forces "brigs" use ordinary bread in their bread-and-water sentences, but there are limits on how long it can be served without being supplemented with other foods. Even so, such a sentence today is literally inhuman.

Why the rolling mills, and the feeling that fine white flour was somehow "better"? It has nothing whatever to do with nutrition, although for some time there were those who argued that it did. It has a great deal to do with class differences in earlier cultures, and something to do with another semantic problem that has been with Western man at least since the written word and certainly for some time before that.

Since long before the Bible, cultures whose languages derive from the Indo-European (and those who have closely associated with them) have equated "white" with good and "black" with evil.[3] White traditionally symbolizes birth, and purity, and nobility.

Black is associated with death and disaster. The worshippers of Satan celebrated the Black Mass, a frightening warrior was called by his enemies the Black Prince, and magic used for evil was black magic as opposed to the white magic of those who fought for good.

This imagery has not disappeared, of course—far from it. We not only speak unthinkingly of the Dark Ages (or "darkest Africa") and usually clothe the devil in black; we do it much more casually, as when too many planes stack up over Kennedy Airport and the controllers immediately dub the day "Black Wednesday." Throughout Western history and to the present day, this pervasive semantic habit has had tragic consequences for relations among the races. Even today, "black is beautiful" is used as a necessary slogan to remind those who are black that it is not necessary to accept the standards set by those who are white.

Although it may seem relatively trivial, exactly the same silly identification seems to have had a great deal to do with the history of bread, and for those concerned with health and nutrition, the result is not trivial at all.

We should perhaps digress here, to talk for a moment about the wheat berry.

The wheat berry is shaped something like a tulip, or a mango. Its outside is a husk, something like the husk of an ear of corn although of course much, much smaller. Three layers of this husk are given separate technical names, but taken together they are called the bran.

Inside the bran is a more or less solid, white, starchy mass called the endosperm. Down at the thick end of the endosperm is a hard, nutty kernel called the germ. When the berry functions as a seed in the ground, it is the germ that is the actual seed (as you might guess from its name); the bran protects it from the outside, and the endosperm feeds it until it is able to draw its own nutrition from the soil.

Other grains—barley, oats, rye, corn—have analogous constructions, and bread can be made from all of them and sometimes is. If the whole grain is used, however, wheat makes the best bread, nutritionally, because the germ of the wheat is one of a very few places in nature in which the entire Vitamin B complex is found.

That means, in case we have allowed any confusion on this point, that it is one of a very few places *anywhere*, because nobody has ever synthesized the entire Vitamin B complex. In fact, nobody is even sure yet what's in it.

Grind up the wheat berry, and the coarsest and heaviest part (and thus the easiest to remove) will be the bran; next heaviest will be the hard, oily germ; lightest of all is the white endosperm, which can also be ground into the finest powder. Bakers love the endosperm powder because it can be kept almost indefinitely without spoiling. The reason it can be kept almost indefinitely without spoiling is that it has so little nutritive value that even bacteria won't eat it. Bacteria do not read advertisements about growing 12 ways, and they don't care what color things are.

Nobody knows how old bread is. The consensus seems to be that our ancestors first made porridge from their grains, then unleavened bread, then the yeast-rising cereal bread to which we usually give the name; but the evidence is far from clear. In the ruins of Swiss lake dwellers, investigators have found bits of bread baked at least 10,000 years ago.

In any case, bread as we understand it quite clearly existed among the Egyptians (some 3,500-year-old loaves were found in Egypt in 1936), the classical Greeks, and the Romans. The Bible and classical literature are filled with references to bread, and the art of the time shows that grains were not only ground with stone mills but bolted: *i.e.*, the coarser and darker parts of the grain were separated from the finer and lighter parts. A kind of sieve shown in a drawing from ancient Thebes is almost identical with a sieve used in parts of Africa today.

One reason for bolting the ground wheat (or other grain) in early times was that all of the methods of stone-grinding then in use were apt to result in chips of stone breaking off and mixing with the meal or flour (flour is the same as meal, but ground more finely). The chips would be sorted out along with the coarsest part of the bran.

Obviously, the more the grain was bolted, the more work would be involved—and the whiter and finer the flour that resulted. It would also be increasingly less nutritious, but nobody knew that until relatively recently; because of our peculiar Indo-European tendency to equate whiteness with purity, it was long believed that the whiter bread was somehow the "cleaner" and more desirable bread.

And since it took more work and consequently cost more (even if the work was done by slaves, the slaves' time was valuable to their owners), white flour and the bread made from it was strictly an upper-class food as long ago as we know of its existence. Abraham, who was no peasant, is represented as demanding for a guest cakes made from "fine meal,"[4] and Greek and Roman literature makes it quite clear that whiteness in bread was as sure a mark of class distinction as purple trim on a toga.

At first, either because it was comparatively rare or because a type of wheat was predominant which required even more work to mill, any "wheaten" bread served this class purpose; but as more wheat of a better variety became available, it was the whiter product which was demanded. The problem here is that the language of the translated classics can be confusing. It was called "white bread," to be sure; but it had little in common with what we call by that name. Almost certainly it was nowhere near white.

Some people have even tried to calculate the degrees to which flour was bolted in the times of classical antiquity, and while they are not in perfect agreement, it's evident that even the Roman emperors got all or almost all of the germ and some of the bran in their "white" bread. Unfortunately, lesser citizens did not simply get what we call whole wheat bread and the British call a wholemeal loaf. Sometimes they got a very branny product made out of what was left after bolting; more often they got other grains mixed in with the wheat.

But the bread on which the armies of Rome conquered what was then virtually the known world was something very near wholemeal wheat bread, possibly with some corn mixed in. And whether they took the concept with them or whether it already existed in some other form, we know that after that conquest, and up until very recently, the same prejudice in favor of whiter bread existed throughout Europe.

Peasants are invariably represented as having to subsist on "black bread" which made them objects of pity for the softer-hearted among those of the upper classes who noticed them at all. Little did they know that black was beautiful. White bread became for many of them a symbol of the better life to which they aspired, so that the more one moved up socially and economically, the whiter the bread one ate.

Happily for the health of Europe, only the very rich could afford bread made from a flour thoroughly enough bolted so that the nutrition was efficiently removed; and when you're that rich, the proportion of bread in your diet goes down, so that the vitamins and minerals you aren't getting in your bread are more likely to come from somewhere else. Nonetheless, people wanted white bread—and millers and bakers were determined to give it to them.

Because it took more work to produce whiter flour, millers and bakers (sometimes and in some places the same people, sometimes not) could legitimately charge more for it. Because there was a demand, they more or less legitimately charged still more. And very like their successors today, they made even more profit by throwing into their flour as much as they could get away with of anything that would stretch it, or make it whiter, or both. Where bakers today rely heavily on air and water, their ancestors had to make do with chalk and alum.

Not that things have changed all that much: "bleached" flour today may contain as one optional bleaching agent (without mention on the label) one part of benzoyl peroxide mixed with six parts of potassium alum. But today, of course, it isn't a fraud.

In England, should the inspectors of the bakers' guild find one of the guild members cheating, it was a common punishment to drive him through the streets with a heavy ball of dough around his neck. Some of what was then called cheating is now, of course, common practice in making "bread." Possibly after the Nutrition Revolution, an angry populace will enforce such a punishment on the president of General Mills or on industry apologists like Dr. Stare, except that the dough *they* use is so full of air that a ball of it probably wouldn't be much of a burden to carry.

The European conviction that "white is right" took hold so strongly that by the eighteenth century, when poverty and hunger were so widespread in England that bread riots took place repeatedly, the populace still insisted that the bread for which they fought must be white bread, and no amount of monarchical persuasion could persuade the workers and the peasants that brown bread was a sufficient substitute (no one then argued that it was superior, except a few scientists regarded as crackpots). From an older England, in fact, comes the rhyme that many of us learned as children:

> Ride a cock horse
> To Banbury Cross
> To see what Tommy can buy—
> A penny white loaf,
> A penny white cake,
> And a two-penny apple pie.[5]

The "white is right" psychology didn't only apply to bread. It was in the seventeenth century that European manufacturers finally developed a process by which, after eight weeks of hard labor, sugar could be refined to something approximating whiteness. The same class distinctions applied: the result, further whitened with chalk, egg whites, and dyes, was wrapped in expensive violet paper and sold at fantastic prices under the trade name, "Emperor's Sugar."

Queen Elizabeth I was a customer (one visitor suggested it as a cause for her blackened teeth).[6]

Napoleon's army was famous for marching on white bread, but even that wasn't today's bread. As noted earlier, it wasn't until the 1870s that a Frenchman invented the steel rolling mill that makes it possible to take *all* of the bran and germ out of the flour, leaving nothing but the completely devitalized (but white) powdered endosperm. Hungarians first turned the French invention to mass production, and almost immediately Governor Washburn of Minnesota, a miller, brought the Hungarian process to Minneapolis and devitalized flour to America.

If what you're after is profit, the result is great. In America as in Europe, Rome, classical Greece, and possibly Egypt, people welcomed white bread as a status symbol. Millers can sell white flour for more money, and sell the germ and bran as animal food for more money yet.[7] They *have* to take the germ out, as all will piously assure you, or else the flour won't keep; there are oils in the germ, and they turn rancid and spoil the flour.

Well yes—if you use steel rolling mills. You see, they are more efficient and much faster, but they don't actually grind, they crush. Stone-grinding relatively small amounts of wheat so distributes the germ, and its oils, that spoilage is minimized; the flour will keep quite well enough in tight containers for distribution and merchandising.

While British and European folklore rather assumes that millers and bakers are crooks (an assumption we seem unfortunately to have forgotten), it should be noted that many quite sincere people, including some scientists, genuinely believed that there was danger in eating wholemeal wheat bread. For one thing, the bran, if it is all included, has a mild laxative effect on some people and is relatively difficult to digest. Its inclusion also tends to make bread drier.

The effect on digestion, particularly, still bothers enough people so that many authorities recommend a flour from which at least part of the bran has been extracted; doing so doesn't affect too seriously the nutritive value, as far as we know right now.

A more recent argument against wholemeal bread, still echoed by some white-bread apologists, has to do with phytic acid.[8] Flour which contains the germ and some of the bran is high in this chemical, and some experiments showed that phytic acid in wholemeal bread interfered with the body's calcium intake. This led for a time to the "enrichment" of some wholemeal breads with calcium, but later and more thorough experiments showed that it was all unnecessary.

What happens, it seems, is that if you've been eating white bread for a while and switch to wholemeal bread, your calcium intake *does* go down for about six weeks, at the end of which time it's back to normal again (or at least back to wherever it was). People who normally eat wholemeal bread don't have any problem with calcium intake. The phytic acid scare wasn't a case of anyone's trying to fool anyone, but sometimes it is still dragged out by people whose motive is exactly that.

What most authorities recommend, then, is what is called an 85 percent extraction flour. If everything were left in from the wheat berry, that would be a 100 percent extraction flour. An 85 percent flour has only the toughest part of the bran removed; up to there, at least, the higher the number the better. The common white flour of America is about 70 percent, which means that you get 70 percent of the bulk and virtually none of the nutrition. Almost all the Vitamin B, the Vitamin E, the minerals (including iron), the essential unsaturated fatty acids, and equally essential amino acids like methionine are removed.

But the flour may not be quite white enough. And there is yet another problem: flour, to be useful in baking, must age. Aging, however, takes time, and there's no profit in that. Besides, all of that crushing and bolting may have left a little of that germ oil so that an insect or two might get some food value out of it. That of course wouldn't do.

The Mad Scientists, you'll be pleased to know, have met all these problems and conquered them. It happens that chemicals will artificially age the flour, and that the same chemicals will also artificially bleach it—and carefully kill off any lingering nutrition that remains.

By far the most common bleaching and aging additive used in flour for years was Agene (a trade name for nitrogen trichloride), mentioned in an earlier chapter. You'll recall that it was found in 1946 that Agene causes running fits in dogs. Since chlorine dioxide has not yet been shown to cause running fits in animals, it has become one of the more popular substitutes. Bicknell finds this to be the final blow:

> Chlorine dioxide destroys all the remaining Vitamin E in flour, destroys or forms a toxic product or a perverted one with the oil's E.F.A. [essential fatty acids] and destroys or may form a toxic product with the methionine. Add to this graveyard of nutrients a lingering

miasma of chlorine dioxide and the millers have achieved a flour as nearly non-nutritious as is possible and as covertly, as insidiously corrupting to the body as food well can be.

Of course Bicknell wrote that in England. In America you don't *have* to use chlorine dioxide. The standard for bleached white flour is much more generous:

> [O]ne or any combination of two or more of the following optional bleaching ingredients may be added in a quantity not more than sufficient for bleaching or, in case such ingredient has an artificial aging effect, in a quantity not more than sufficient for bleaching and such artificial aging effect:
> (1) Oxides of nitrogen.
> (2) Chlorine.
> (3) Nitrosyl chloride.
> (4) Chlorine dioxide.
> (5) One part by weight of benzoyl peroxide mixed with not more than six parts by weight of one or any mixture of two or more of the following: potassium alum, calcium sulfate, magnesium carbonate, sodium aluminum sulfate, dicalcium phosphate, tricalcium phosphate, starch, calcium carbonate.
> (6) Acetone peroxides [as described in another CFR Section].
> (7) Azodicarbonamide [as described in another CFR Section].[9]

Calcium carbonate, in case it slipped past you back there in (5), has another name when you run into it in a schoolroom. It's called chalk. What once earned a miller a ball of dough hung around his neck is now a federal standard.

Chlorine, another of the bleaches, has a characteristic of its own that makes it dear to the millers: it also causes starch to swell. Since what's left of the wheat berry by the milling process is almost all starch, this is a delightful turn of events from their point of view. Same amount of flour, but more volume.

If you care about bread, you should know that the standard quoted is for "flour, white flour, wheat flour, plain flour." Note that "wheat flour" is *not* a different product from white flour. Watch those labels. The only label requirement is the product name itself, plus the word "bleached" if any (or all) of the above chemicals are used.[10]

Bleaching bothers other people besides Dr. Bicknell. In 1954, a reporter for a German weekly visited Otto Warburg, who was already at that time a Nobel Laureate in medicine, and found in his laboratory a small mill for hand-grinding flour. It turned out that Warburg and his associates made their own flour and from it their own bread; naturally, the reporter asked why.

"In the bleaching process," Dr. Warburg explained to the reporter and thus to much of West Germany, "the vitamins which are essential for life are destroyed just as are the protein building blocks which are important, as for example, methionine. Everyone, without his knowledge and certainly against his wishes, is constantly being fed substances such as boracic acid and bromine [there is a separate U.S. standard for bromated flour, which is widely used in commercial bread with no label requirement]. Constantly used, these substances are dangerous."

Whether because of Dr. Warburg's eminence or not, the German Federal Republic banned the use of *all* bleaching agents in flour 14 years ago.

So far we have been talking mostly about what has happened to flour, not bread. If bread has any nutritional value, of course, it comes from flour; but as we have already seen, that has been crushed, bolted, and bleached out before the flour ever gets to the baker.

About the missing Vitamin B we shall have more to say. Vitamin E is known to have something to do with animal reproduction and is believed to have something to do with human reproduction; its absence may also be related to heart disease (whether the AMA likes it or not) and to various bodily phenomena that we lump together under the word "aging."

As mentioned earlier in the book, some doctors still insist that it is difficult to imagine an American diet deficient in Vitamin E. This stems from their having been taught, erroneously, that there is Vitamin E activity in all of a class of chemicals called tocopherols. It now appears that only *alpha*-tocopherol, one of the group, has this activity. Thus there can quite easily be a deficiency, and flour-pounding helps in building one up.

The fatty acids, says Holum, are "the richest source of energy of all foodstuffs," and we cannot do without them, though we are forced to do without them in bread if we eat the standard American loaf.[11] Amino acids are the relatively simple substances of which all proteins are built; a very few amino acids (fewer than 3 dozen) make up most protein matter, varying in the order in which they appear along a chain. Among the absolutely essential amino acids are methionine, lysine, and tryptophan, all of which are in wheat but not in American white flour after it's been through the hands of the Mad Scientists.

Bread, says the Food and Drug Administration, is the name for a product which is "prepared by baking a kneaded yeast-leavened dough, made by moistening flour with water" or with another liquid ingredient or some combination thereof.[12] Fair enough, if you understand what we've already told you about what the FDA means by "flour."

But alas, that is not all there is to bread ("bread" and "white bread" are the same thing according to federal law, which is a bit of a semantic fast shuffle in itself). Bread is the prime example of a product whose standards permit all sorts of "optional ingredients" but do not require them to appear on the label. Maybe it's because there wouldn't be room on the label. The only label requirements are that some of the preservatives permitted must be listed if used, and the words "spice added," or similar phrase, must be used if spice is added.

In the first place, bread can be made with plain old flour, or with bromated flour (plain flour with potassium bromate added), or with phosphated flour (plain flour with monocalcium phosphate added). Beyond that, listed below are only *some* of the things which are likely to be in anything called "bread, white bread, rolls, white rolls, buns, white buns":

Lecithin; mono- and di-glycerides (already discussed earlier); milk or milk solids in all kinds of forms (probably an improvement when used); carragheenan; cheese whey; eggs in various forms; sugar or molasses or various related substances; enzymes or bromelain preparations of particular types; "harmless preparations of *alpha*-amylases, obtained from *Bacillus subtilis*"; lactic-acid-producing bacteria.

And/or:

Any of a number of other flours or starches besides wheat flour, up to 3 percent of the total; ground soybeans; calcium sulfate, calcium lactate, calcium carbonate (that's more chalk), dicalcium phosphate, ammonium phosphates, ammonium sulfate, ammonium chloride; potassium bromate, calcium bromate, potassium iodate, calcium iodate, calcium peroxide; azodicarbonamide; tricalcium phosphate; monocalcium phosphate; vinegar; calcium propionate, sodium propionate, sodium diacetate, or lactic acid (the last four are preservatives which must be listed if used).

And/or:

Calcium stearoyl-2-lactylate, lactylic stearate, sodium stearyl fumarate, succinylated monoglycerides, ethoxylated mono- and diglycerides.

If you remember the previous chapter, you'll have recognized the emulsifiers, including the worrisome "poly compounds." And then there's *L*-cysteine, which may be added in the form of the hydrochloride salt, including hydrates thereof, if that makes you feel any better. All in all, there are 93 different ingredients which you can get in a ham sandwich even if somebody forgets the ham, none of which have to be listed on the bread label. Almost none of them have any nutrient value whatever.

There is no standard for anything called "wheat bread." There are products called "wheat bread," however. They call for careful label-reading. They may be standards-avoiders because they're not even as good as the horrible stuff the standards describe. They also may be standards-avoiders because they're much better.

There is a standard for "whole wheat bread," however. And if white bread is that bad, then of course we should all buy whole wheat bread. Right?

Wrong.

Everything listed above that can go into white bread can also go into whole wheat bread. Every single additive. The only difference is that wheat bread is made with white flour, and whole wheat bread is made with whole wheat flour—*as whole wheat flour is defined in the standards.*[13]

It is whole wheat flour—"the proportions of the natural constituents of such wheat, other than moisture, remain unaltered"—and is therefore much, much healthier than white flour. It is preferable; but good it ain't. For the same aging-bleaching process is allowed with whole wheat flour as with white flour; the only difference is that the bleaching agents listed as (1), (5), and (6) for white flour are not permitted for whole wheat. And it is steel-rolled, not stone-ground.

Chlorine can still be used to bleach and "age" the flour and to swell the starch; chlorine dioxide can still be used to go through its reactions with the fatty acids and with the methionine in whole wheat flour as it does in white flour; everything Dr. Bicknell warned about would apply, perhaps with even greater force. It's even true that, like white flour, whole wheat flour can be bromated and so used in whole wheat bread without your being told about it.

You'll recall from the previous chapter that the principal purpose in adding emulsifiers to bread, according to testimony before the Delaney Committee, was "to increase the absorption of the dough," *i.e.*, to get more water into it. You'll recall also that there was testimony to the effect that those emulsifiers react with the starches in the wheat to produce what appears to be an indigestible product. Now you know that the starches are about all that's left of the wheat in the first place: starches already, in some cases, swollen by chlorine. More air, more water, no food—that's the name of the American bread game.

Before going any further we have to call attention to one of the most common, most insidious tricks from the bag of the Mad Scientists, the PR men, and their profit-hungry bosses. Once you've learned to spot this one, you'll run into it over and over again whenever there is a controversy about the nutritional value, or lack thereof, of anything sold as food—whether it's bread, breakfast cereal, or imported frozen eucalyptus sawdust.

We call it The Old O.A.D. Trick.

"Old" because it is old. We have a prime example from the U.S. Public Health Service (dealing with bread) dating from 52 years ago. The "O.A.D." stands for "otherwise adequate diet," and the trick is well explained by that 52-year-old episode.

As knowledge of nutrition advanced, more and more scientists raised questions about the wisdom of refining flour to such a lovely whiteness. It was noticed, for example, that men from northern England and southern Scotland were large and powerful men during the Napoleonic wars. However, that same region produced short, frail men, many unfit for military service, at the time of the Boer War. A commission, appointed to investigate this odd fact, concluded that it was due to the fact so many of the men had moved to cities and were eating white bread (and sugar) while their fathers had eaten plenty of whole wheat bread.

During World War I, Denmark simply stopped refining flour. Later it was found that the death rate had dropped, and there had been a marked decline in cancer, heart disease, diabetes, kidney trouble, and high blood pressure. No other marked change in diet or living habits had taken place.

Right after World War I, in the American South, grains began to be highly refined, and there was a wave of deficiency diseases, most notably beri-beri and pellagra. In April, 1919, the Public Health Service announced that there was a definite connection between those diseases and the overrefining of grain.

The millers went immediately into action, not to change the flour and do anything about the diseases, but to apply political pressure to get the Public Health Service to shut up. Within six months the Public Health Service abjectly issued a "correction" to its bulletin. White bread, they said, was perfectly wholesome—if it was eaten in conjunction with an otherwise adequate diet of fruit, vegetables, and dairy products.

So is cardboard. But that's The Old O.A.D. Trick.

Anything, so long as it isn't actually poisonous, is wholesome when it's eaten in combination with an otherwise adequate diet. It's simply another semantic fast shuffle.

The food industry knows that no one food is adequate for life by itself. They know that you know that too. So you're supposed to read quickly over the O.A.D. formula, or half-listen to it, and you're supposed to think that it means: *this food can reasonably be considered to be part of a balanced diet.* But that is not what it says. All it says is that it probably won't hurt you if you eat an "otherwise adequate diet."

There are only two things wrong with The Old O.A.D. Trick, aside from its use to sell worthless food. One is that the food in question, what with all the gunk we needle into it, may not be wholesome after all. The other is that it works very well for bureaucrats and Mad Scientists whose incomes are in five figures a year, or for food industry executives whose incomes are in six. It does nothing whatever for the poor in those parts of the South who went right on getting pellagra and beri-beri.

Not everyone uses bread as an incidental adjunct to otherwise adequate diets. The less fortunate of us must use it as a staple, both a nutritional base and a filler to overcome simple hunger. Stone-ground whole wheat bread without a bunch of gunk in it will not, by itself, make the malnourished hale and hearty, but it will do more than any other single improvement would to improve their nutrition and their general health.

The poor do not have, and cannot afford, an "otherwise adequate diet" in the United States, possibly because nobody has yet figured out how to make a profit on it.

The bakers' determined effort to sell America on nutritionally worthless white bread, with the addition of contemporary advertising techniques to sell lies and in part to play on those ancient class and semantic prejudices, pressed on through the 1920s and 1930s. For reasons best known to the oligarchy that runs it and to the people who controlled the advertising policies of its publications, the American Medical Association jumped on the bandwagon and endorsed all sorts of pseudoscientific hogwash about white bread.[14] And of course, as always, the federal regulatory agencies went where the "regulated" told them to go.

On May 7, 1930, the Department of Agriculture, for example,

issued a statement signed by several of industry's and government's best Mad Scientists:

> Bread, either white or whole wheat, is always an economical source of energy and protein in any diet. The form may be left to the choice of the individual when the remainder of the diet is so constituted as to contribute the necessary minerals, vitamins, and any necessary roughage.

Recognize The Old O.A.D. Trick? With a little practice you can spot it in your newspaper at least once a month, and in all sorts of textbooks and free give-away food industry materials being used in almost all of our public schools. But bread is a source of damned little energy and very little protein, no matter how widespread the propaganda.

By 1938, however, The Old O.A.D. Trick wasn't enough. There was a war about to start in Europe, for one thing—a war in which America might well become involved. Wars have been known to turn on the quality of the diet fed to a nation.[15] For another thing, even the American Medical Association is sometimes unable to hold down the integrity of some American doctors for more than ten or 12 years.

At any rate, a couple of the AMA's committees took a look at white bread at about that time, and decided that maybe, at the very least, some synthetic vitamins should be added (vitamins were beginning to get to be a Big Thing at around that time). They directed Dr. George Cowgill to make an investigation of the subject, and in 1939 he reported back. During the year of his investigation, others, too, were looking at the same situation, particularly with regard to what we know as the B vitamin complex.

It must be understood that we didn't know much about the complex then (we still must deal with more areas of ignorance than of knowledge about this subject; nobody even knows how many B vitamins there are yet). Investigators had learned, though, that a B_1 deficiency was behind beri-beri, and in 1938 beri-beri was still endemic in the United States.

"A large fraction of the population," Dr. Norman Jolliffe wrote in that year, "particularly those spending per capita less than $2 a week for food, subsist on diets of borderline adequacy in Vitamin B_1."[16] Lest that sound to younger readers as though it describes only the very poor, we might note that those were Depression dollars; a comparable figure today would be somewhere around $10, possibly more.

Dr. Cowgill, in his report to the AMA, noted that paupers in London, in 1838, got more Vitamin B_1 than did the highest-income groups in the same city a hundred years later, a deficiency clearly traceable in large part to the bread they ate. Obviously, he said in his report, something would have to be done about the nutritive quality of bread.

An increasing number of doctors and nutritionists began to call attention to the failures of American white bread. First the Axis countries and then Britain forbade the baking of anything but wholemeal bread, recognizing the simple fact that it was the one enormous improvement in nutrition that could be made, not only at no extra cost, but at an actual saving. The American milling and baking industries went into panic. No American corporation ever, ever admits that it has been conning its customers for decades.

They would, they said, "fortify" their bread.

It was called "fortification" then; it is called "enrichment" now. It's the same thing. As we shall see, it may be better than nothing, and then again it may not; but it has nothing to do with nutrition really. It was a propaganda effort to stop the government from forcing the Mad Scientists to leave the food alone and let it be healthy.

But there was one turning point: the 1939 convention of the American Institute of Nutrition, at which the industry proposed a formula for adding thiamine, riboflavin, niacin, and iron to bread. It may be wiser, though, to digress from the chronological and first cover a little of what is known about B vitamins. Most people (including ourselves) tend to get them mixed up.

In the chapter on genetics, we talked briefly about enzymes and their role: they are substances that make it possible for processes to take place in the cells. Coenzymes are substances that work with enzymes to make those processes possible.[17] For our purposes, this is close enough.

Holum's excellent elementary chemistry text defines vitamins and minerals as "substances the organism needs but cannot itself manufacture from organic raw materials in its diet." The key word is "needs," as these further quotations from Holum show:

> The importance of vitamins in daily living cannot be exaggerated. There are about sixteen independent vitamins, and several variations of these are known [this quotation is nearly ten years old, by the way]. The chemical roles played by many of the vitamins . . . are still quite obscure. B vitamins, however, are parts of enzyme systems. A large number of coenzymes either are identical with certain of the B vitamins

or are simple derivatives of them. This relationship makes clear why the B vitamins are essential. . . .

All vitamins must be supplied continuously in the diet, because the body cannot make them from simpler substances. Unless they are on hand, several enzyme systems fail to materialize. When this happens, many important chemical reactions in the body go too slowly or not at all. These are critical matters, for all the B vitamins are used in every cell in the body every day.

The last sentence is worth reading over again at least one more time.

You may wonder why there is a Vitamin A, a Vitamin C, a Vitamin K—but a Vitamin B *complex*, with little numbers and odd names and all sorts of confusion. The answer is the key to the section that follows, and the key to understanding many of the claims and counterclaims not only about enriching bread but about other foods, about vitamin pills, about a number of nutritional controversies.

The complex is called a complex because that's what it turned out, on examination, to be. Once it was thought that there was a Vitamin B; then it was found that Vitamin B is actually a number of separate substances which work together. As more is learned about what they do and how they work, it can be seen that there must be parts to the complex that we have not yet isolated.

But, although the various parts of the B complex do different things, they work together. This is extremely important, because it means that if you take extra B_1, for example, your system will need more of each of the other B vitamins—*and in the same proportions as though you got them naturally.*

We did not evolve into such intricate forms by taking in nutrients in random amounts over the last few hundred thousand years.[18]

For some reason, only four of the B vitamins are commonly known by numbers: B_1, B_2, B_6, and B_{12}. These all also have names, and the others have names, and even people who really want to try to get it straight have their difficulties. A guide seems like a good idea—not a guide to all the B vitamins, not even to all of those known, but at least to the ones you're most likely to run into.

First, though, you'll recall Dr. Holum's saying that "all the B vitamins are used in every cell in the body every day." That means that one can study human tissues and see what the proportions are of the different parts of the complex. One can also study human waste and see what isn't used. From these two kinds of studies, one can get a pretty good (though tentative) idea of what the proper proportions are of the various parts of the B complex. That information is included in what follows; but it is tentative.

—Vitamin B_1 is also called thiamin, or thiamine. This relatively simple (fewer than 30 atoms) molecule, in the human system, becomes thiamine pyrophosphate (also called cocarboxylase), and as a coenzyme system works to help in the process of breaking down carbohydrates. For purposes of setting up proportions, let's say that a natural B-complex intake has one part thiamine.

—Vitamin B_2 is also called riboflavin (it was once called Vitamin G). A little less simple than thiamine, it also does much of its work by joining with a phosphate group to form something variously called riboflavin phosphate (a logical enough name), or flavin mononucleotide, or FMN, or—among people who like things simple—"yellow coenzyme." This also works in the breakdown of carbohydrates, but with a different, though related, function. If you have one part thiamine, you should have one part riboflavin.

—Vitamin B_6 is also called pyridoxal or pyridoxine. It is also simple; it also combines with a phosphate group to form (predictably enough) pyridoxal phosphate, but its coenzyme function is related to the utilization of amino acids, from which proteins are formed. If you have one part thiamine, you should have one part pyridoxal.

But now look at your bottle of vitamin pills. If your pills are from one of America's large corporations, you have just discovered the relationship between nutrition and economics. Look in the breadbox and you will also find that there is no pyridoxal added to "enriched" bread. Pyridoxal costs about sixteen times as much as thiamine.

—Vitamin B_{12} is also known as cyanocobalamine. As a quick look at that name might tell you, it is notable for the presence of a cobalt atom in its extremely complex (hundreds of atoms) structure, an example of how "trace amounts" of an element can be vital in your diet. B_{12} is known to be essential. It acts, for example, in a process that prevents pernicious anemia, but its workings are as yet little understood; even the coenzyme structure isn't certain. If you have one part thiamine, then you need very little cyanocobalamine; somewhere between one- and three-one-thousandths of a part appears to be enough.

—*p*-Aminobenzoic acid (the "*p*" can be read as "*para*") is more often called PABA. It is itself fairly simple, but it goes into the middle of a much more complex coenzyme which we have met before: folic acid (also called pteroylglutamic acid). Like pyridoxal, PABA has to do with using amino acids.

There is a problem about amounts of PABA. Folic acid, itself, is often classed as a B vitamin, and can be obtained separately. You may recall from an earlier section on the setting up of the GRAS list that folic acid was originally on the list, was taken off after FDA scientists protested, and has since been put back on. One reason for the complaint was that it tends to throw the B complex out of balance.

According to the kinds of cell and waste studies described earlier, one part of thiamine should be matched with one part of folic acid *and* 20 parts of PABA. This indicates that we ought to get our PABA in its natural form, and not just as a constituent of folic acid.

—Niacin and niacinamide are closely related B vitamins which are usually regarded as interchangeable. Niacin is also called nicotinic acid, niacinamide is also called nicotinamide. Also a very simple substance, it becomes a part of something called nicotinamide-adenine dinucleotide (happily known as NAD most of the time). NAD forms a part of a number of enzymes having different functions. You should have ten times as much niacin as thiamine.

—Pantothenic acid, usually called just that, forms part of a much more complex substance known simply as Coenzyme A. You may see it written somewhere as CoA-SH (there's a sulfur and a hydrogen atom out at one end of its structure). It's important in the body's use of fatty acids, and it, too, should be ten times as abundant as thiamine.

—Biotin (once called Vitamin H) is itself a coenzyme for at least two enzyme systems, probably more. It's one of the B vitamins about which not much is known; it may be necessary for the manufacture of other enzymes, and it seems to be involved in the body's handling of carbohydrates, proteins, *and* fatty acids. If there is a study of how much you need, we can't find it.

—Lipoic acid, also called thioctic acid, is also itself a coenzyme, and not much is known about it either, including the amounts.

There are two other B vitamins—choline and inositol—which are known to be required in large amounts as a part of the complex. Choline has at least something to do with the function of the liver bile; inositol is known to act as a check on cholesterol. For every one part of thiamine, the system should have 500 parts of each of these two substances.

You won't find these proportions in a lot of vitamin pills. Some vitamins are cheap, some are expensive; and manufacturers would much rather charge you a lot of money for cheap vitamins. Needless to say, millers and bakers feel rather the same way.

The formula proposed by the industry to the American Institute of Nutrition convention in 1939 called for adding only three of the cheaper B vitamins to bread—thiamine, riboflavin, and niacin (they also proposed to add iron). Some of the people on the panel that talked about it said it would be better than nothing. Dr. Lydia J. Roberts of the University of Chicago, for example, said that it might be okay "*if* it could be done at no additional cost to the consumer—for the lower-income groups are the ones that most need to have it provided in this way—and *if* all advertising for the resulting product could be properly controlled [her emphasis]."

Some others didn't like the idea at all. Dr. Alonzo Taylor was there as Research Director of General Mills (he was previously head of the Food Research Institute of Stanford University, another example of the back-and-forth travel among food technologists), and even he said that retention of "native vitamins" should come first. But Dr. W.H. Sebrell of the Public Health Service had nothing good to say about the proposal at all:

> [I]t does seem a little ridiculous to take a natural foodstuff in which the vitamins and minerals have been placed by nature, submit this foodstuff to a refining process which removes them, and then add them back to the refined product at an increased cost.

Even that, of course, is not what the millers were proposing. They were proposing to add only a couple of them back, not all of them. As you might guess, by the way, Dr. Sebrell and his agency, within a year, were enthusiastic supporters of "enriched" bread. By 1941 the AMA was back in line, too; its Council on Foods and Nutrition announced that enrichment was the greatest thing since the thermometer, noting that "thiamine is the component which makes whole wheat most significant in the diet at the present time."

This is total nonsense, as is much of what the medical profession says about nutrition when it speaks more or less "officially." But the AMA went on to say that the other components of whole wheat bread were something called "plus values"—nice but unnecessary. The standard for "enriched bread" went into the law in May, 1941, and was slightly revised (only the amounts, not the ingredients) in July, 1943; it still calls for only three of the least expensive B vitamins and for iron, with the baker having the option to add Vitamin D, or calcium, or both.[19] All the other standards for white bread and white flour (the optional bleaches, the emulsifiers, the chlorine, the 93 bread additives) are the same for "enriched" white bread and white flour.

There have been a lot of experiments dealing with "enriched" bread and whole wheat bread, but many of them are hard to use in argument because of the presence of milk solids in uncontrolled amounts. Since the milk solids contain some of the same elements used in "enrichment," they confuse the results.

Of course you can still make and sell good bread, as has been demonstrated by Dr. Clive McCay at Cornell University. He came up with an *unbleached* (but not whole wheat) flour to which wheat germ, soybean flour, and a lot of milk solids were added. The effect is to put back most of the wheat except for some of the bran, and to add some other food values too (bran is not only tough on some people's digestion, but not incidentally contains more strontium-90 from fallout than other parts of American wheat). Dr. Estelle Hawley of Rochester University then tried the McCay-Cornell bread on some rats, and tried ordinary commercial "enriched" white bread on some other rats.

The rats on the McCay-Cornell bread lived happily and healthily until the fourth generation, when the experiment was stopped. The rats on our everyday "enriched" sandwich bread got sick and produced stunted offspring (their bodies didn't grow twelve ways, or even one). The reason for terminating the experiment was that the rats on "enriched" bread were extinct by the fourth generation.

This, of course, the bread interests couldn't have. Obedient as always, the FDA moved in to say that, what with federal standards and all, McCay's bread couldn't be called bread. If someone wanted to sell it, he would, the FDA said, have to call it cake!

Fortunately, that attack was fought off, and the product can be found in some parts of the country; all its ingredients must be listed on the label. It's called "special formula bread" where we shop,[20] and can sometimes be identified by a label reference to the "Cornell formula."

But in the meantime, most Americans eat "enriched white bread," which is as "enriched" as you would be if somebody gave you 15 cents after robbing you of $100. Refined out of the wheat and never returned, even to "enriched" bread, are not only the remaining B vitamins and Vitamin E but a number of minerals and at least three amino acids, lysine, methionine, and tryptophan. Studies repeatedly demonstrate that all of these appear to work together—certainly the B vitamins do—as an ecological whole, so that to remove one of them may inhibit the operation of all the others.

Despite this, we still hear, from among the eminent, such phrases as Dr. Stare's previously quoted bit in *Life* about how "key vitamins" are added to bread. Clearly, there can be no such thing as a "key vitamin." It's a contradiction in terms, a meaningless phrase in the English language.

While we're on the subject of vitamins, it should be noted that these are the food additives the industry most likes to brag about. Look at the Vitamin D we put in milk, they say; it's practically wiped out rickets in America.

It has, too. But while a lot of doctors don't want to talk about it, the Vitamin D added to milk and to some "enriched" bread may have something to do with heart disease, with damage in blood vessels, and with birth defects. The vitamin speeds up the body's production of calcium, and it doesn't appear that there's a limit to this: too much Vitamin D, too much calcium, is the way it looks. It also passes across the placenta if taken in by pregnant women.

In Britain a few years ago, it was found that a number of babies were being born with unhealthy calcium deposits, and with subnormal intelligence that seemed to be related. On investigation it turned out that their mothers, during pregnancy, had been eating bread and drinking milk both high in added Vitamin D. They cut the vitamin by half in the milk, took extra Vitamin D out of the other foods to which it had been added, and found that children born later didn't have the same birth defects.

In America, some dentists are arguing that we get a lot more malocclusion than we used to, and that excess Vitamin D is probably involved. It may be more complex than that (especially if we remember Dr. Price traveling around the world looking at teeth and old skulls and relating the results to total diet rather than to a single component). But the dentists aren't just making wild guesses either. In addition, children have been found with arterial lesions, caused by excessive calcification of the blood vessels, caused in turn by too much Vitamin D in their mothers' diets during pregnancy.

Then there's Vitamin A, discussed briefly in an earlier chapter. In 1968, Canadian scientists reported that in that nation there are a surprising number of cases of Vitamin A deficiency, despite the fact that the people affected seem to have enough Vitamin A in their diets. According to T. Keith Murray, head of the nutrition division of Canada's Food and Drug Directorate, there is no connection with any particular disease, so that it must mean that something is interfering with the body's use of Vitamin A. "We suspect," he says, "that some environmental factor—drugs, pesticides, food additives—may be reducing the utilization, or increasing the metabolism, of Vitamin A."[21]

Probably the biggest American problem about the utilization of vitamins (and minerals, and other essential substances) lies in the attitude of the organized medical profession. This attitude is based on underlying assumptions which are never defended outright; on these assumptions, however, rests much of what is still being taught in medical schools. Much of medical education is based on the idea that your genetic inheritance is fixed, determined by a sort of luck of the sperm-and-egg draw, and that "health" means, simply, staying that way. This idea was accepted in many of the sciences dealing with the nature and development of the human body until recently, but it is now given less and less weight by biologists and others.

Put a little differently, this older assumption would mean that your inner ecology, however complex, is simply a mechanism for maintaining a *status quo*—a somewhat different *status quo* for periods of growth, maturity, and aging, perhaps, but a *status quo* nonetheless. "Health" is thus a process of staying the same (or, in case of illness, "getting better" until you're again the same, which is called "well").

From this assumption naturally proceeds the idea that in order to maintain the *status quo*, you have to have a certain amount of this, and a certain amount of that. Then you can stay "healthy" and will be able to reproduce during the appropriate years. Well, of course, you do need those certain amounts; but that's far from the whole story. For by that line of thinking, health simply equals a kind of functioning, walking-around survival.

Thus we arrive at such common ideas as that a human being has a "minimum daily requirement" of Vitamin B_1, and that as long as you get it, and get your MDR of everything else, you won't get sick. That's true—in the sense that if you don't get a certain amount of B_1 you will get sick in certain predictable ways.

But even assuming that the government's "minimum daily requirements" are accurate (nobody really knows), the concept assumes that if you get one milligram of B_1 every day, you won't have any trouble for lack of B_1. It doesn't say anything whatever about the necessary interplay of that milligram of B_1 with ten milligrams of pantothenic acid, 20 milligrams of PABA and 500 milligrams of inositol. You may be getting the B_1 from bread—but none of the others. It doesn't say anything about your B_1 requirements going up as your sugar intake goes up. It doesn't say anything about how too much of one B vitamin can give you a "deficiency" in another, even if the amount of the second one is well above the minimum.

Far beyond that, however, the whole concept assumes that once you are healthy, you cannot be improved by eating better. Organized medicine tends to insist on this. Yet there is no evidence to support this assumption, and a considerable body of evidence to the contrary.

You may recall Dr. Price's discovery, as he examined the diets of people who had not yet come into contact with the processed foods of contemporary civilization, that while their diets were always "balanced" in terms of modern nutrition, they often took in far more of some substances than is usually regarded as "adequate" in America. Dr. E.V. McCollum and others argued for decades that what is *adequate* is by no means what is *optimal* in human diet; that the human body is, in fact, in a state of potential and that what you eat can make a difference in what you become. McCollum's phrase was that there is definitely a "principle of the nutritional improvability of the norm."

McCollum is not, by the way, a faddist or a quack. He was the first to isolate Vitamin A (in 1913), the first to isolate Vitamin D (in 1922), and toward the end of World War II first isolated one of the water-soluble components of Vitamin B.

Thus, from a number of experiments it at least appears that the human body is improved (it becomes "healthier than normal," if you like) if its Vitamin C intake is well above the amount that is considered adequate. It appears to be healthier with at least twice as much, perhaps even four times as much, Vitamin A as it can be shown to "need." With other substances—cholesterol is a well-known example—the human body would probably be better off with less than it usually gets (and not only in this fat-prone country).

Obviously this is a thought to be treated with care; you can get too much of something, even something that most people should get more of than they do (we've talked about the dangers from too much Vitamin A or D, for example). But so long as we operate on the assumption that an *adequate* amount is the *right* amount, we are unlikely to pursue the scientific question of finding the right amount: the amount that would not only keep us in our present state of "health" but improve us.

No doctor, pressed in an argument, will deny the existence of deficiency diseases: anemia, beri-beri, pellagra, scurvy, and the rest. But there is enormous resistance to examining the assumption that our inner ecologies are fixed processes, always the same. An improvement in the diet of pregnant women has been shown to affect, very possibly for life, the learning abilities of their children. Except for scattered experiments, outside the principal stream of medical research, such indicators are not pursued.

It is understandable; nobody likes to have his assumptions challenged, and some people are incapable of dealing with such a challenge. But it is not excusable. In World War I, as noted, Denmark changed its bread and very little else, and cancer and heart disease dropped. England had a similar experience in World War II, when the government forced use of a wholemeal bread. But our bakers have piously put back only four cheap substances into the flour from which they have purged dozens. Cancer and heart disease problems in America have risen in a correlation to changes in our eating patterns that is almost as precise as the correlation between smoking and lung cancer.

But the thrust of our cancer and heart research is not at all toward diet, with the single exception of the investigations of cholesterol and saturated fats. The idea of dealing with total diet is unbearable to the sachems of medicine.

It was that way for a long time with the common ailments now known as "deficiency diseases," too; most of organized medicine refused to deal with them as having anything to do with the diet, and insisted on regarding them as being caused by microbes. It might be thought that this experience would convince medicine that other present-day illnesses might be due to combined deficiencies, or imbalances; but no. They defend food processors and their additives instead.

In Manila, for most of 1901, three months in jail meant three months in jail. By 1902, thanks to some thoughtful white missionaries' wives, it was a death sentence.

What the wives had accomplished was to get "the very best rice" introduced into Manila prisons. Prior to the change, there had been two known cases of beri-beri, neither fatal. In December, 1901, there were 52 cases. By October, 1902, there were 4,825 cases of beri-beri and 216 deaths. The only thing that kept it from getting worse was that one doctor discovered an 1883 report about the Japanese Navy; beri-beri had disappeared when the Navy switched from its rice diet to the English Navy diet. On a hunch, the doctor persuaded his colleagues to change the Manila jail diet away from rice. The beri-beri disappeared.[22]

This experience did not convince doctors in America or elsewhere in the world that an epidemic disease might be caused by the lack of a substance in the food; they insisted that it had to have something to do with a microbe.

But it doesn't. The "very best rice" (which is what they had in the Japanese Navy too) is polished, and of course white. If you polish rice, you polish the thiamine off it.[23] One of the things that B_1 does is to help break down sugar in the liver, the muscular system, and the nervous system. In its absence, the sugar isn't broken down correctly, and the residues, if they're present in strong enough concentration, are toxic to the human system.

Notice that "if" in the last sentence. A slight thiamine deficiency may result in a slight concentration of those residues. With such a concentration, you don't get beri-beri. You're just not healthy. Maybe you "don't feel good." Maybe you shrug it off. But you're not healthy—and your system is being hurt, almost certainly in some ways that we don't yet know about.

While medicine refused to recognize deficiency diseases as such, the people who most had to deal with them often understood the idea quite well. An example is scurvy, which results from a deficiency which is not completely understood but involves at least two vitamins—C and P—and can be "cured" (*i.e.*, the patient can be restored to where he can get up and walk around) by Vitamin C alone, though he won't be quite so "healthy" as he would have been if he'd eaten natural food substances which contain other elements as well.

Scurvy was serious enough at sea to cause Vasco da Gama, on his celebrated voyage around the Cape of Good Hope, to lose 100 men out of a 160-member crew. But it was already known in the sixteenth century that fresh fruits and vegetables would cure it, and prevent it. The "uncivilized" had known it for much longer than that. American Indians on the St. Lawrence River cured Jacques Cartier and his party of scurvy with a pine-needle broth in 1536, and other Indians traditionally divided among all family members the adrenal glands of a slain elk, a good Vitamin C source. Eskimos traditionally eat certain layers of skin of baluga whales and get their Vitamin C that way.

One "food" long believed to be a protection against scurvy is beer. It was, once, but not any more. Like bread, beer has become a different product while retaining the same name. A few hundred years ago, it was a headier brew which often contained tiny "sprouts" rich in Vitamin C. Today's sterile product has that nutrition removed.

But scurvy and pellagra became widespread in more "civilized" cultures by a sort of back-door method. Refusing to recognize that food deficiencies cause diseases, Westerners proceeded with the processing of foods: pasteurization of milk, for instance, which eliminated one evil and created another. The boiling of milk, it was later discovered, either destroyed the Vitamin C in the milk or, at least, seriously impaired its effectiveness. The result was "Barlow's Disease," in which babies grew fat and ate well, then lost their appetites, bled from the skin and mucous membranes, and underwent painful changes in the structure of their bones. Barlow's Disease, it turned out, was infantile scurvy. Milk is now pasteurized by a different method.

Pellagra, as we noted earlier, was directly associated in the South with the refining of grains; but its earliest epidemic was at the turn of the century, when corn, which grew easily and cheaply, became increasingly a staple grain in that part of the United States (from which corn bread comes as a delicacy to all of us, and corn liquor to a few).

Like refined flour, corn happens not to have any niacin in it. Pellagra is a niacin deficiency. While the makers of "enriched" bread now put back at least some of the niacin they mill out of the wheat flour, they do not put back the amino acid tryptophan which functions (in part) as a precursor from which more niacinamide is made, and which is also absent in corn.

The same pellagra epidemic that hit the South at the turn of the century (it rose to 100,000 known cases in Mississippi by 1913, and if you know much about Mississippi you can guess at how many unknown and unrecorded cases there were) also hit Europe, but in Europe corn consumption dropped off, and the epidemic went away. Still nobody really put it together.

Not until 1937 was it discovered that niacin (more widely called nicotinic acid then) cures many cases of pellagra. It does not, however, cure them all. Many need more of the B-complex vitamins in addition to niacin, and attention to other parts of their diet, before their disease can be relieved. None of this is by any means completely understood even yet, as Albert von Haller carefully noted:

> Since 1937, pellagra can be quickly cured. But the condition which precedes it has no specific name; it is not marked by specific symptoms; it rarely impels the patient to consult a doctor. There are a good many indefinite complaints that may put in their appearance: lack of appetite, loss of weight, digestive disturbances involving the stomach and the intestines, insomnia, headaches, great irritability, loss of

memory, lack of concentration. . . . The prevention of this preclinical stage of pellagra . . . remains an unsolved problem to this very day.

It might—it just might—be solved if Americans could break the assumption that "normal health" cannot be improved by diet.

There is so much that we don't know, as in the study of *any* ecological problem. And, as with the study of the other ecological problems that confront us, we keep blundering on as though our ignorance were not important. Instead of not doing something until we know what its effects will be, we insist on doing anything that occurs to us until we find out that it's dangerous, even lethal—and sometimes, even then, we continue. Certainly there must be risk attached to life; but just as certainly there is a clear distinction between risk and heedless folly.

For years it was believed that calcium and phosphorus were necessary to the formation of bones in the young; only much more recently was it discovered that all the calcium and phosphorus you can eat won't do a thing for your bones unless two-one-thousandths of a milligram of Vitamin D is taken into the system every day. That's enough to prevent rickets. Without it, the complex inner-ecological process that is bone formation just won't happen.[24]

Vitamin C not only prevents scurvy (in the right combination with other substances). It is to some extent involved in the healing of wounds, in protection against infections, in the formation of teeth; it has something to do with whether other vitamins do their jobs; it plays a part in the action of certain hormones, and possibly in the body's formation of those hormones.

And it dies from oxidation—as, for example, when food is heated in an uncovered pot. Or quite often when the Mad Scientists process it.

It is interesting that the deficiency diseases followed civilization, so to speak. Europeans, sometimes with the best of intentions, proceeded eastward with their ideas that white equals pure, and their "very best rice," and spread beri-beri all through Asia in a kind of "nutritional colonialism." And our treatment of what some of us, distressingly enough, still think of as second-class human beings has not grown all that much better; there are still acute deficiency diseases, not in the remote corners of the world, but right here in the United States.

In 1970, for example, a team of doctors sponsored by the New York Field Foundation visited migrant workers in Texas and Florida. They were quite sure about the reasons for what they found:

We believe the deplorable state of health and welfare of the migrant farm workers to be consciously determined by those who use them. We find the lives of hundreds of thousands of our fellow citizens manipulated and managed in such a way as to reduce them to sub-human status.

And they were quite sure about what it was that they had found: scurvy, pellagra, and rickets.[25]

It may seem a little far-fetched to link deficiency diseases to racism, but it's hard to know how else to regard this statement by a Mayo Clinic doctor (quoted by Rorty and Norman) during the period of the 1930s when organized medicine was allied with the determined bakers of (not yet enriched) white bread:

The most progressive races, those most sound in mind and body, have voluntarily selected white bread as their main diet by the exercise of natural biological laws.

He was an expert.

Nutritional colonialism goes on. As the milling industry once followed the missionaries across Asia (supplying highly polished rice for the poor at a cost comparable to their previous cost for home-pounded rice) and brought beri-beri with it, so today it is pushing the growing use of highly refined white flour (not even "enriched") in Africa.

In East Africa, commercial millers have gone to work on corn, already deficient, as we have seen, in niacin and tryptophan, and are producing a flour with only .05 milligrams of thiamine per 100 grams. The rice that produced the beri-beri in Asia had about .06 milligrams of thiamine per 100 grams. The danger is quite clear; happily, African governments are trying to do something about it, although they have the same trouble with rich and powerful citizens that our government has.

Because of the determination of American manufacturers and processors to sell their products to as wide a market as possible, there is constant pressure from the United States to get the United Nations, and the governments of other countries, to adopt *our* standards for foods (and food additives). It may not be, precisely, "nutritional colonialism," but the metaphorical political phrase, "Coca-Cola imperalism," has a more literal content than is usually thought.

And it's more than Coca-Cola: it's refined sugar, white flour, and our general determination to make and sell nutritionally worthless (and whiter) foods, plus our dogged push to get the whole world to throw all kinds of unhealthy junk into its dinner pail.

CHAPTER TWELVE

Who Dies in the Desert?

In the propaganda of the Mad Scientists, carcinogenic food colors and fit-causing flour bleaches are slurred over, and food additives are discussed strongly in terms of "nutrients." Nutrients, usually, turns out to mean vitamins and minerals. But vitamins, as we hope has been made clear, are not in the strictest sense nutrients at all; they are substances without which our bodies cannot make use of nutrients.

Nor are they individual substances each of which we need for "health," nor nature's "remedies" for scurvy or beri-beri or pellagra.[1] They are parts of an extremely complex, intricately interrelated whole—the whole of what we eat, all of which functions together not only to keep us alive but to keep us growing. When we don't get it all, in the right proportions (not "adequate amounts" of each), and in the right forms, we go downhill as individuals and as a species.

We need to know the importance of these interrelationships more than we need to know most details about nutrition. It is a subject which could easily be taught in junior high school (when we have much less nonsense to *un*learn) but it is not, in fact, taught to any real degree in many medical schools.

No one believes (the idea would make a doctor laugh) that if we separately eat 30 parts of hydrogen, 20 parts of carbon, and one part of oxygen, our stomachs or intestines will somehow magically turn them into Vitamin A. Yet we fail to understand the implications of the fact that the foodstuffs we need come to us in a naturally arranged form.

There is no logical difference between the fact that we must ingest the various elements of Vitamin A in an arrangement that takes place in advance outside our body, and the fact that we must ingest the various parts of the B-complex in a proportion that exists outside our body; in each case we have evolved in such a way as to make use of a pattern preexisting in nature.

The distinguished physicist, Erwin Schrödinger, is the author of a famous essay titled "What Is Life?" What he wrote in that essay still comes closer to answering the title's question, in scientific terms, than anything else ever written. It is far too long and too technical to summarize here, but it is essential to his essay that what we call "life" is deeply involved with organization.

Nonliving things, left alone, tend to disorganize, to deteriorate, to fall apart into their various components all the way down to individual atoms.[2] Living things, at least temporarily, do the opposite: we organize those components, from the bacteria that "fix" the free nitrogen in the air, and the plankton that fix the free oxygen into carbon compounds, all the way to the animals (like us) at the tops of the food chains.[3]

Man, in the course of his evolution, has never until the last insignificant blink of time eaten lysine, or cholesterol, or Vitamin A, or casein. He has always eaten grains, meat, vegetables, fruits, fish. All of them in turn had done their work of organizing substances still less organized—by eating them, or by bacterial or chemical action, or in some other way.

This entire process of organizing the elemental materials of the planet (including constant releases and uses of energy) is what we call life; and since before we can imagine, as an inherent part of what it is, it has continued to evolve. What does not evolve becomes extinct—excepting only a very few forms of life which have achieved a "final" balance, which (for the evolutionary moment at least) continue to exist substantially without change, frozen in place in the middle of a flowing stream.

It is the basic flaw in the common medical view of health and nutrition that it sees health as a state, a condition, rather than as a process. Essentially, doctors abandon the concept of evolution as a part of life, and treat your body and its functions as though you were a coelacanth—or a cockroach.

Obviously if you are suffering from beri-beri and can be kept alive by massive doses of thiamine, you should have it. The point is that most of us are not suffering from beri-beri (although we might be suffering from subclinical problems—headaches, lethargy, etc.—which are related to thiamine imbalance). We are suffering from being treated like cockroaches, frozen in the same form for countless generations, when in fact we are evolutionary processes.

We must learn all we can, of course, about the particular effects of particular deficiencies; but more than that, we must learn that we don't know what a deficiency means. What we do know is that nothing we ingest works in isolation. Beyond certain broad limits there is no such thing as too much sugar or too little something else. Given enough food in general, what must be maintained is a natural balance of foods in natural form. Given that, we can have a glass of possibly useless beer or a bar of possibly harmful candy with some confidence that the strength of our bodies can handle any momentary problem.

In whole wheat bread and butter can be found, in their natural form, sizable amounts of Vitamins A, B, D, E, and F. Some years ago, scientists at the Lee Foundation, a nutritional research institution at the University of Wisconsin, undertook what is called a "literature search": they went through all the journals and scientific papers they could find, to see how many health problems they could find that had already been related to deficiencies in those vitamins. They found 150 separate diseases, illnesses, or metabolic disturbances.

In a recent *Nutrition Review*, Oswald Roels described symptoms which probably sound familiar to every American:

> Fatigue, malaise, and lethargy are common complaints, frequently accompanied by one or more of the following symptoms: abdominal discomfort, bone and/or joint pain, severe throbbing headaches, insomnia and restlessness, night sweats, loss of body hair, brittle nails.

Such symptoms might be brought on by any of a number of causes; but the subject of the paper was what happens when you get a relatively small overdose of Vitamin A.[4]

Folic acid, usually described as a B vitamin, can hide the existence of pernicious anemia (as previously noted, an absence of B_{12} may be involved here too). Or, used as a treatment, it can cure the anemia but mask other bodily degeneration that accompanies it.

Dr. Stare, in a pamphlet distributed by the Manufacturing Chemists' Association:

> Don't let any food faddist or organic gardener tell you there is any difference between Vitamin C in an orange and that made in a chemical factory and added to grape or apple juice. . . .

But there is an obvious difference—though a chemist may not be able to find it after the Vitamin C is isolated. The difference is that one is in an orange, quite naturally, and the other is not where it is naturally; just as the various parts of the B-complex are together in natural proportions in wheat germ (along with other substances), but are not when they're taken in separately.

It does matter. It matters with Vitamin D, for instance, which has been found, when its various known functions are taken into account, to be eight times as effective in its natural form as a part of butter than when it's given separately in the same amounts. In its role as a substance involved in combating rickets, Vitamin D is at least five times as effective as it naturally occurs in milk than it is when given in other forms.

Obviously there is something else, something we don't yet know about, that works in some complex and interrelated way with Vitamin D and that occurs alongside it in nature, but not in the laboratory. That's the difference.

An excellent example is the sugar beet. There is nothing wrong with sugar *per se*. In the refined form in which we get it in America, there is nothing much good about it, but it isn't bad in itself. What makes "too much sugar" dangerous is that every one of a large number of different substances must be present, in the right proportions at the right times, to break down the sugar. Otherwise it leaves residues which are dangerous, sometimes toxic.

One of them, in fact, can cause beri-beri. The presence of thiamine enables the breakdown process to take place so that the residue does not build up. In nature, the thiamine is in the sugar beet with the sugar—and that is what makes the sugar beet an edible food for man over the period of evolutionary history (that and a lot of other things, of course).

Does Vitamin E, also present in the germ of the wheat, have anything to do with the body's use of the Vitamin B complex? Nobody knows for sure, but it's probable. Scientists think so because in the absence of Vitamin E in their diets, rats suffer from a deterioration of their nerves and muscles, so that obviously the Vitamin E is necessary to make possible *some* general metabolic function.

And to do other things too, of course. Without Vitamin E, rats become sterile. In humans it plays a role in the utilization of iodine by the thyroid gland. It has something to do with the repairing of scars both inside and outside the body. It is available to almost all of us, in a natural form and in its natural combination with other and possibly crucial substances, only in the germ of the wheat (it is *not* restored to white bread by "enrichment") and in a few green and leafy vegetables, including lettuce.[5] We must stress again that many doctors still contend, on the basis of out-of-date nutritional knowledge, that there is plenty of Vitamin E in everybody's diet, and that they are wrong.

Another word about Vitamin E. From its apparent effect on scarring, and from some experiments on rats by one scientist, there has arisen a theory that Vitamin E will slow down the wrinkling of the skin and at least postpone the aging process—which has led to something of a fad involving Vitamin E pills and ointments among middle-aged people. We don't want to get into this argument, but so far it's only a possibility, and it's one that involves the risk, again, of an imbalance with other nutritional substances, known and unknown. Adelle Davis, who is often accused of getting pretty far out with her nutritional theories, is eminently sensible about Vitamin E ointments and pills in relation to wrinkles and aging:

> My conviction is that scars will remain unless every nutrient is supplied so that healthy cells can be formed to replace them. Because Vitamin E is only one of the . . . essential nutrients that work together, to expect results by adding it alone is like trying to play chess with nothing but the bishop.[6]

The Mad Scientists won't tell you, either, that all the synthetic vitamins are not the same as the originals, not even the same chemically. Vitamin K, which among other functions plays an important role in blood clotting, is normally fat-soluble; but some vitamin-happy experimenters came up with a water-soluble variety which they immediately started feeding to pregnant women to prevent hemorrhaging at birth, or in the baby. It turned out to cause anemia in the fetus—specifically, hemolysis (the separation of the hemoglobin).

Not only vitamins, of course, cause mischief when they get out of balance. Americans, and particularly women, are constantly being bombarded with newspaper and magazine advertising about their need for iron; but iron, too, has to be in a fairly precise balance with everything else. Adelle Davis has been known to go on about this also, and we were impressed with one paragraph in which she used no fewer than ten separate footnotes (which we checked), every one of them to the original technical-journal report of the experiment in question:

> Other so-called "safe" drugs are ferrous sulfate and various iron compounds, though as early as 1928 they were found to destroy Vitamin E.

Later studies show that they tremendously increase the need for oxygen, pantothenic acid [one of the B-complex, which in turn should be in balance with the others], and several nutrients, that they harm the unsaturated fatty acids, and destroy carotene and Vitamins A, C, and E. When little food can be eaten during illness and the protein intake is low, iron compounds can cause serious liver damage. During pregnancy, taking iron salts, which increase the need for oxygen, already undersupplied the fetus, can bring about miscarriages or premature or delayed births, and may cause the infant to be malformed, mentally defective, or susceptible to anemia and jaundice.

The anemia, just in case your eyebrows went up too quickly, would come from a fetal deficiency in Vitamin E. As you can see, it all fits together.

One of the most dramatic demonstrations of the interrelationships of even seemingly insignificant substances was an experiment with two groups of rats, fed identical diets except that, as nearly as possible, all the manganese was removed. Then, for one group, it was "added back" in the proportion of .005 percent of the diet—one part in 20,000.

All went well for quite a while, until another generation of rats was born. The ones whose parents had had no manganese were a little more listless, but that was all. Then, suddenly, a startling thing happened: the mother rats refused to nurse their young.

While the group which had received the tiny amount of manganese appeared to be perfectly normal, the females in the other group—97 percent of them—were missing what has been regarded as one of the basic characteristics of all mammalian females. They were not *unable* to feed their young; they *refused* to. As a further test, the experimenters put the young rats from the healthy group in with the reluctant mothers. Because the young rats themselves were livelier, they were able to get some milk from the mothers—but only in two cases out of ten.

Here is a case, not of a physical change from a nutritional deficiency, but of a serious behavioral change. Nothing like that has yet happened in experiments with humans (not counting the uncontrolled experiments with additives on humans that the Mad Scientists are conducting all the time), but there are effects from similar tiny amounts of other elements.

Dr. Walter Mertz of the Department of Agriculture, for instance, has found various forms of malnutrition associated with the absence of things like copper, zinc, and chromium.[7] Factory processing of carbohydrate foods, it must be noted, often destroys these minerals.

In studying the ecology of a forest or a prairie, a group of animals in the wild, or a colony of algae, the scientist must deal constantly with the concept of the limiting factor: the population of a given species is limited by the amount of whatever, among its needs, is in the shortest supply. If there is enough food and territory for ten prides of lions, except that trace amounts of chromium are present only in enough supply to keep alive nine prides, then there will be nine prides and no more.

The same is true of our bodily metabolism, of what we have called our inner ecology. Drop the thiamine intake below a certain level while everything else is all right, and you'll get beri-beri.

But nutrition, and health, are not simply matters of staying ahead of one limiting factor at a time. Health is not the absence of illness; it is an optimally operating process over time.

About all of this, Bicknell wrote that "we have exchanged easily recognizable diseases for the insidious dry rot of degenerative diseases, for long-lasting chronic illnesses," and then added: "The only all-embracing cause that I am able to find lies in our nourishment." This immediately brought replies, most of them taking the form, "If that's true, how come we aren't all sick all the time?"

It's a good question. We tried to answer part of it in our brief section on statistics. But for those who are not comfortable with that kind of demonstration, perhaps an anecdote told by a research scientist about a crossing of the Sahara (quoted by von Haller) will make the point:

> On our journey to the south, our two wagons went out of commission, one after the other. We were forced to return to our starting point on foot. There were eight of us in the party, and our water supply was most inadequate. Naturally we rationed what water there was and divided it equally among us—but it was interesting to note that thirst had a different effect on different individuals. On the fifth day three of us reached an oasis and were able to go back and pick up those of our comrades who had fallen by the wayside. For two of them, any help that we might have brought came too late; they had died of thirst. The others recovered, some quickly, some more slowly.
>
> Do you mean to say that the cause of the deaths was not the lack of water of all, because every single one of the group did not succumb to it?

We do not even "ration our water and divide it equally." The poor start from further behind with each generation (for life is a continuing process), and we need not go beyond the borders of our own wealthy nation to see it. But those whose parents refuse the knowledge of the interrelationship of nutrients and the substances which allow us to use them properly: the vitamins that become parts of enzymes, the minerals that in tiny amounts are vital catalysts, and the organizing principles that put them together before we eat them and demand that we eat them in the right forms and in the right proportions; they also start further behind.

We did not evolve by eating yogurt and wheat germ cookies and sesame seeds to the exclusion of balance; nor by eating vitamin pills with added minerals. If that were the reason for the FDA's relentless pursuit of "health food stores," *in the presence of a truly natural diet in America*, then that pursuit might be justified, and they might even be right: the purveyors of those foods might indeed be perpetuating frauds.

But allowing for the occasional fraud among them, or the occasional faddist whose remedy is as bad as the ill prescribed for, they are, by and large, simply trying to give us a way to get back into our inner ecologies what has been stolen from them. And the prosecutions are not to curtail the frauds but to protect the real thieves.

There have of course been sick people, and poorly fed people, since there have been people. The "primitive" diet was imbalanced more often than balanced, and no one can ever count the branches of our race, the individual villages even, that simply didn't make it because of some limiting factor in diet. But the species has survived, and evolved, and it is tough; it got that way by eating, to put it simply, what came naturally. It cannot change that in two generations, and spread the change over the planet in two more, and survive.[8]

We have come a long way from the bread in your breadbox, or so it might seem. But it is not very far, really. As elsewhere in this book, we have found ourselves unable simply to quote scientists who refute other scientists, or to say flatly and loudly that "white bread is bad." The questions behind these arguments, the logical assumptions that underlie the public positions of "authorities," the basic knowledge of the workings of our systems—both bodily and politically—are an essential part of understanding what the controversies are about.

We are, however, talking all this time about bread—or, more accurately, about grain and about what happens to it between the field and your breadbox that destroys it as a meaningful food. It would be a relief if we could say that most Americans have other and better ways to get their cereal grains. There are other ways, but they aren't better.

There are cakes and pies and cookies and things, of course, but the flour is the same; there are just more useless additives, none of them "nutrients" even by the Mad Scientists' definition. And then there are the good old breakfast cereals, two-thirds of which are cold cereals.

That's not two-thirds of the cereal *portions;* it's two-thirds of the weight of cereals sold. If you count portions, the cold cereal proportion goes up (it's 85 percent of total sales). It isn't by any means a small business, as you can guess from the amounts spent on advertising every year. Every American averages out at about ten pounds of the stuff a year—but of course we don't eat "on the average," and in practice kids eat a lot more of it. The national average is about 1 percent by weight of all the food we eat—but again, the percentage goes up for children.

Just as what we now call "bread" is not what our great-grandparents called bread but is a completely different "food," and just as the same is true of beer, so what we now call "cereal" is different. Cold cereals are a fairly recent invention; the practice for almost all of man's history has been to eat hot cooked cereals, "mush," "porridge," "gruel."

And like bread, it was made out of the whole grain and was a far more nutritious food, because it had in it all of its own natural constituents (and because it didn't have food additives). Nowadays, what we call breakfast cereal almost invariably has most or all of its vitamins, minerals, and high-quality protein puffed, shot, crunched, snapped, crackled, and popped completely out. You would be much better off eating shredded beetles.

If you have to have something for breakfast (for yourself or your children) that is quick and easy to prepare, tear up a couple of hunks of whole wheat bread in a bowl and pour the milk over it. If you use breakfast cereal as a sneaky way to get the kids to eat fruit, throw the fruit in too. If you're using real whole wheat bread without the industry's bleaches and preservatives (or something like the McCay-Cornell bread), that breakfast is more nutritious than any you can buy in a box from a major American manufacturer.[9] Leave the sugar off; the bread has enough taste without it, and the sugar is nutritionally useless.

The above paragraph is literally accurate. Every major cereal manufacturer has one cereal (a couple have more than one) that he likes to trot out on display: Kellogg's Product 19, General Mills' Total, Quaker's Life are all good foods nutritionally—as breakfast cereals go. They are all just about as good as an equivalent amount by weight of bread—but not the best bread. The rest of the cereals (there are six or eight others that might be in the above class) go downhill from there, most of them a hell of a long way downhill.

If you must have a breakfast cereal that comes in a box, have a hot one, preferably one made from oats which are milled as little as possible.[10] Oats have the largest percentage of digestible protein and fat, and they retain more of their amino acids and protein during milling than do other grains (they are short on lysine, but most grains are). It isn't the "hot" that matters in having hot cereal (although every mother we've ever known has believed that hot breakfasts are better than cold, and maybe we ought to trust folk wisdom one more time); it's the fact that cereals prepared for cooking at home generally have fewer additives as well as having had fewer things done to them.

Not that all are good. Cream of Wheat, for instance, is mostly made from farina (the most highly refined and nutritionally useless white flour there is) with a little wheat germ thrown in—not much or it wouldn't be the color it is—and enough vitamin content so the farina can be called "enriched." It is not nearly so good nutritionally as "enriched" white bread—and that makes it pretty bad.

But please don't let your children (or yourself) eat any of the breakfast cereals whose names have become household words. They are not manufactured in order to feed anybody. They are manufactured solely because they are by far the most profitable "foods" produced in America despite the fact that those profits come after an astonishing 19 percent of the total sales is spent on advertising and sales promotion.

Difficult as it is to believe today, cold cereals actually began as "health food," though their origins are mixed up with controversies not only about foods but about the early days of the Seventh-Day Adventists. In 1876 Dr. John Harvey Kellogg took over the church's Western Health Reform Institute in Battle Creek, Michigan, and soon renamed it the Battle Creek Sanitarium. In 1891, the sanitarium treated a patient named Charles W. Post. In 1895, distressed by the fact that some patients with tooth trouble couldn't eat zweiback (the sanitarium was big on zweiback), Dr. Kellogg invented the idea of a cereal flake.

The first were wheat flakes. A previous food product, consisting of crumbled bits of baked wheat, was being marketed under the name Granula; when Kellogg called his produce Granola,[11] he was sued, and changed its name to Granose. At about that time, elsewhere, Henry Perky invented what he thought of as a new way to eat whole wheat in biscuits with soup or other foods; it was baked into shreds and he called it Shredded Wheat (it was once offered to Kellogg for $100,000, but he turned it down).

Kellogg also made a coffee substitute out of grains, called Caramel Coffee at first and later Minute Brew. In 1892 Post, Kellogg's patient, offered to market the substitute. Kellogg refused, and Post went off to make his own. In 1895, when Kellogg was discovering the cereal flake, Post was beginning to market his coffee substitute—Postum.

Happy with its success, Post came up with another health food, a wheat cracker that he broke into crumbs and for no apparent reason called Grape-Nuts. He went on to apply Kellogg's cereal-flake idea to corn, inventing what he called "corn flakes." The rest is history: Battle Creek became the health food center of the world, with a great deal of charlatanism thrown in (though both Kellogg and Post got rich, no one has ever demonstrated that they did not genuinely believe in the value of what they were doing; some of the fly-by-night competitors they had for a while were something else again).

Gradually, of course, it went from a sizable small business to a giant business. Kellogg's (taken over and turned into a monster by the doctor's younger brother, W.K.) stayed independent; Post's company became a subsidiary of General Foods. But one thing didn't change—the underlying purpose of the whole thing.

Only a few years after the battle of Battle Creek was under way, old C.W. Post himself uttered for all time the philosophy of the breakfast food business:

> It's not enough to just make and sell cereal. After that you get it half-way down the customer's throat through the use of advertising. Then, they've got to swallow it.

Only a short time ago, a Kellogg official was quoted in *The Wall Street Journal*: "With television, we can almost sell children our product before they can talk." Nothing has changed since old C.W.

During 1970, Dr. Robert B. Choate, a nutritionist, attacked breakfast cereals and made the nation's front pages. A short time before, the Federal Trade Commission (not the FDA) announced an investigation into the cereal industry to see whether domination by a small number of firms leads to antitrust violations (we're still waiting for the results). During that month, somebody at United Press International in New York and Washington put together a lot of data on the economics of the industry, and columnist Milton Moskowitz soon added more (the month was July). From those sources, we can see what that stack of Junk Flakes and Sugar Frosted Armadillos in your supermarket means in the American economy. If all the figures are right (none of them are incompatible with each other), Kellogg has 40 percent of the market, most of it accounted for with Corn Flakes, Rice Krispies, Sugar Frosted Flakes, and Special K. Post (General Foods) and General Mills, together, split another 39 percent. Quaker Oats is good for 6 percent, a little ahead of National Biscuit (Shredded Wheat), with about 5 percent, and Ralston-Purina, with about 3 percent. The other 7 percent is everybody else, including all the private brands in supermarkets.

You mustn't think that that's how big the companies are. Nabisco and Ralston-Purina are both bigger than Kellogg. But not in breakfast cereals, which was all Kellogg did until very recently, when it acquired Salada Foods (tea bags).

Put together, they sell $900 million worth of this junk a year.

In 1969, Kellogg sold $542 million worth of goods (very little of it tea bags) and made $44.6 million in profits after taxes.[12] The year before, they made $42 million after taxes on sales of $466.6 million. The company manufactures 855 million pounds of cereal a year—and that's less than half of the total of what's made.

And they all get bigger. The retail price for breakfast cereals, the price you pay in the store, went up 45 percent between 1954 and 1964, a greater increase than in the price of any other "food." According to *Chain Store Age*, a supermarket trade magazine, aggregate supermarket sales of all products jumped 5 percent between 1969 and 1970, but dollar sales of cereals jumped 10 percent.

Those figures led UPI to wonder whether Dr. Choate's highly publicized charges had hurt cereal sales, so they checked a few chains around the country. "Supermarket chain officials," according to the report, "say it is still too early to tell if there will be any lasting effect and the bulk of cereal buyers are not that worried about nutrition anyway."

That's you they're talking about. You know how the manufacturers feel about you; now you know how the chain store executives feel about you: you're dumb.

Since you're paying for all this, you might want to put some of those figures together. If they're putting 19 percent of sales back into advertising, and if their after-tax profit is something like 9 percent, that means that nearly a third of what you pay is either total profit for the manufacturer or goes to persuading you to buy more, so that nearly a third of what you then pay can go to—etc. The profit on breakfast cereals is twice the average profit on food products generally. And we aren't including the profit factor for the distributor, the store, and so on.

The next time you hear some businessman talking about passing savings along to the consumer, by the way, you might remember that in the same period in which these astronomical profits are being made, the price of breakfast cereal is going up.

What is it that you're paying for them to tell you? Well, in C.W. Post's day it was all about "stomach comfort" (Grape-Nuts was sold as an alternative to surgery for an inflamed appendix!). In the 1930s, it was "bulk" to cure constipation (Kellogg came in strong with All-Bran). In the 1940s, it was "vitamins." Today, it's all about protein. In testimony before a Senate subcommittee, Dr. Michael Latham, a Cornell nutritionist, recommended:

> . . . hopefully a restriction, of the subtle, often false, insinuations made by breakfast cereal advertisers which lead the public to believe that these foods have nutritional qualities which are superior to many cheaper and often more nutritious ordinary foods such as bread.

The Federal Trade Commission says it is investigating antitrust questions, but it can't do anything about nutrition claims, because cereals are foods and they're regulated by the FDA. The FDA, of course, says it has nothing to do with advertising or false claims as long as there's no mislabeling. There it sits.

In the meantime, the manufacturers grow because they've got the best deal there could be: a product that is cheap to make and for which an overwhelming demand has been created. All the grains widely used in cereal manufacture are products that can be grown over wide areas in mass production, with a maximum use of machinery and a minimum of hand labor. The manufacturing processes are simple (they would be less simple if there were a genuine attempt to retain nutritive values).

As a result, Puffed Wheat and Puffed Rice—two relatively small sellers which provide the smallest number of calories per bowlful because they're mostly air, and which are manufactured through a process using so much heat that virtually all the nutrients are destroyed—were themselves doing $200,000 a month in sales more than ten years ago.

They all use heat processes, of course, and the heat processes damage protein at best, and usually destroy the lysine, if any. A few of the Mad Scientists have suggested "enriching" the stuff with lysine—whether to avoid attack or to increase sales appeal with a new gimmick, we don't know—but even white bread champions are unwilling to go along with that further step in ecological nonsense.

Are cold cereals actually that *bad* nutritionally? Yes. Dr. Choate came in for some criticism of his methodology and for his apparent approval of "enrichment" as a remedy, but the essence of his criticism of the cereals was quite correct. You could guess this from the cerealmakers' first responses, none of which contained any nutritional arguments. According to *The Washington Post*, "The cereal companies said Choate's testimony contained numerous technical errors and that he totally ignored children's taste preferences."

What do "children's taste preferences" have to do with it? What do you do if your six-year-old decides he loves the taste of ant poison? The Post people, put out because Dr. Choate made the obvious comment that sugar-coated cereals aren't good for teeth, said that those cereals "provide a measurement of control over children's sugar intake." You only hear that kind of nonsense when they can't think of anything else to say.

In a few days, however, the cereal industry wheeled up its big guns—and its "big guns" turned out to be Dr. Stare, ready with The Old O.A.D. Trick, cereal industry variation:

> Breakfast cereals *with milk* contribute to the nutritional quality of the total breakfast [our italics].

Sure they do, if you don't forget the milk, and especially if the whole thing is only part of "the total breakfast," whatever that is. Usually the industry's defenders casually drop a little fruit into the cereal, too, which is why we call this variation The Old Milk and Fruit Trick. Watch for it.

Yes, there is some protein in some cereal. Some of the protein that is in some cereal is even usable. Some of the "high-protein" cereals like Total or Product 19 might even be considered as light meals (although most of the value will still come from the milk). But compare the cost of those products with the cost of Shredded Wheat or corn flakes (and compare it by the pound).

One reason for the large sale of breakfast cereals is that poorer parents can sometimes afford them for the kids. Those parents don't buy Total or Product 19; they can't afford to. And if more affluent parents do, they're silly. If you're trying to make your money stretch, you're coming very close to buying nonfood; if you're trying to buy food, feed the kids bread. *Consumer Bulletin* put it with perfect accuracy ten years ago:

> When breakfast cereals are of the best quality and type (as a few are), they are nutritionally about equivalent to bread; when of the usual over-processed and sugary sort, they are more nearly equivalent to a starchy dessert, which few would regard, under that name, as a satisfactory breakfast food.

You get "calories" from a starchy dessert, too. Even a little protein once in a while.

There was one experiment in which, again, two groups of rats were measured against each other. One group got a mixture of corn, oats, and rye, pounded. The other got the same grains, "puffed" as in breakfast cereal. The first group stayed healthy. The growth rate of the second group was depressed by from two-thirds to three-quarters.

Dr. Philip Jeans of Iowa State University's College of Medicine did some looking around, and discovered that in our "well-fed" society, a great many children have average weight for their height and age, but they lack good muscles and good posture. He went so far as to say that it is extremely uncommon for even four-year-olds to have good posture.

The kids get plenty of calories. But not enough of them come from animal-protein foods nor from high-protein vegetables (animal protein and vegetable protein differ considerably, and the body can make much better use of the vegetable protein if there's a little animal protein present). Muscles don't achieve what Dr. Jeans called "the expected normal growth"—and you'll notice that he didn't say the *optimal* growth.

The latest twist of the cereal scientists is to add phosphates to their products to help curb the cavities caused by presweetened cereals. They blandly tell us they have, once more, "enriched" their cereals for us. These selfless philanthropists of course don't mention that the cavities wouldn't be there in the first place if they hadn't presweetened their cereals. And we aren't even mentioning the fact that adding phosphates might have damaging inner-ecological effects. Phosphates, for instance, can interfere with the body's adsorption of iron—one of the other minerals added in "enrichment"!

Because it seemed to follow logically from talking about bread, we have talked about breakfast cereal and stressed that it, like bread, has most of the nutrition and all of the nutritional ecology hammered out of it. We should note that it also has all kinds of additives in it. We won't go through another catalogue, since most of them are discussed in earlier chapters, but don't ever forget that they're there. Give the kids something else for breakfast. At least give them *something* for breakfast, even bread if you can find some that's any good.

Bread was, once, the staff of life. It was literally the nutritional base of the existence of whole populations, and was most of what they ate, and they were able to live on it. You couldn't now, of course—and as this increasingly became true, people ate increasingly less bread.

That itself is probably good; the less the better. But unfortunately, at the same time that the diet was becoming more varied, the Mad Scientists were doing to everything else what they were doing to bread. With the decline of bread, people all over the world eat, among other things, more and more fats and sugar. But at the same time, these things, too, have been increasingly refined, denatured (think about the word!), and processed.

There are fats and there are fats—depending on links to their carbon atoms. An atom of carbon is said to have four valences, or a valence of four. What that means is that each carbon atom can connect with, or "bond to," as many as four other atoms. They can be other carbon atoms, or oxygen atoms, or hydrogen atoms, or atoms of other elements; but there can be no more than four of them.

If there are fewer than four of them, however, the carbon atom will connect with, or bond to, one of them twice, or maybe even three times; this is called a double bond, or a triple bond. In drawing some kinds of chemical diagrams, these bonds are represented by lines. You could have, for example, a string of carbon atoms, with hydrogen atoms attached, that would look in part like this:

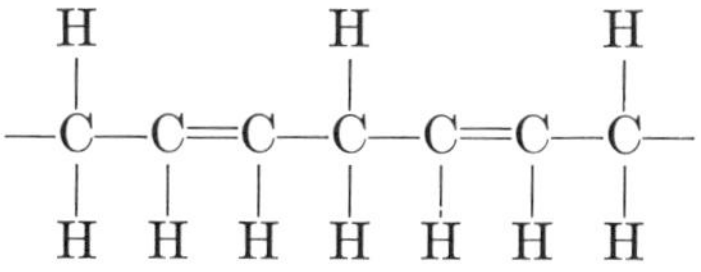

```
  H         H         H
  |         |         |
—C—C=C—C—C=C—C—
  |    |    |    |    |    |    |
  H   H   H   H   H   H   H
```

You'll notice that from every C, there are four lines radiating, representing each carbon atom's four "bonds." In some cases every bond is used up with a separate atom. In other cases the carbon is doublebonded to the next carbon.

One thing may not be clear if you're not used to these diagrams: each of the doublebonded carbons has one bond "available." One of the lines between the two Cs is a bond available to both—it can swing up from the middle like a drawbridge, and each of the two Cs can accept, for example, another hydrogen atom.

That could be the diagram for part of an oil, or fat. Because the possibilities for bonding are not all used up, that kind of oil or fat is called "unsaturated." A "polyunsaturated oil" is simply a long chain of such structures, with many double bonds.

If you added four more hydrogen atoms to our diagram, so that you would use up all the available carbon bonds, you would then have a "saturated" fat:

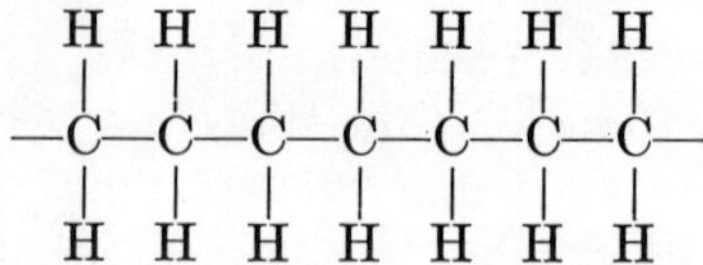

The process of adding hydrogen to unsaturated fats in order to get saturated fats is called "hydrogenation." That's all there is to those terms—"saturated fats," "polyunsaturates," and "hydrogenated vegetable oils"—that are so much kicked around these days.

Vegetable oils are liquid at room temperature. Animal fats are solid at room temperature. These are characteristics of unsaturated and saturated oils respectively. The makers of commercial shortenings like Crisco or Spry make them by taking vegetable oils and hydrogenating them. The smooth modern peanut butter in which the oil doesn't separate is made that way by hydrogenating the peanut oil.[13]

When makers of margarine say that their products are "high in polyunsaturates," it is, therefore, important that it's a relative claim. Margarine is hydrogenated vegetable oil, emulsified with milk and with other additives (dyes, things to make it smell like butter, and all sorts of other junk) thrown in. And it is solid at room temperature—a direct measure of how saturated it is.

But there is (as there always is in dealing with the manipulations of the Mad Scientists) a serious catch. The parts of vegetable oils which contain these long carbohydrate chains are acids—fatty acids. And it turns out that the fatty acids in vegetable oils have important functions in our metabolism—so important that some of them are now generally referred to as "essential fatty acids."

The *essential* fatty acids are in *unsaturated* fats, always. They are "essential" in the same way that vitamins are. We cannot make them in our bodies; our bodies cannot function without them; and they seem to have a multiple importance, so that we cannot pinpoint a single "function" that each performs. It was first discovered in 1929 that rats got sick if essential fatty acids were not present in their diet. The progress since is summarized by von Haller (but the italics are ours):

> Gradually it grew to be a certainty that certain unsaturated fatty acids—above all, linoleic acid and arachidonic acid—have qualities resembling those of the vitamins. The human organism cannot produce them. These unsaturated fatty acids play their part in preserving the healthy state of the blood vessels and the skin, in promoting normal growth and in protecting the organism against infections. *This many-sidedness of their functions points to the fact that they do not merely have a limited local significance, but that, like the vitamins, they are absolutely essential for the proper functioning of the body's metabolism as a whole.*

Linoleic acid, to use just one example, makes up about 55 percent of soybean oil, 44 percent of cottonseed oil, 38 percent of corn oil, 20 percent of peanut oil, and smaller amounts of linseed or olive oil (the composition may vary by several percent).[14] This assumes, of course, that the oils are not hydrogenated.

By 1957 it was well known to nutritionists and to some doctors that there is a decided correlation between the rate of heart disease in a given geographic area and the consumption of hydrogenated (or "hardened") fats. While the correlation is known, however, the mechanism is far from clear—exactly the same situation that exists with cigarettes and lung cancer. In completely unscientific laymen's terms, this means that we know damned well that it's true but can't prove it conclusively enough to break through the bonds that tie the food industry to the regulatory agencies.

It is generally believed by nutritionists and, now, by a sizable number of doctors that saturated fats lead to the body's overproducing a substance called cholesterol, a kind of "solid alcohol" that is found in the liver bile but that also makes its way to a great many other parts of the body in the performance of functions still not well understood.

What is understood is that too much cholesterol is one of the things that hardens arteries, which means that the insides of the arteries get "caked," rather like the insides of drainpipes eventually get caked by solid material. What damages the arteries is quite likely to damage the heart.

One of the substances which helps to counter the action of cholesterol in the arteries is one of the B vitamins that is taken out of our bread and not restored—inositol.

By 1969 the Food and Drug Administration was under pressure (as it still is) to ask—indeed, to demand—labeling regulations that would allow consumers to figure out how much of what kind of fat is in their food. The pressure was from such quacks and food faddists as the National Research Council of the National Academy of Sciences, the American Heart Association, the American Diabetes Association, and even the American Medical Association. Medical authorities were so alarmed that in its issue of January 10, 1969, *Time* was able to print this item from one of its corps of excellent science reporters:

> Medical researchers studying heart disease are coming reluctantly to a revolutionary conclusion. The Federal Government, they suggest, may have to intervene and decree a radical change in the prevailing American diet. This would involve taking most of the fat out of those marbled steaks and from those billions of gallons of milk, as well as altering the chemical constitution of cooking oils and fats.

In 1959, however, in hurried protection of the industry it's supposed to regulate, the Food and Drug Administration had placed into the law a new regulation, which is still there, reading in part:

> . . . (b) The role of cholesterol in heart and artery diseases has not been established. A causal relationship between blood cholesterol levels and these diseases has not been proved. . . .
>
> (c) [Therefore] any claim, *direct or implied*, in the labeling of fats and oils or other fatty substances offered to the general public that they will prevent, mitigate, or cure diseases of the heart or arteries is false and misleading, and constitutes misbranding within the meaning of the Federal Food, Drug and Cosmetic Act [emphasis added].[15]

If the FDA had confined itself to an order that prohibited direct claim of a cure, the regulation could certainly be justified, even with present knowledge. But they made it quite clear then, and they have consistently reiterated since, that by their "direct or implied" clause they meant that no one may label a product as "fat-free," or "high in polyunsaturates." As indicated in an earlier chapter, some people, particularly margarine manufacturers, do it anyway, and the FDA is afraid to sue, lest the evidence prove to be too strong (though margarine is surely a saturated fat itself; it doesn't make any difference that corn oil is "higher in polyunsaturates" than some other oil, if you're going to hydrogenate the oils anyway).

But the point is that the FDA is doing the opposite of what it's supposed to do. By its regulation, it is preventing many food processors (including small processors from whom innovations often come but who don't have the resources for a long legal battle) from developing relatively fat-free versions of a number of products like cheese.

The FDA's rationale, according to Turner, is that permitting such labeling would give the manufacturers of the probably healthier products an unfair advantage! In the meantime, the Justice Department refuses to prosecute violators because the violators' case looks too good.

The quotation from *Time*, with its reference to milk and "the fat in marbled steaks," may be misleading. Man has always eaten meat and drunk unskimmed milk, and while some cultures make do with very little (though always getting adequate amounts of animal protein), our problems probably come much more from the deliberate hydrogenation of unsaturated fats, and a national fascination with foods fried either in such shortenings or in animal fats.

If we could get rid of the hydrogenated shortenings (most of what most people do with Crisco can be done with soya oil) and feed our families bread good enough so that we wouldn't have to spread butter or margarine on it to make it edible, we could probably go on having steak every now and then if we can afford it. And it may be that there are ecological reasons why we should take our meat and milk, like our vegetable oil, the way it comes.

A lot of what we do is habit, of course. But we have to learn, too, that we aren't born with food habits.

That's important, because people rear children according to what they themselves believe. Food habits are not all that well understood, but it is certain that they are *learned*. Nobody is born with a "sweet tooth" (nor even a taste for salt); we learn our tastes, and we learn some of them very, very young. It is amazing what you can do for your children by remembering for the first two or three years that you are manufacturing almost every food preference he or she will have for the rest of his or her life. You can indeed make that life a good deal longer.

Of course you can make it longer yet if that growing child somehow acquires the understanding of his or her body as an ecosystem—if, indeed, the whole idea of ecology as a way of looking at things becomes a part of the consideration of every question, from the importance of salmon and the value of trees to imperialist wars and the social cancer of racism.

In the meantime, though, there isn't much else that you can do about all this. Clearly, deep changes in our entire economic and distributive system are called for, but there are a few things we can do to help. Unfortunately, these things are more available to the middle-class people who usually read books. It isn't so easy for the less fortunate.

Most people, rich or poor, don't know much about nutrition. But if you're rich, or even fairly comfortable, you eat by preference enough of a variety of foods, some of them of quality, so that there is some value to your diet almost in spite of yourself. It is not, of course, nearly so good as it could be (for about the same money) if you avoided ballooned white bread, presweetened puffed or flaked or otherwise phonied-up cereals, excessive sugar, and so on, and if what you ate weren't pumped, stirred, and needled full of all the junk we've been talking about. But it might be okay.

When you're poor, it is not okay. Those with less money eat a lot of starches because they have to, and they're all "purified" of food value. They eat the cheapest cuts of meat when they eat meat—the cuts most likely to be prettied up with some additive or other (and the fattiest). They eat a lot of sugar; check, for example, the soft drink consumption of poor Southerners. Canned instead of frozen foods are not so good in quality and have more additives. Fresh fruit and vegetables are too expensive for every day. Milk intake is usually inadequate. And so on.

One of the things you can do, if you are in a reasonably middle-class bracket, is bake your own bread.

We know: it's a lot of work. But stay with us a minute on that one.

Fifty years ago, 95 percent of the bread eaten in the United States was baked at home. Today, it's less than 5 percent. Somewhere during those 50 years, a whole lot of other laborsaving devices have come along as well, and you'd still be a lot better off than your grandmother even if you did bake your own bread.

We wouldn't dream of telling you how; any decent cookbook will do that. But we will tell you to get some stone-ground whole wheat flour, unbleached, and with as few additives as possible, to start with; it's a good idea to buy it in relatively small amounts and use it quickly (another problem for the poor, who cannot afford the slight upward difference in cost for good flour). It can be found in most cities and towns of any size.

About your time: you'll have noticed (at least anyone paying attention to the women's movement will) the reference above to your grandmother making bread. We didn't mention Granddad.

But we think "women's liberation" has some relevance here if we can get away from the TV and press stereotypes about it.

One of the great arguments by the PR men for the food industry has to do with how liberated the American woman is. Thanks to this industry, prepared foods, mixes, and other "convenience food" items have cut by hours the daily time spent in the kitchen. Well, yes. But isn't it a little silly to liberate yourself by killing yourself? Possibly it's time to think ecologically about the household too.

The figures about home-baked bread a few paragraphs back came from Morrison Wood's newspaper column, which is always about food and cooking. The column is called "Mainly for Men."[16] As anyone knows who has ever laid eyes on a functioning outdoor barbecue, men love to cook. So do kids.

The other side of the argument is that the horrors of the American diet affect male stomachs as much as female ones. If he won't help, bake the bread for yourself and let him eat the stuff that comes from the store. You can bet on how long that will last.

Seriously, our stereotypes about "women's roles" and who belongs in the kitchen are a little silly. If you're a man who worries about your masculine image, there's an easy way out: cook one meal, or one major dish for one meal (inside, in the kitchen, and not something simple like steaks), and then invite for dinner the person you're most afraid of as far as kidding is concerned. Take the kidding if it comes but watch him (or her) eat what you cook. That'll take care of your ego.

If you're a woman who is terrified at letting the ape loose in the kitchen, just remember that it's mostly washable and that you don't have to be a tyrant. Your way may be more efficient, but let him find that out for himself; otherwise you take the fun out of it.

And if neither of you can stand the idea, then divide the rest of the housework up differently. Just because added cooking happens in the kitchen, it doesn't have to be done by the female in the house on top of everything else she already does.

In any case, try this experiment. Regardless of how you work out your domestic relationships, try baking your own bread, as we have suggested, for a month. Then you decide whether to keep it up after that.

All right. Yes. We've used another stereotype: everybody has married and lived happily ever after.

In fact, some women have to cook and work and bring up their kids without a man. Some men have to do the same without a woman. Some people live alone. Some have other obligations, or things they'd rather do with their time (like maybe fighting for causes related to the world's ecology). In that case, at least hunt around until you find a place where you can buy decent bread and flour, stay out of the packaged cereal department, and cut down on the sugar. *That* at least you can do.

And while you're at it, try to eat plenty of fresh fruit and vegetables.

If you can find any that aren't impregnated with DDT.

CHAPTER THIRTEEN

"What If We Do?"

If the invisible insecticide residues on and in all of our foods were in fact visible (if they showed, for instance, as an ugly yellow-green slime that got worse as the concentration increased), most of us wouldn't be able to bring ourselves to pick up anything in the produce department. A great many of us wouldn't even be able to walk through that part of the store without becoming physically ill.

The same is true of the meat department, and of any other section in which the food shows through transparent packaging. The slime would show in bottled milk and fruit juices—and at home, it would be there whenever you opened a box or a can.

It is there. It just doesn't happen to be an ugly yellow-green slime.

If those residues were in fact visible, the wave of shock and horror that would by now have swept America would have run the insecticide manufacturers out of business, and driven into exile the government bureaucrats who let it happen and who are still letting it happen.

Eight years ago, a Public Health Service toxicologist told a Senate subcommittee that DDT has been found "in every complete meal analyzed in this country." Today, it would be extremely rare not to find it in every component of every meal.

DDT is only one insecticide.[1] There are more than 500 basic chemicals used as "pesticides," and they are used in more than 50,000 formulations (what we usually describe as "pesticides," when we talk about this environmental problem, includes herbicides, fungicides, and some other poisons with related functions).

We have used the term "pesticides" too up to now, because it's the word most people know; but as we see what semantic confusions do to our understanding of ecological problems, it is time to note that the word "pesticide" is itself a con job, an attempt to confuse us about what these compounds do, and to make us regard them favorably. They do not kill *pests;* they kill *insects.*

Some of the insects are pests at some times and in some places; some are harmless; some are beneficial; many we just don't know about. The compounds don't know the difference; the people who use them often either don't know or don't care.

Herbicides kill *plants,* not *weeds.* Some are specific for certain kinds of plants, some less so, some not at all. And of course when we talk about their effects (or what little we know about their effects), it is necessary to talk not about insecticides or herbicides but about classes of chemical compounds (chlorinated hydrocarbons or organic phosphates or carbamates) and ultimately about specific compounds: DDT and chlordane, Carbaryl and 2,4,5-T.

There are perhaps a million varieties of insect; more than half that many have been catalogued. They have inhabited the earth for something like 250 million years, having attained a marvelous variety of specialization long before another evolutionary branch developed stereoscopic vision and the opposable thumb. If we were to make one of those metaphorical evolutionary compressions for comparative purposes, and say that insects have existed for the equivalent of a year, then man was literally born yesterday.

We can say that insects have been here since before the beginning of our evolution as humans, or even as primates. Everyone knows something of their natural ecological functions: they pollinate plants, they aerate the soil, they feed on organic material and turn it back into soil nutrients, or they play some role in the life of other insects, or other life forms, that have other functions.[2]

Until extremely recently in human history, we have managed to live with insects. Not without difficulty, of course: the *anopheles* mosquito carries malaria, and no one who has read the book (or who remembers the movie, for that matter) can forget the impact of the locust plague on Chinese peasants in Pearl Buck's *The Good Earth.* But even locust plagues have their ecological function, and if we are to eliminate them (or malaria) we must do so in a way that fits the ecological facts—not in a way that turns out to do more harm than good.

We managed, too, to develop agriculture itself, and then to refine it, in company (if not always in harmony) with insects. Of course "the engineering mentality," as applied to agriculture, far predates modern times; vast barren areas, once fertile, in Pakistan and in the Middle East remind us of the folly of intensive irrigation with insufficient attention to the effects of salinity. Today, the same problem haunts the San Joaquin and Imperial Valleys of California. Much of the American West has suffered from man's unwillingness to learn from his ancestors the lessons of erosion or of overuse of the land.

But only recently has agriculture become truly commercial, truly a big business; and only recently has the urban resident forgotten completely what most men and women once knew about the appearance of food, so that we can be conned into wanting (and eventually "demanding") apples or spinach or tomatoes that simply look pretty—unblemished, uniform in color and size, even shiny—instead of nourishing.

We have been conned into calling this "quality."

And we still think of "the farmer" as he exists in literature and in the movies, and nowhere else: the nice, friendly man with calluses on his hands, a pitchfork somewhere in the background, an amiable and extremely competent wife who puts up endless amounts of fruit in mason jars and takes care of the chickens, barefoot children who help with the milking before going off to a one-room schoolhouse.

He is long dead, that nice man. The "farmer" is a giant corporation, run from a plush office in an urban financial district somewhere, interchanging directors with banks and utility companies and obscure holding companies whose functions are known only to highly trusted accountants.

It also interchanges directors, of course, with can companies, food processing companies, and insecticide manufacturers. Very often these are the same: in California, for example, the Dow Chemical Company, Hunt-Wesson, Purex, and the California Packing Corporation (the Del Monte label) are "farmers," as are the people behind Safeway Stores (who do their own canning, processing, etc., under private store labels).

The DiGiorgio family, in the 1930s as today, major targets of farm union organization, are behind the S&W label and are represented on the boards of directors of the Bank of America and other major firms. Since the Steinbeck days, the role of can companies as a part of this corporate and financial complex has been clearly understood, at least within the state.[3]

Twenty percent of the nation's insecticides are used in California, where in the most fertile areas a tiny percentage of the number of landowners own the overwhelming preponderance of the farm acreage. The use of insecticides is no accident. It is related directly to the profit factor in agriculture.[4]

California also provides another example of a phenomenon too little understood even by the "informed" consumer. The vast profits of what Californians call "agribusiness" (five of the eight largest governmental farm subsidies for 1970, all of them in the millions of dollars, went to California corporations) are made possible, in part, by the corporations' complete control of the vast state university, which provides in effect a free (*i. e.*, taxpayer-supported) research and development arm for the state's largest industry.

The university thus invents and licenses (at fantastically small fees) many of the insecticides, does what research is done on them, provides the "experts" to justify their use, and lends enormous prestige to the careers of men (they are mostly men) who in effect go along with the game.[5] The University of California, then, is as much a creature of the major corporations as is the Food and Drug Administration—perhaps even more so, since most of its employees, at least in those areas concerned with agriculture, are willing participants in what's going on.

It is this that makes it difficult to take seriously as an "expert" such a person as Emil Mrak, just because he is a U. C. professor with "stature" and chancellor emeritus of the campus at Davis (which used to be called the "ag school" and which actually does much of the free research and development for the state's agribusiness corporations). He could not have attained his eminence in those surroundings if he were often critical of what the big growers do.[6]

California's is not, of course, the only state university in thrall to giant agricultural interests; we use it as an example only. The point is that when you put together the giant agricultural corporations, their captive Mad Scientists on public payrolls at universities, and such manufacturing leviathans as Olin, Shell, Dow, and DuPont, you have an elephantine conglomerate of such power that it is no wonder that they sometimes seem unstoppable.

They care not at all, of course, about the history of man's evolution, and his development of agriculture alongside the insect world; they care not at all about the planet on which they live, nor about the health of the people they brag about feeding. They care only about yield per acre—more money from the land.

About that they care a good deal. One book, Rachel Carson's *Silent Spring*, stirred enough concern about insecticides that the propaganda to counteract it from The Bugkillers' Conglomerate has never stopped. Their line is, of course, that it's all because you demand it, that you insist on "quality," and on produce and other food that is "pest-free" (as though every potato or head of lettuce that our grandparents bought was swarming with bugs).

The government, predictably, goes right along. Here, for example, is H.L. Haller, assistant chief of the Agricultural Research Service of the Department of Agriculture—in a statement made nine years ago:

> Consumers will increasingly demand higher-quality farm products free from pest damage and contamination. Farmers, processors, and those who market farm products will demand the tools to furnish these higher-quality, pest-free products. People everywhere will demand freedom from the nuisance, discomfort, and the hazards to health that are associated with insects and other pests. In short, chemicals are here to stay.

He was certainly right in that last sentence. They may be here to stay long after all the people are gone. And so, surprisingly, may the insects—especially if DDT kills all the birds that eat them.

A now famous example of how that might come to pass is the gypsy moth fiasco of 1957, a $5 million Department of Agriculture program that led to a lawsuit and a great deal of fuss. In that year, the department set out to rid New York, New Jersey, and Pennsylvania forever of the gypsy moth, via the spraying of 3 million acres, including some densely populated areas, with one pound of DDT per acre.

As an insecticide, it didn't do too well. It did, however, kill animals, fish, and birds in enormous quantities. It killed a lot of bees. It ruined the paint on a number of automobiles, poisoned millions of gallons of milk, and contaminated untold quantities of food (including the food of dairy cattle, so that milk went on being contaminated for some time afterward).

It did not kill the gypsy moths. And the citizens who sued lost the suit.

It was done, of course, with no research whatever as to the effect on other forms of life, on the ecology of the area, or on the soil. And since it caused a fuss, the Department naturally did it again—this time spraying 27 million acres in nine Southeastern states with two pounds per acre of heptachlor and dieldrin, to kill fire ants.

Both are far deadlier than DDT, and they did indeed kill a lot of fire ants. They also killed a whole lot of other things, including livestock. The bird kill was fantastic. After the spray, the Audubon Society managed to get some people to listen to a minor fact: the fire ant is in fact not a pest, but a benefactor to agriculture in the area, since it preys on a number of other insects that are pests.

As most of us know by now, chlorinated hydrocarbons tend to persist in the soil, some for a long time, and to spread literally all over the earth. DDT has been found in the fat of Antarctic penguins, and the Bermuda petrel—which lives on the ocean and nests on a remote island far from any human contact—is endangered by DDT's effects on its egg-forming mechanism.

None of this bothers the giant "farmers." Although parathion, an organic phosphate that will kill you if a drop of it gets into your eye, is permitted for use only under rigidly controlled circumstances, a California farmer was found putting parathion dust into the crowns of plants by the teaspoonful during the picking season. A lettuce farmer was spotted by the FDA using (in sequence) DDT, chlordane, endrin, dieldrin, toxaphene, malathion, cryolite, and rotenone on his crop.

Every new insecticide is hailed as though it were the answer to all of man's problems. Never does it have drawbacks, never is there danger. "Harmless for humans" is always the battle cry, as it was for benzene hexachloride. They were selling 98 million pounds of BHC a year when it was found that it turned up in the brain tissues of laboratory animals and produced tumors elsewhere in their bodies. Since then it has been found to be associated with blood abnormalities in humans (so has lindane).

As a result BHC was given a "tolerance" for use on citrus fruits and vegetables. It, and chlordane, have been found active in the soil after 12 years, and they are absorbed into the crops that are grown in the soil. It has turned up in the milk supply, and for some time was used as an insecticide in restaurants and in home vaporizing devices.

Yet it was known, during all that time, that BHC, along with whatever other effects it may have, destroys inositol—the B vitamin that helps to check the action of cholesterol in "hardening" the arteries. You don't hear much about the suggestion that the way we use insecticides may have something to do with our "rising" cholesterol levels—but it's true. We are ecosystems.

With The Bugkillers' Conglomerate firmly in control, very little has ever been done, by any government agency, to control any insecticide (or herbicide, or fungicide); and what has been done has come always as the result of pressure from private citizens or organizations. It was, for example, a scientist not connected with any government agency who found excessive mercury in swordfish and tuna, and (luckily hitting a time when public concern about food contamination already existed) was able to raise enough fuss so that some steps were taken.

Private citizens, organized in such groups as the Environmental Defense Fund and the Audubon Society, have pioneered the approach of the "environmental lawsuit" to force recalcitrant government agencies into action. After grower-employers, aided by state agencies, fought for years with every weapon they could command (including "Red smears" and hired armed thugs), farm workers represented by the United Farm Workers Organizing Committee in California have achieved some measure of insecticide control through collective bargaining.[7]

A very few senators and representatives in Congress have shown concern—most notably Senator Gaylord Nelson of Wisconsin, who has tried to have DDT and other chlorinated hydrocarbons banned outright for years. A few state legislators have also taken action, usually after being pushed by groups of concerned citizens.

But virtually nothing has been banned yet, not even DDT, although there's been a lot of talk about it. The only bans, and they have been few, have been against specific compounds, whose extremely close chemical relatives were ignored in the edicts; even those prohibitions have been made only when evidence has been overwhelming and pressure prolonged. Most insecticides are hardly touched by the new regulations. Some are more toxic in the short run than DDT (like the organic phosphates, directly related to nerve gases), some equally long-lasting (picloram), some known beyond a doubt to be teratogenic (2,4,5-T).

There are, now, rules for their "safe" use. But the rules are usually ignored, except on unionized farms in California, where the farm workers themselves enforce them. Numerous studies, including the government's own, have repeatedly shown this.

Even if the rules were followed, there would still be residues on crops. The FDA says there is no adequate testing procedure to find out how much contamination there is on or in fruit and vegetables. When the FDA does make a spot check, which is all it ever does, it confiscates the produce, tells the insecticide user never, never to do it again, and otherwise lets the whole thing go.

Somewhere (sources vary) between $700 million and $900 million a year is spent on agricultural "pest" control in the United States.

Nearly 200 million pounds of insecticide is spread every year in this country alone. Almost all of it is poisonous by definition.

The chlorinated hydrocarbons—DDT, aldrin, dieldrin, endrin, lindane, chlordane, methoxychlor, toxaphene—are all polycyclic aromatic hydrocarbons and as such automatically suspect of carcinogenicity.[8] The organic phosphates—chlorothion, Malathion, parathion, phosdrin, schradan, tetraethyl pyrophosphate (TEPP), thimet—are direct analogues of nerve gas and work like nerve gases on the human system. The carbamates, among their other effects, appear very likely to be mutagens (one of them, urethane, is known to be). The herbicides 2,4,5-T and 2,4-D are known teratogens.

These substances are present in everything we eat. A few of us can avoid some of the ingestion that the rest of us cannot avoid; and that is the best we can do. Most of the substances are present in us (using federal standards for edible meat, none of us is fit to eat). The only hope we have, literally, is to change our practices all over the world and hope that we haven't already gone too far with what we've done.

In the 23 years before its functions were taken over by the new Environmental Protection Agency, the Pesticides Regulation Division of the Agriculture Department registered about 60,000 separate "pesticides" (including herbicides and fungicides), and by 1970 was registering about 5,000 new ones a year. Of the 60,000, some 20,000 were eventually withdrawn by the manufacturers (in favor of newer and more profitable versions); there are still about 40,000 approved for use, some of which can be used in more than one formulation. Most, of course, are variations of the same basic chemical, as 2,4,5-T and 2,4-D are virtually the same thing except for one atom in a highly complex molecular structure.

In those same 23 years, although it has always had the power to do so at will and although evidence about the dangers of many insecticides has piled up frighteningly, the pesticide regulation division has removed exactly *three* of those 60,000 products from sale as "imminent hazards" to public health.

One was 2,4,5-T, which was removed only from some nonagricultural uses, although its primary use has always been in agriculture in this country. Even this removal turned out to have been mostly one of those verbal fast shuffles with which we're getting so familiar. In Vietnam 2,4,5-T is used to starve noncombatant Vietnamese and bring about defective births.

Another considered an "imminent hazard" was a group of compounds containing mercury; that move was clumsy and inept, and hasn't yet successfully resulted in the banning of the products. The third to be removed was a product of the Aeroseal Company, and the ban, as will be seen, appeared to be for the protection of a corporation, not of the people.

The confusions surrounding "regulation" of "pesticides" are nowhere better illustrated than in the case of the most famous insecticide of them all—the one that may yet destroy the ecology of the planet.

In 1874, in Strasbourg, Germany, an obscure graduate student in chemistry—Othmar Zeidler—put together some chloral hydrate (the "Mickey Finn" of countless adventure stories) and some chlorobenzene, in the presence of a catalyst, and created a new and—as far as anyone knew—useless organic compound. He wrote it up and forgot about it. Its only name was the long and technical one derived from its molecular structure: dichlorodiphenyltrichloroethane.

The only thing Zeidler ever got out of it was a niche in the history of what may turn out to be an abruptly temporary and final stage in the story of humankind. It was not until 1934 that DDT was rediscovered, in a deliberate search for effective insecticides, in a Swiss manufacturer's laboratory by a team of technicians under Dr. Paul Müller. Five years later it was tried experimentally on the Swiss potato crop when that crop was threatened by an alien American invader, the Colorado beetle; it worked. In 1942 it came into the hands of the United States, and since then it has spread all over the world.

Today, when most of us have acquired at least some working knowledge of the principles of ecology, the glowing descriptions of the miracle insecticide of the 1940s and 1950s read like a horror story. Hailed because it is "triple-functional" (it is a stomach poison, a contact poison, and a respiratory poison), DDT was almost literally sprayed into every corner of the world, with virtually no testing for any of the dozens of disastrous side effects that have since been discovered. We will never know, even by orders of magnitude, how many necessary and beneficial microbes and insects, how many kinds of wildlife, or even how many human beings were wiped out.

The only information given to the public (and indeed, it was all that was known, since no one was even trying to be skeptical) was that a new and delightfully cheap miracle substance had been discovered, one that would surely rid the world once and for all of pestilence and probably of famine as well.

Possibly the most chilling book we have read in years is a 1946 publication written by two chemical engineers, O.T. Zimmerman and Irvin Lavine, and published by the Industrial Research Service of Dover, New Hampshire.

It makes incredible reading today, and not because it was a blatant example of flying in the face of fact (almost nothing factual was known). Rather it is simply a somewhat eager version of what we were all being told in magazine and newspaper stories over and over again. Consider this quotation:

> Although everyone should have sense enough not to use an insect spray where it might contaminate food, it is highly improbable that even the most careless person could possibly get enough DDT on his food to endanger his health.

Even more unbelievable is the comment on ecology—a word not used in the book. Some people, it seems, had expressed concern about "upsetting the balance of nature." The authors take care of that by pointing out that it's never static anyway.

> And yet, in spite of the fact that there is nothing static in nature, many alarmists start shouting whenever a new insecticide is introduced, and their cry is always the same: "You will destroy the balance of nature!" To which we can only reply: "What if we do?"[9]

As if that weren't enough, the authors go on to argue that if there is a change in "the balance of nature," there is a 50-50 chance that it will be for the better! And even if all of the insects on a continent could be killed, they say "it would be interesting to see what would happen."

It is, of course, quite true that "the balance of nature" is always changing, that in the long run even whole species come and go, that even a climax ecosystem like a redwood forest is "permanent" in only a relative sense.[10] The difference is that when ecological change is *natural*, it is not *arbitrary:* the change itself is in balance. When it is arbitrary, on the other hand, it cannot happen as it does naturally; things get out of whack and sometimes cannot repair themselves.

The evolution of our inner ecology follows this rule; we, too, have changed over time (consider for example the appendix). But we have changed in a natural way, in the direction of continued survival and adaptation; the changes to which we have adapted have been natural, not arbitrary. Arbitrary changes do not have a 50-50 chance of being improvements, any more than do sledgehammer blows on a television set.

What happened with DDT was understandable enough. Stupidity is always understandable. You sprinkle some powder on some plants, or spray a liquid on some cattle (or on people's heads). You don't expect the substance to turn up substantially unchanged ten years later and 10,000 miles away. And that, in effect, is what happened.

As early as the 1950s, Dr. L.G. Cox of the Beech-Nut Packing Company was telling the Delaney Committee that Beech-Nut was spending $100,000 a year in attempts to control the problem of residues on the foods it bought. He told the committee of a number of cases in which Beech-Nut had rejected foods because the insecticide residues were too high. The rejected food, of course, was simply sold to someone else.

Longgood's book, now more than ten years old, mentions "recent" tests in New York showing 50 parts per million of DDT in egg yolks, 70 ppm in cheese, ten in butter—and a report from one investigator who had found 150 ppm in cheese and believed that butter might go as high as 2,000.

One part per million, by the way, is only about one teaspoon of DDT (or whatever) in about ten tons of food. It might seem that even a couple of hundred parts per million might get lost in the metabolic shuffle. We hope we've made it clear by now that it doesn't work that way.

Also in the 1950s, a Public Health Service survey at a federal prison found 69 ppm of DDT in stewed fruit and 100 ppm in bread. An investigator from the Texas Research Foundation checked out some supermarkets at about the same time and found up to 13.8 ppm in milk and a range in meat from 3 ppm (lean meat) to 68.5 (in the fat).

A photograph in the Zimmerman-Lavine book shows stockmen cheerfully spraying DDT directly on the backs of cows (and presumably inhaling the spray). The caption reads, "Cows kept free from flies by direct application of DDT will give from 3 to 8 percent more milk than their untreated sisters." But back in 1949 the Department of Agriculture suddenly decided to advise *against* using DDT, not only on cows but in dairy barns. It had been found that even if DDT was used in the barns while the cows were kept outside, and even if the feeding troughs were covered, the poison would show up in the milk 24 hours later anyway. Today, of course, you can't buy milk without it; nor, if you're a woman who has just had a baby, can you *produce* milk without it.

By now, of course, as everyone knows, DDT is everywhere. And as most people have learned, it concentrates upward in food chains—that is, it is present in living organisms in higher concentrations than in the biosphere as a whole, and the concentrations get higher as one gets nearer the top of a food chain.

This is particularly true among life forms that live in the water (salt or fresh), because—to oversimplify it—their "concentrator" mechanisms are much more efficient. They have evolved that way because the distribution of vital substances in the water is much more diffuse than on land.

In a now famous experiment (one of the first to attract widespread public attention) scientists discovered in 1965 that in Clear Lake, California, a sizable body of water in an agricultural area, plankton had concentrated DDT to ten parts per million, but by the time the insecticide moved up the food chain, the largest carnivorous fish had concentrated it to 2,690 ppm.[11] It was at about the same time that the public began to realize that there was DDT in the fat of Antarctic penguins, Arctic caribou, deep-sea whales, and many other species who spend their lives far from any use of the insecticide.

In 1968, in the United States alone, 125 million pounds of DDT were manufactured—and sold for $20 million; so there was ample reason for a few people to fight against a ban, and for a lot of pro-DDT propaganda to be issued.[12] But in 1969, as the "ecology fad" hit and public concern expressed itself, government agencies began, albeit sluggishly, to move. This was the visible result of some years of invisible work by scientists like Charles Wurster, lawyers like Victor and Carol Yannacone, and groups like the Audubon Society and the Environmental Defense Fund.

One contribution among many by the Audubon Society was the publicizing early in 1969 of the fact that unless the use of DDT is sharply curtailed or banned, the American national bird, the bald eagle, will become extinct. Examination of dead eagles disclosed concentrations of DDT not only in their fat (where it's expected) but in their brains.

In March, 1969, the FDA seized 28,150 pounds of Coho salmon from Lake Michigan—salmon brought a few years earlier from Oregon and now found to contain DDT in concentrations of from 12 to 30 ppm. In April, the tolerance for fish was set at 5 ppm—if, and only if, they move in interstate commerce, the only fish federal agencies can control. Like most such tolerances, this was not so much a determination of what is "safe" as a matter of making legal what was already fact. If a lower tolerance were set we would virtually have to ban fish entirely.

In that same month, the FDA estimated that the average daily consumption of DDT by humans in the United States was seven-tenths of a milligram. This was already one-tenth of what an international organization had set as a "permissible" level.

The Food and Agriculture Organization of the United Nations, which set the seven-milligram maximum, is not all that disinterested; its chief concern, too, is the increase of agricultural yield.

Even so, they admitted that they set the "permissible" level on extremely shaky grounds and with the full knowledge that the figure is tentative. They made no effort to pretend—and neither does anybody else, any more—that there is any such thing as a "harmless" ingestion of DDT.

In December, 1970, for the first time, the FDA seized a load of salt-water fish, and the news reports of the seizure are interesting indeed.[13] Eight thousand pounds of kingfish, having been found to contain 14 ppm of DDT, were seized from the State Fish Company at San Pedro, California. The FDA tried to seize an earlier batch of 1,260 pounds from the same company, but they could not find the missing fish, which presumably had been sold.

Naturally, there turns out to be nothing illegal about selling off the poisoned fish before it's seized. If it makes you feel any better, the president of the State Fish Company said that it was all sold for pet food.

Not only DDT is present in food. In November, 1969, shortly before Thanksgiving, the Department of Agriculture inspected 150,000 turkeys at Little Rock, Arkansas, and seized 90,000 because of heptachlor residues. Heptachlor is another chlorinated hydrocarbon, and there wasn't supposed to be any residue in those turkeys. This insecticide used to have a tolerance of one-tenth of a part per million for use on more than 30 crops, but after we had been eating the residues for more than ten years, somebody paid close enough attention to discover that heptachlor breaks down into a more potent poison. The tolerance was then cut to zero, but nobody told the turkeys.

The massive use of DDT continues to show new effects—and there is no way, as yet, to assess the consequences of some of them, nor to know what they might ultimately indicate as far as humans are concerned. One effect whose ecological consequences are still largely unknown is the development of new, resistant strains of "pests," a problem that did not arise with prewar insecticides. As long ago as 1952, the then Secretary of Agriculture, Charles F. Brannan, noted with puzzlement, "We have more insect pests although we have better insecticides and better ways to fight them."

In 1956 the World Health Organization listed 36 "pests" of public health importance, all of which had developed insecticide-resistant strains. Ten years before, said the WHO, there had been one. "The world," said the statement, "may soon face an emergency because the resistance problem is intensifying day by day."

As an example, there are flies that can take without effect a thousand times as much DDT as once killed their direct ancestors. The fish-ladder response of the Mad Scientists, of course, is stronger and stronger insecticides; but it is quite clearly not working.

On July 28, 1971, a story by Marshall Schwartz appeared in *The San Francisco Chronicle* under the chilling headline, "The Mosquito That Can't Be Killed." The insect is the *culex tarsalis*, a carrier of a disease called Venezuelan equine encephalomyelitis. During the first four weeks of July, the disease killed 3100 horses and infected 21 people in California.

Schwartz quotes entomologist Robert van den Bosch: "We created this monster, and now we can't kill it. We're at the end of the line as far as chemicals are concerned." *Culex tarsalis*, by the way, became immune to DDT in 1953—and in that year 50 Californians died of encephalomyelitis and many more suffered serious brain damage. Control efforts switched to other chemical families of insecticides; the mosquito is now immune to them all, and the disease is still with us.

Another effect on insects is the transformation of once innocuous varieties into problems because the other insects who once preyed on them and held their populations in check have been destroyed. Mites are an example. In fact, they have become so much of a problem in some areas that the deadly carcinogenic insecticide Aramite, whose story was told in an earlier chapter, was developed just to deal with them.

There are other and stranger effects, however, whose possible meaning for humans is chilling. One change, almost certainly due to DDT concentration, has been in the behavior of California sea lions. We have seen it. When these animals bear young, the females live for relatively long periods of time off the energy stored in their fat—which means that DDT is released in large amounts into their systems from the fat in which it, too, is stored. After the pups are born, the mothers normally respond to the pups' cries, and to their tendency to wander a bit, simply by reaching out and pulling the pups back.

In recent seasons, however, a great many sea lion mothers have responded abnormally and tragically: by picking up the pups, dragging them around the beaches, dropping them on rocks, mauling them, or even smashing them to death on the rocky beaches. Scientists are still studying this phenomenon; the prevailing belief is that the suddenly and massively released DDT affects the hormones which in turn affect maternal behavior (DDT is known to affect hormones in laboratory animals). We can simply note without comment that hormones affect maternal behavior in human beings also.

Lester Brown, a scientist long with the Department of Agriculture, commented recently on DDT in a roundup for *Scientific American* on ecological issues; his remarks are worth quoting at some length:

> It is now clear that the use of DDT and other chlorinated hydrocarbons as pesticides and herbicides is beginning to threaten many species of animal life, possibly including man. DDT today is found in the tissues of animals over a global range of life forms and geography from penguins in Antarctica to children in the villages of Thailand. There is strong evidence that it is actually on the way to extinguishing some animal species, notably predatory birds such as the bald eagle and the peregrine falcon, whose capacity for using calcium is so impaired by DDT that the shells of their eggs are too thin to avoid breakage in the nest before the fledglings hatch. . . . Concentrations of DDT in mothers' milk in the U.S. now exceed the tolerance levels established for foodstuffs. . . .
>
> This illustrates how little man knows about the effects of his intervening in the biosphere. Up to now he has been using the biosphere as a laboratory, sometimes with unhappy results.

Not only eagles and falcons, but pelicans, petrels, and other species are threatened by the thin-shell phenomenon. It affects especially those birds which are at the top of relatively long food chains—either those that feed on animals or are sea feeders.[14]

As long as 20 years ago, the American Medical Association, through its Council on Pharmacy and Chemistry, issued a warning still pertinent (and still little understood) today. The Council cautioned against the use of then new insecticides in household vaporizers, on the ground that anything that kills insects in a few hours is extremely likely to mean danger to humans who are repeatedly exposed. Most important, the Council noted that the effects of the insecticide might easily go unnoticed.

The resultant injury may be cumulative or delayed, or simulate a chronic disease of other origin, thereby making identification and statistical comparison difficult or impossible.

It's still true. Doctors, as a rule, don't connect a lot of their patients' complaints to the pervasiveness of insecticides and herbicides in the environment; they aren't taught to. But there is some evidence all the same.

Also 20 years ago, the Public Health Service issued an even more precise warning about DDT as a "delayed action poison":

> Due to the fact that it accumulates in the body tissues, especially in females, the repeated inhalation or ingestion of DDT constitutes a distinct health hazard. The deleterious effects are manifested principally in the liver, spleen, kidneys and spinal cord.
>
> . . . Children and infants especially are much more susceptible to poisoning than adults.
>
> . . . A recent report from the National Cancer Institute discloses that the inhalation of petroleum mists or fogs (used as carriers for DDT in spray) has been incriminated as a major cause of lung cancer and laryngeal cancer. These types of cancer occur with overwhelming predominance in men.

And they can occur, you'll remember, after a long time, or as a result of the cumulative effects of ingestion over time. Early in this book, we discussed the concept of the "totality of toxicity," and talked of cumulative effects. Those concepts directly apply to the consideration of the effects of insecticides.

In 1958, Dr. Malcolm Hargraves, a Mayo Clinic blood specialist, testified in a court case that he was absolutely certain of the role of DDT as a causative agent in leukemia, aplastic anemia, Hodgkin's disease, jaundice, and other disorders. It should be noted in passing that the liver and/or spleen are involved in all these maladies. There has been a sharp increase in the incidence of all of them since DDT came into use after World War II—and leukemia has gone up most sharply in the states that use the most DDT.

Dr. Morton Biskind, who was on the AMA Council, told the Delaney Committee that DDT might also be involved in the postwar rise of "virus-like" diseases. He cited the case of the "X disease" in livestock, which was believed to be viral but which turned out to be caused by the ingestion of chlorinated hydrocarbons.

About the long-term effects of DDT on humans, not much is known. As Longgood wrote in 1960, "It is an acknowledged delayed-action poison, so everything about it has been studied except its delayed effects." But about shorter-range effects more is known than the Mad Scientists like to have publicized. Some significant experiments were done in California in the 1950s by Pottenger and Krone, who found that rats fed chlorinated hydrocarbons died earlier than control rats—and that it would be difficult to demonstrate it medically:

> [T]he tissues of the animals showed no changes at autopsy, demonstrating how easily such toxicity might escape notice. In short, one could not prove the diagnosis of insecticide toxicity in these animals by clinical or laboratory methods. But their lives were shortened by insecticides. We shall never know how many humans suffer similarly.[15]

By that time a sizable number of experiments with all kinds of animals *had* shown effects, most of them degenerative, not only on the liver and spleen but also on the skin, gall bladder, lungs, kidneys, thyroid, adrenal glands, ovaries, testicles, muscles, blood, peripheral nerves, gastrointestinal tract—and the brain, the spinal cord, and the heart. Liver injury was shown with rats if the DDT in the diet was five parts per million.

Think about that for a minute. Five parts per million is the present federal tolerance for fish.

And it is the *same* concentration that produces liver damage in rats. It is *not* set off by the 100-to-1 safety factor usually used in interpreting animal experiments. In other words, according to normal scientific practice, we are getting at least 100 times as much DDT in our food as is known to cause liver damage to an extent considered medically significant.

As we've indicated before, DDT, in humans as in other animals, is stored mostly (though not entirely) in the fatty tissues; women, who have a little more such tissue in proportion to their weight, tend to store a little more DDT than men. But "stored," it should be noted, does not mean that it is tucked out of harm's way like a container of leftover peas in the refrigerator. The cells of the fatty tissue are involved in a constant turnover, and the fat is involved in a number of inner-ecological processes all the time. The cumulative poisons that fatty tissues may contain are involved in cumulative poisonous activity.

Another effect of this storage (remember the sea lions) is that a relatively sudden use of the fat, as in any sudden weight loss, releases the DDT into the blood stream and can even cause an attack of poisoning. This is particularly important to children, who store a lot of fat relative to their weight and who often suffer sudden losses in childhood illnesses.

It may also be important to pregnant women who attempt to lose weight (or who do so without attempting it), because DDT passes across the placenta. It's just one more danger to the poor fetus, who throughout the course of human evolution has had to cope with none of the new problems of the twentieth century.

In the late 1950s the Public Health Service estimated the average concentration of DDT in human tissue at 5 ppm. Longgood says that "objective investigators" he consulted thought it was closer to 9 ppm. A 1958 statement by Dr. Pottenger, who often tested his patients, said that concentration in the fat of one patient was 5 ppm as early as 1949, and by 1953 had gone up to 23 ppm; another had 173 ppm in 1955 (but he didn't say whether that patient had had unusual contact with chlorinated hydrocarbons).

In 1970, while he was Secretary of Health, Education, and Welfare, Robert Finch said that the national average was 12 parts per million. The national average, however, is clearly a figure to be regarded with some suspicion; 12 is the average of 11 and 13, but it is also the average of one and 23. With 200 million numbers to average out, you can get a pretty wide spread. Other evidence makes it clear that some Americans have many, many times 12 ppm of DDT in their tissues.

The thin-shell mechanism in birds is believed to involve the action of DDT on the liver, and the most convincing evidence of damage to human beings is also concerned with that organ. It is the job of the liver, in effect, to "strain" poison out of the system—to "detoxify" as far as possible the substances in the diet. Additional poisons force the liver to work harder (which sometimes enlarges it). The detoxification function of the liver, in turn, is dependent, in ways not at all clearly understood, on several vitamins and several enzyme systems (some of which themselves involve vitamins).

Hepatitis (the word simply means "inflammation of the liver") has risen sharply in America, and in many other parts of the world, since World War II. In one two-year period (1955–7) the number of cases in America doubled. The Army has worried for more than a decade about the sharp rise in hepatitis among soldiers.

Cases of hepatic necrosis (the death of liver tissue) have risen in a fashion some authorities consider alarming. And it was in 1957 that cirrhosis of the liver became one of the top ten (recognized) causes of death in America.[16] Contrary to American mythology, cirrhosis of the liver is not necessarily a disease reserved for us drinkers.

"Monkeys," said the World Health Organization in 1966, "given

DDT orally in a dose of 0.2 mg/kg body-weight for 7 to 9 months developed symptoms of hepatitis. After a year, the animals showed liver enlargement and hyperglycaemia." Hyperglycaemia means that the blood-sugar level is too high, which may not be dangerous in itself but which may indicate other and more serious disturbances.

The experiment, incidentally, shows the danger of "national averages." Remembering that the FAO's tentative "permissible" ingestion of DDT is seven milligrams a day, we might wonder how a human dose would work out in the same proportions—0.2 milligrams per kilogram of body weight—as those given to the monkeys. For a 150-pound adult, it would be between 13 and 14 milligrams. But for a 40-pound child, it would be only about three-and-a-half milligrams, half the "permissible" dose.

And that does *not* allow any 100-to-1 safety factor.

As with the experiments of Pottenger and Krone, however, we are sure of the effects in the WHO experiments only because they were deliberately brought about; we can't necessarily find them in the general population, where they might go unnoticed, particularly where doctors are not trained to look for them. At one time in the late 1950s, the FDA examined at random the bodies of 20 Californians who had died of cancer, heart disease, or pneumonia. They found DDT or comparable insecticides in the systems of 19 of them, and liver damage in 16.

The 1966 WHO report which told of the experiment with monkeys also discussed possible applications to humans:

> [T]he possibility that [the storage of DDT] might have deleterious consequences later in life cannot be ruled out. In addition, it has not been demonstrated that metabolic changes in the liver cells comparable to those observed in . . . rats do not take place in man. . . . At the present time, [we remain] concerned about the storage of DDT which occurs in all species and about the cellular changes produced in the liver of rats by DDT and by other compounds chemically related to it.

Other investigators found that DDT caused liver tumors (hepatomas), and that they were also caused by other chlorinated hydrocarbons—by aldrin and dieldrin in mice, and by methoxychlor in rats. Those were all feeding tests.

Because of the liver's role as a "poison strainer," or detoxifier, DDT has strange effects in conjunction with other chemicals. Dr. Kenneth DuBois of the University of Chicago, for example, found that insecticide residues get in the way of the action of barbiturates. The enzymes that normally do the detoxifying are stimulated by the presence of the DDT, then act on the barbiturates to rob them of their intended effect; in one experiment the amount of barbiturate that normally put a rat to sleep for an hour worked only for ten minutes.

This effect, DuBois pointed out, would generally be disregarded by physicians, who would conclude that the patient had an unusual resistance to the drug at that time, and that's all. In other experiments with other drugs, the DDT, through its complex effects on the liver, actually increased their effectiveness, which of course might not be so good.[17]

By damaging the liver, says Dr. W. Coda Martin (once president of the American Academy of Nutrition), DDT contributes to the "slow, progressive, internal deterioration of the tissues and cells of the body"—and ultimately to results that would never be diagnosed as insecticide poisoning. But it isn't only the liver. Many authorities are convinced that DDT interferes directly with the functions of a number of vitamins (it has been shown to destroy vitamins in laboratory tests) and enzyme systems; its presence in fatty tissue can influence endocrine, enzyme, and nerve functions, and interfere with the ability of the body to manufacture hormones.[18]

"DDT," says Dr. Martin, "affects the enzyme system which controls the supply of oxygen to the individual cells. The greater the concentration, the greater the damage. There isn't a cell in the body that isn't affected by DDT."

Which means that there isn't a human cell anywhere in the world that isn't affected by DDT—a sentence certainly worthy of meditation.

In 1967, three scientists from the University of Miami did some more studying of dead bodies, this time in Florida. They compared patients who had died of liver or central nervous system disease to those who had died in accidents, and found that the first group had in their fat twice as much DDT (or other chlorinated hydrocarbons). That's not so surprising, perhaps, but they also found concentrations twice as high as "normal" in *all* cases of cancer of the lung, stomach, rectum, pancreas, prostate, and bladder (many of which, they determined, were complicated by liver disease). The concentrations were at least twice as high and usually higher in high blood pressure cases.

No one autopsied had had any connection with insecticides in any occupational way; but the correlation was definitely there between body storage of insecticides on the one hand and, on the other, cancer, cirrhosis, and high blood pressure.[19]

Still another DDT effect, which has received almost no public notice, was first suggested by a 1965 study at the University of Arkansas by Dr. Douglas James. He used quail, of all animals, and found that adding DDT to their food resulted in a decline in learning ability. Since then there have been a number of studies showing a decline in mental abilities in humans that correlates with exposure to insecticides (and the worse effects, incidentally, seem to come from DDT and the organic phosphates in combination, which shouldn't surprise anyone who has mastered the idea of the cumulative and/or synergistic effects of poisons).

Occupationally exposed men have been shown to have relatively slow reaction times, losses in vitality, high irritability, and poor memories. These are considered by some investigators to be "soft signs" of biological damage. One San Francisco psychiatrist said eight years ago that all his colleagues should check on insecticide exposures of their patients as a matter of course.

Dr. Douglas Campbell referred to a number of patients whose doctors could find no organic cause for loss of coordination, blurred speech, or other, similar symptoms, and had referred the patients for psychiatric treatment. "On a hunch, I had a pharmacologist study fatty tissue from one patient. It contained six times the average amount of DDT." He later found that most of the patients in this classification were regular users of chlorinated-hydrocarbon household sprays.

Recently a PBS television documentary focused on the problems of farm workers, and of workers in chemical plants that manufacture insecticides and herbicides. In the plants, it was found, headaches, irritability, and sleeplessness are chronic; the divorce rate is measurably higher than usual.

One worker pointed out that a week's vacation clears up the headaches and irritability, *and* a troublesome skin condition which he and his fellow workers often had. Back at work, it comes back. A visit to the plant nurse results in a squirt of hand lotion and an offhand attribution of the problem to the soap used at home, the children's measles, or some other non-job-related cause.

Electroencephalographs taken of the workers, however, according to a doctor who took part in the program, resemble those of average Americans six to ten years older than those tested. Memory lapses, similar to those found in other tests, are not uncommon. The Shell Chemical Company, one of the firms investigated, has been asked to put occupational safety provisions into its union contracts, but has so far refused.

Farm workers, said a doctor interviewed on the documentary, must "accept a chronic state of ill health as the price you pay to earn a living." This is, of course, the situation against which Cesar Chavez and the United Farm Workers Organizing Committee has been fighting with its demands that insecticide regulation be a part of every contract negotiated. Otherwise they must face the fact, as one farm worker noted on the program, that "laws regulating economic poisons are in the hands of the California Department of Agriculture—which is like putting civil rights laws in the hands of the Ku Klux Klan."

Finally, DDT is beyond any doubt known to cause both cancer and mutations in test animals. In addition to the effects on humans we've already mentioned, it is widely believed to cause panmyelophthisis—a wasting away of the bone marrow. And there is a very high probability that it causes leukemia.

DDT is still not banned by the government of the United States.

The past two years have, however, seen some action against this one insecticide, but less against its "chemical neighbors" in the chlorinated hydrocarbon neighborhood. Sweden banned DDT in March, 1969, Denmark three months later, and the USSR in May, 1970 (Russia also announced that restrictions would be placed on the use of other insecticides which had been found responsible for wildlife deaths).[20] New Zealand followed suit the month after that.

Several states have also acted independently, Michigan becoming the first to ban the sale or use of DDT within its borders as of April 16, 1969. Illinois, Wisconsin, and Maryland have done so since. Maryland bans all chlorinated hydrocarbons "except for instances when the public health may be threatened or where severe economic losses would occur without their use." Wisconsin allows for emergency use if the emergency is certified as such by a three-man board.

Washington has announced a ban on "two-thirds of the registered uses," and since mid-1970 has forbidden DDT use for homes and gardens, shade trees, mosquito control, or the dusting of farms. Other states have similarly limited controls.

In California, DDT has been banned for use on 35 food and forage crops, 12 seed crops, and in homes and gardens (there are 200 food crops grown in California). In April, 1970, the United Farm Workers Organizing Committee signed the contracts that broke the nationwide grape boycott, and in those contracts are clauses forbidding the use of "hard pesticides," including DDT, on union farms. But neither cotton nor citrus crops are covered, and there is no way to prevent a grower from buying DDT for use on cotton and then using it on something else (it isn't unusual for a big California cotton grower to grow food crops as well). In July, 1970, the California legislature killed for the second time a bill by State Senator John Nejedly to ban DDT and other chlorinated hydrocarbons completely.

It is on the federal level, however, that something must happen if any real progress is to be made in the United States; and on that level there have been, over the past two-and-a-half years, a number of "actions" and "announcements," all of them intended to give the impression that something important was being done about insecticides, but none of them actually achieving much. This is what has actually happened.

Until late in 1970, the Agriculture Department (USDA), through its Pesticides Regulation Division, was "in charge of" insecticides, herbicides, fungicides, and related substances. The procedure for registering a new one involved a ruling by the division on its effectiveness, followed by an FDA advisory opinion on safety (no tests were done or required on mutagenicity, teratogenicity, or synergistic effects). The division then decided (usually ignoring the FDA opinion if it was negative) whether to register the new product, and the FDA set the tolerances that could remain on or in food, and theoretically enforced them (residues in meat remain the responsibility of USDA).

With all the fuss going on in state legislatures and around the world, Secretary Finch in 1969 appointed a Commission on Pesticides and Their Relationship to Environmental Health. At the time he noted that "each American body" (an unfortunate distortion of the meaning of an average) contained 12 ppm of DDT in its fatty tissue—5 ppm more than is allowed in the meat we eat. The commission's chairman was Emil Mrak, long-time academic champion of the DDT users.

The pressure was more than a commission could remove, however. Senator Nelson had been trying to get DDT banned, and in the same month that the commission was appointed, Nelson introduced a bill that would ban interstate shipment of dieldrin, aldrin, chlordane, endrin, heptachlor, DDT analogues, lindane, and toxaphene. It was to lose, but it attracted more attention than any other such bill had before.

In November, 1969, Secretary of Agriculture Clifford Hardin (ex-chancellor of the University of Nebraska) canceled the registration of DDT for four uses: around the home, on shade trees, on tobacco plants, and in marshes. He left a loophole: it could be used to control disease carriers.

That's a pretty big loophole, since it theoretically includes both flies and mosquitoes. But that's not the *only* loophole. For under the law, "canceling" the registration of an insecticide is not the same as "suspending" it. The government can "suspend" a registration if it finds that the insecticide is an "imminent hazard"; that means that distribution stops *right now,* and if the manufacturer wants to fight about it, the insecticide stays off the market while the fight goes on.

"Canceling" a registration, however, is good for headlines and not much more. The manufacturer files an appeal (if he wants to, of course, and he usually does) and the product stays on the market and stays in use while the appeal and subsequent appeals go through administrative channels and through the courts. The "canceled" registration of DDT for four uses, loophole and all, had precisely *no* effect on the use or distribution of DDT.

In 1970, however (after Nelson made another try with his anti-chlorinated-hydrocarbons bill, still unsuccessful), there entered on the scene the Environmental Defense Fund (EDF), an organization which has been trying for several years to develop a new concept called "environmental law." Joined by the Audubon Society, the Sierra Club, and others, the EDF filed two lawsuits, one against the Department of Agriculture and one against the Department of Health, Education, and Welfare (as the parent of the FDA).[21]

Early in 1970, USDA canceled the registration of aldrin and dieldrin for "aquatic" uses against insect larvae. This had the same effect as the DDT cancellation had had.

EDF had begun by petitioning the FDA to set a zero tolerance for DDT because its presence on food is banned by the Delaney Amendment, DDT being a proven carcinogen in animals. The FDA turned down the petition on the practical, but not necessarily legal, ground that there's no way to get the DDT out of the food.

But EDF argued that according to the law the FDA has the power to set tolerances "to the extent necessary to protect public health," and that that doesn't mean that the FDA can go around legalizing whatever it can't prevent. There are some things the FDA can do: set a zero tolerance for residues from applications of DDT after the ban, for instance, or allow some time limit.

The USDA fought the other suit on the ground that only registrants (the people who make insecticides) can challenge USDA decisions. This was, of course, one of the major issues of the suits:

can the public be represented in court actions to force regulatory agencies to regulate?

At the end of May, 1970, the U.S. Court of Appeals for the District of Columbia handed down decisions in favor of EDF and its co-plaintiffs in both DDT cases. Chief Judge David Bazelon wrote the opinion, holding that "the interest of the public in safety" is just as good a reason for court action as the "economic interest" of the insecticide industry. It was time, he wrote, for beginning "a new era in the history of the long and fruitful collaboration of administrative agencies and reviewing courts."

That was pretty language, but it's doubtful whether the regulatory agencies in question felt as though they had been collaborated with. They were, in fact, told off, and good. In detail, the court's opinion meant that the Secretary of Agriculture was given 30 days in which to decide what he was going to do about DDT. If he wasn't going to ban it, he had to give good and detailed reasons. The legal language took a little longer to say it:

> If [the secretary] persists in denying suspension in the face of the impressive evidence presented by petitioners, then the basis for that decision should appear clearly on the record, not in conclusory terms but in sufficient detail to permit prompt and effective review.

In the case against the Department of Health, Education, and Welfare, the court threw out the Delaney Amendment argument, on the ground that under the law, insecticides are not technically food additives within the meaning of that amendment (see the earlier section on food additive law). But it picked up the phrase about the FDA's setting tolerances "to the extent necessary to protect public health," and decided that that was authority enough. The court ordered HEW to start a proceeding right now on the "zero tolerance" petition.

The court also noted that the argument about "we can't-get-it-out" wouldn't do, noting the alternatives mentioned above, and took a major legal step forward by saying, in effect, that the FDA doesn't need a Delaney Amendment to tell it that a carcinogen is unsafe. If the HEW secretary decides not to change the tolerances, the court said, despite the evidence of carcinogenicity, then the secretary will "be required to explain the basis on which he determined such tolerances to be 'safe.' "

Late in July, 1970, Secretary of Agriculture Hardin said that he wouldn't suspend DDT registration because there was no scientific evidence that DDT "constitutes an imminent hazard to human health," and because it has some beneficial uses. He did, however, cancel (not suspend) its registration for use on 50 types of fruits and vegetables, forest trees, lumber, livestock, and buildings—leaving only cotton and citrus growers as major users. Like the previous cancellations, this one was promptly appealed and had no effect on the actual sale or use of DDT, though it was headlines in the newspapers.

Hardin's refusal to suspend DDT, however, was probably more a grandstand play for agribusiness leaders than anything else, for by then he must have known that he wasn't going to have to play insecticide games much longer. President Nixon had proposed earlier in the month creation of a new governmental body, the Environmental Protection Agency (EPA), to which a number of regulatory functions would be transferred—among them insecticide control. The EPA came into legal existence in October, 1970, and took over the USDA insecticide control functions a short time later.

In the meantime, Secretary of the Interior Walter Hickel, who had progressively been seeing so much of the light that he eventually blinded himself out of a job, made his fellow cabinet members look a little shabby. He ordered a ban on the use, not only of DDT, but of a total of 16 insecticides and herbicides on 534,000,000 acres of federal land controlled by his department. He went further than that, ordering 32 other chemicals put on a "restricted list," to be used only under limited circumstances and with prior approval. That ban, as far as it went, was real.[22]

The switch in responsibility from the Agriculture Department to the new EPA meant new court action, and this time it was tougher. On January 7, 1971, the same court ordered EPA administrator William Ruckelshaus to cancel all registrations of DDT immediately, and to examine immediately the evidence as to whether the registrations should be suspended on imminent-hazard grounds. On January 15 Ruckelshaus announced that the agency would not appeal and that cancellation orders would be issued without delay.

As they were. But while the court order was tougher, there is still, at this writing, no DDT *ban*. Cancellation still means that manufacturers can appeal, and keep selling the stuff while the appeal procedure goes on.

On January 18, however, the EPA began what it said would be a 60-day study, embracing both DDT and the herbicide 2,4,5-T, to see whether they should be declared "imminent hazards" and suspended immediately.[23] This passage is being written in the midst of that 60-day period, but we'll try to get an update either into the text at this point or, at least, into a footnote.[24]

That's the situation at this writing. Except for very limited uses (the loophole sizes vary from place to place), DDT is wholly banned in four states and partially in many others. Despite all the headlines (and what is in fact genuine progress in one sense), the only federal action that has actually stopped any DDT use was former Secretary Hickel's order against its use on Interior Department lands.

One more note: if you're a student of which states have the most clout in federal politics, you should know that the biggest users of chlorinated hydrocarbon insecticides are cotton growers. They account for two-thirds of the DDT used agriculturally in America. The biggest cotton state is Texas. Approximately tied for second is California.[25]

As of now, there are no FDA "zero tolerances" for any chlorinated hydrocarbons except heptachlor; some is permitted on or in all your food, though this is clearly not a decision about safety but a decision to go along with what has already happened. The USDA still has zero tolerances for some insecticides for some meat and poultry, but it ignores its own rules unless the contamination is unusually high. If it didn't, it would have to ban meat.

Otherwise, tolerances are supposed to be (as a rule of thumb) 1 percent of the "no-effect level" on humans. This is obviously nonsense too as applied to insecticides. The figures are in fact arbitrary, or might as well be. The DDT tolerances vary from 3.5 ppm in or on some foods (avocados, cherries, and others) through 5 ppm for fish to 7 ppm for apples, beans, and other items. There is no attempt to explain why a concentration twice as high is safe if it's on or in an apple instead of an avocado.

For milk, the legal tolerance for DDT is five-one-hundredths of a part per million. If you'd rather call it 50 parts per billion, it's the same thing. It would mean something like five ounces of DDT for ever 10 million gallons of milk—if anybody believed that it really meant anything at all.

Mothers' milk, by the way, regularly contained DDT in amounts more than ten times the legal limit for cows' milk as long as 15 years ago; in 1960 one reporter mentioned a study in which mothers' milk was found to contain 116 ppm of DDT—2,320 times the legal limit for cows' milk.

No wonder biologist Charles Wurster says that mothers' milk couldn't be taken across state lines in any other container.

CHAPTER FOURTEEN

Birth Defects at Home and Abroad

By now you have the idea, we hope, that things are pretty cozy between the regulators and the people who are supposed to be regulated. Usually it doesn't degenerate into outright corruption, but it has happened. And sometimes it expands to something far, far worse.

There was a brief mystery a few years ago when the Pesticides Regulation Division suspended an insecticide strip made by Aeroseal Company—suspended, not merely canceled—but allowed Shell's competing and almost identical No-Pest Strip to stay on the market. The Public Health Service has been screaming about these strips for eight years: dangerous in the same room with food, in the same room with babies, in the same room with the ill or aging—and probably just plain dangerous anyway.

The mystery was solved when three PRD employees were found by a Congressional committee to be on Shell's payroll at the same time. Shell's No-Pest strip is, of course, still on sale—but whatever you do, don't buy one. Don't allow one in the house, and if you've got one, throw it away (in a sealed container).

That sort of corruption, though, is mild in its effects compared to the FDA's coziness with the Big Three of 2,4,5-trichlorophenoxyacetic acid. Even Turner and such a veteran Washington reporter as Robert Sherrill treat this relationship as just more of the same industry-agency bedfellowship; but it is much more. The fact appears to be that we are all eating 2,4,5-T and 2,4-D residues, despite their being known (certainly in one case, probably in the other) as teratogens, because it serves the high policy of the United States in Southeast Asia and in world politics.

While one group of scientists worked frantically before and during World War II to develop a new and awesome bomb that was to change the politics (and to some extent the ecology) of the world, others worked just as hard in other fields—notably a group at Camp (later Fort) Detrick, Maryland, the center of U.S. research in chemical and biological warfare. Most Americans, when they hear that phrase, think of poison gas or of evil plans to spread deadly disease among the enemy; what a lot of those scientists were working on, though, were herbicides.

To quote an actual document from 1941, they were specifically investigating "the toxic properties of growth-regulating substances for the destruction of crops or the limitation of crop production." The products they developed were never used (not then, anyway) and after the war some were released for commercial use. Among them were two closely related acids (they differ by the presence or absence of a single chlorine atom in each molecule) called for convenience 2,4-D and 2,4,5-T.[1]

It was then claimed that the direct toxicity levels for animals and men are low for both products, and that they are short-lived in the soil. The USDA, the FDA, and the Fish and Wildlife Service all smiled happily on the new products. 2,4-D, the cheaper, became a popular household product, what was often called a "weedkiller" (though in fact it is not so discriminate in its action). Both were used for commercial purposes, with 2,4,5-T being favored for clearing woody brush out of pastures. There were *no* tests for carcinogenicity, for teratogenicity, for mutagenicity, for ill effects in combination with other substances, nor for long-range effects.

But the commercialization of these products didn't mean that Fort Detrick was giving up on them. They refined them, tested their uses (there was a full-scale defoliation test in Puerto Rico), and finally sent them off to Vietnam. Defoliation and crop destruction started there in 1961.

What the Army talks about doing is clearing the vegetation, for example, along the side of a river, so that snipers can't hide and fire on boats. That is done, of course. What the Army does not like to talk about is the deliberate destruction of crops in order to starve people.

They will admit that they do that in the northern part of South Vietnam, where the "enemy" is, and claim that it's okay because the National Liberation Front soldiers, when they see them, do look undernourished. What is happening to children, to the elderly, to the sick, they do not mention. The Army does not admit that it is guilty of deliberate crop destruction in the Mekong Delta area and other areas to the south; it also denies that many crops are destroyed by "defoliants" drifting from other targets. Independent observers, and even South Vietnamese biologists, say that the Army is lying.[2]

Before 1965 defoliation was a small operation and probably most often devoted genuinely to reducing enemy cover. Between 1965 and 1968 it shot upward in terms of acres of land involved; criticism of the practice rose also. Beginning in 1967, the American Association for the Advancement of Science (AAAS) tried to obtain data from the Department of Defense on ecological damage, on the precise agents used, and other information, and tried to arrange its own study in Vietnam; they got either stony silence or a glad-handed runaround.

At about the same time, however, the thalidomide scandal broke. As noted in the discussion of cyclamates, the federal government wanted no more of those; so the National Cancer Institute set out to do a little research. As a part of it, they awarded a contract for more than $2.5 million to the Bionetics Research Laboratories Division of Litton Industries, to test more than 200 products for carcinogenicity, teratogenicity, and mutagenicity. Most of them were insecticides or related compounds.

Back in 1963, President Kennedy's Science Advisory Committee had recommended toxicity studies on all pesticides (they used the term to include herbicides), involving studies of at least two generations of at least two warm-blooded species. This was, needless to say, a recommendation that went almost unnoticed. One man who noticed it was Emil Mrak, who remarked that there was no evidence whatever that any such substances "may eventually affect human health."

In the fall of 1966, word began to come out of the Bionetics study; 2,4,5-T was definitely teratogenic, possibly much more so than thalidomide. It didn't come out far enough to reach the general public, however, and even most scientists, many concerned with the more visible danger of DDT and its relatives, were unaware of the finding.

And if they didn't find out then, they didn't find out at all. The Bionetics study of 2,4,5-T literally disappeared. No progress papers were published; no preliminary data (as far as anyone knows) ever reached USDA or the FDA. But the use of 2,4,5-T in Vietnam continued to escalate until 1968, and has slacked off only slightly since.

Somebody, quite obviously, put the lid on that report, and it is ridiculous to think that it was the National Cancer Institute. As a matter of policy, the U.S. government was using a teratogenic spray as a defoliant and crop destroyer in Vietnam; thalidomide scandal might be nothing compared to deliberate biological warfare.

But scientific findings are well-nigh irrepressible. In February, 1969, for instance, Samuel Epstein of the Children's Cancer Research Foundation in Boston seems to have heard about it. We know now that the National Cancer Institute (NCI) had the Bionetics report on 2,4,5-T in that month; an official of NCI's parent body, the National Institutes of Health, says that in the same month the report was made available to officials of the FDA, the USDA, and the Defense Department (DOD). But it turns out that "making it available" to DOD meant that a couple of people from Fort Detrick were called; a bet that DOD already knew about the report wouldn't get you many takers around Washington.

In March, possibly in an attempt to check a swirling rumor mill, some Bionetics results were released—on cancer research. Fifty-three of the substances had some carcinogenic effect, but 2,4,5-T wasn't among them. If the release was supposed to stir a diversionary flurry of attention, it didn't work.

Possibly at the same AAAS meeting where the carcinogenicity report was made, more definite leaks began to turn up in the scientific community about the fetal damage caused by 2,4,5-T; the AAAS, remember, was already concerned with trying to find out more about this chemical and its possible effects in Vietnam.

An official of the NIH has since contended that by this time the Bionetics report was already in the hands of the Mrak Commission—Finch's pesticides commission that had just been appointed and which we mentioned a few pages back. It may or may not have been in the hands of Mrak, but we have the word of several Commission members that the Bionetics report had not been shown to them.

The break came in the summer of 1969, when one of "Nader's Raiders," Anita Johnson, was nosing around the FDA offices.[3] She came upon a copy of the Bionetics preliminary report and wrote to Nader about it. At this point there were two separate threads of discovery going on, because Dr. Epstein, now appointed to the Mrak Commission, demanded in August, 1969, that a copy of the report be provided to the group (this is the report that the NIH said had been in the group's hands for five months already). Ten days later, he was told that a *summary* of the report would be made available, but not the whole thing.

Back on the FDA front, Nader, who gets a lot of mail, hadn't pursued the communication from Anita Johnson. By September, though, she had a copy of the Bionetics report, and showed it to a friend of hers, a Harvard graduate student in biology. At about the same time, Epstein, having done some shouting and table-pounding, finally pried the full report out of the NIH for the Mrak Commission's subcommittee on teratogenicity.

In October, 1969, it all came out. Anita Johnson's friend showed the Bionetics report, or the part of it that she had a copy of, to Professor Matthew Meselson of Harvard, who was already concerned about reports of an unusual incidence of birth defects in Vietnam. Meselson, a scientist with some high-level contacts of his own, tried to get a copy of the full report, and was told that it was classified. The Nader group, now doing their own pursuing, drew a blank at every office door they approached.

Meselson, however, managed to get bootlegged copies of the full Bionetics report and of a statistical summary that went with it. He went to both White House and Army acquaintances to see whether some action couldn't be taken against 2,4,5-T. Whether with Meselson's knowledge or not, some "acquaintances" of his leaked the data on 2,4,5-T as a teratogen to the press.

That was apparently the final push. As with cyclamates, the government was again on the edge of a scandal—but this was a scandal that could put thalidomide on a back page forever where scandals are concerned. This was the story of the apparently deliberate use of a teratogen against a civilian population for more than three years.[4]

This was no matter for a mere cabinet member. On October 29, 1969, a statement on 2,4,5-T was issued by no less than President Nixon's personal science advisor, Dr. Lee A. DuBridge.

It is a masterpiece of headline-grabbing governmental obfuscation. The lead item of the DuBridge announcement was that from then on, the Department of Defense would restrict the use of 2,4,5-T in Vietnam "to areas remote from population." In addition, just in case the folks at home were worried, DuBridge also said these things:

—After January 1, 1970, USDA would cancel registrations of 2,4,5-T for food crops, unless the FDA finds a basis for setting a tolerance.

—USDA and Interior would stop using 2,4,5-T in their own programs in any populated areas, or any areas where the residues could reach man.

—By January 1, 1970, HEW would complete action on a petition asking that a definite tolerance be set for 2,4,5-T.

—Almost no 2,4,5-T was used by home gardeners or in residential areas anyway.

We won't keep you in suspense:

—USDA did *not* cancel registrations of 2,4,5-T for food crops by January 1, 1970, and the FDA did *not* set a tolerance.

—USDA, at least, made no visible change in its 2,4,5-T use program. Hickel's order, described in the chapter on DDT but including 2,4,5-T, did not come until the following summer.

—HEW did *not* complete action on the petition described by January 1, 1970.

—Monsanto, a major 2,4,5-T manufacturer, said that mixtures of 2,4,5-T and 2,4-D were quite commonly used on residential lawns throughout the country.

And one day after the DuBridge statement, the Department of Defense issued its own statement, saying that no change would be made in the use of 2,4,5-T in Vietnam because usage (they said) already conformed to the statement.

This was the *second* version of the DOD statement. The first version, given orally at a press conference, said about the same thing but referred to the use of 2,4,5-T against "enemy training and regroupment centers." A reporter asked how this could be put together with DuBridge's statement about "areas remote from population." When the written version of the DOD statement came out, the reference to "enemy" centers was missing.[5]

In other words, it was all a lie. DuBridge's statement was made entirely for its propaganda effect, in the knowledge that few, if any, of us would follow up to see whether any such changes or cancellations or other actions were ever actually taken. As if that weren't a miserable enough performance for a man once a distinguished scientist, DuBridge topped it off with this:

> It seems improbable that any person could receive harmful amounts of this chemical from any of the existing uses of 2,4,5-T and, while the relationships of these effects in laboratory animals to effects in man are not entirely clear at this time, the actions taken will assure safety of the public while further evidence is being sought.

But behind DuBridge's performance was a fact: the Bionetics information on 2,4,5-T and other substances was finally public. The substances tested for teratogenicity were put into four classifications:

1. "Probably dangerous" and worthy of immediate condemnation.
2. "Potentially dangerous" teratogenic effects—that is, they're really teratogens, but it isn't clear how serious—and "needing further study."
3. Adversely affecting fetal development but "probably not teratogenic."
4. Okay.

Certainly anything in the first two categories and probably the first three should be discontinued immediately; but our concern for the moment is only with the first and worst. Of all the substances tested, only two were found to be in that category. One was pentachloronitrobenzene. The other was 2,4,5-T.[6]

The lowest dose of 2,4,5-T fed to rats caused the number of abnormal births to increase by from three to more than five times. That was not a daily dose. It was given *once*, shortly after the onset of pregnancy. The equivalent dose by weight for a 150-pound human would be about one one-hundredth of an ounce.

This was and is the stuff that the Department of Defense kept right on using in Vietnam, and American growers kept right on using on pasturage and as a weedkiller, with not even a maximum tolerance prescribed for our food. Less than a month after DuBridge's astonishing statement, President Nixon himself had a statement on another subject. He said:

> The United States shall renounce the use of lethal biological agents and weapons, and all other methods of biological warfare.

Did that include herbicides? a reporter quickly asked the White House. Absolutely not, the White House said.

Thomas Whiteside, a nosy type in the best journalistic tradition, looked in something called the Joint Chiefs of Staff Dictionary, which he says "governs proper word usage within the military establishment." Biological warfare, according to the Joint Chiefs, is:

> employment of living organisms, toxic biological products and *plant-growth regulators* to produce death or casualties in man, animals *or plants* . . . [emphasis added].

Whiteside pursued it further through military-legal sources, but we need not. It was biological warfare even *before* they knew that it's a teratogen.

The rest of the world certainly thinks so. About a month after the President's statement, the political committee of the General Assembly of the United Nations (a committee on which every nation is represented) voted on a resolution specifically rejecting the President's definition of biological warfare to exclude herbicides. There were some abstentions; many nations cannot afford politically to condemn the United States, though they would not support us in the use of biological warfare. But the vote against the Nixon view was 58 to 3.[7] A few days earlier, the World Health Organization ringingly condemned the use of defoliants as "possible causes of birth defects in children." Even the Mrak Commission said that use of 2,4,5-T should be "restricted."

The next development in the 2,4,5-T story came in January, 1970, after all those departments didn't do what DuBridge had said they were going to do. There had in fact been no change in the use of 2,4,5-T whatever, and reporters who tried to find out more about it at Bionetics or in government agency offices met a silence beside which the faces on Mount Rushmore are downright chatty. In January, however, came The Great Dioxin Diversion.

By the time we had carried the story this far we were so suspicious of any public statements that we were ready for a while to believe that dioxin didn't really exist at all, but apparently it does. Its full name is 2,3,6,7-tetrachlorodibenzo-*p*-dioxin, but people call it by its last name, for obvious reasons.

The Dow Chemical Company, of Saran Wrap and napalm fame, is also a principal manufacturer of 2,4,5-T. In January, 1970, they announced that the 2,4,5-T in the Bionetics tests hadn't been pure. In the manufacture of the herbicide, they said, it often becomes contaminated with dioxin. They had had that trouble themselves, Dow said, but had cleared it up (after a bunch of their employees came down with a skin ailment called chloracne). But the Bionetics 2,4,5-T had come from another manufacturer and was contaminated with dioxin to the extent of 27 parts per million.

It was the dioxin, they said—not the 2,4,5-T—which was teratogenic.

That would make it pretty teratogenic. The lowest dose to rats, which as we've noted led to from three to five times as many birth defects, was only 4.6 milligrams per kilogram of body weight. If the dioxin was only 27 one-millionths of *that*, it's powerful stuff.

Again, we'll save you suspense. Dioxin is a teratogen, and it is a powerful one. Dr. Jacqueline Verrett (who did both the cyclamate and the MSG tests on chick embryos) ran a test on it and found that, for chick embryos anyway, it might be a million times as bad as thalidomide! But of course tests on much purer 2,4,5-T were made too, just in case Dow was right. The Great Dioxin Diversion turned out to be interesting, but irrelevant. 2,4,5-T is still a powerful teratogen, even without dioxin in it.[8]

Another fast shuffle started in April, 1970, when Deputy Secretary of Defense David Packard announced what was reported as a suspension on the use of the defoliant agent "Orange" in Vietnam. Orange, the press told us all (obviously repeating what DOD briefers told them), is the Army's name for 2,4,5-T.

There are two things wrong with that. One (the important one) is that the Army didn't quit using Orange, as a number of newsmen discovered, for several months. The other wrong point is that Orange and 2,4,5-T aren't really the same thing; Orange is half 2,4,5-T and half 2,4,-D.[9] Packard didn't say, by the way, that the Army would stop using defoliants; he just said that they'd switch to another one.

The pressure about 2,4,5-T use at home was of course going on all this time too; the same scientists who were concerned about spraying a teratogen all over Vietnam were just as concerned about spraying it all over our own food crops (it's used mostly for pasturage, but 75 percent of our corn fields are sprayed with herbicides, too, and at least some of it is 2,4,5-T). Besides, no matter what DuBridge had said, American scientists knew that unsuspecting householders (herbicides don't have to list ingredients on the label) were still using the stuff on their front lawns.

In that same month of April, 1970, Secretary of Agriculture Hardin, for only the third time in USDA history, suspended a product; he banned liquid 2,4,5-T, and banned nonliquid 2,4,5-T for home use or for use in aquatic areas. Hardin said that this was for our safety, and proved it by refusing to ban use on food crops.

He said that he couldn't do that because food crop use is a "hazard" but not an "imminent hazard." *The Washington Post* explained it more fully:

> Hardin's reasoning involved an emphasis on immediacy in defining the word "imminent." Thus he said that a possible hazard in using 2,4,5-T on food crops cannot be "imminent" when the chemical is applied, but only several months later, if it is on food that is about to be eaten.[10]

Pregnant women will undoubtedly be interested to learn that. It was in the same week as Hardin's statement that the experiments were reported showing 2,4,5-T to be teratogenic whether it has dioxin in it or not.

The Center for the Study of Responsive Law, set up by Ralph Nader, promptly petitioned Hardin to declare 2,4,5-T an "imminent hazard." On July 17, Hardin turned the petition down. In the meantime, we heard from Emil Mrak again.

In *The San Francisco Chronicle* for June 16, 1970 (one of our favorite newspaper editions of the past few years), an article headlined "Expert Criticizes Ecology 'Panic' " reports Mrak's remarks on what he described as the "panic-button approach" to "pesticides":

> Emil M. Mrak, former Chancellor of the University of California at Davis, criticized in particular the ban on the herbicide 2,4,5-T. . . .
>
> "Emotions play a big part in such actions," he said. . . .
>
> Mrak criticized the methods generally used to test new pesticides to see if they cause cancer or mutations.
>
> "Right now, we base everything on extremely large doses," he noted, "sometimes thousands of times what a human being takes in every day."
>
> ". . . With these chemicals we are banning, there may very well be a dose effect—where a large amount will cause tumors or mutations, but continued small doses will not."

This is nonsense. For one thing, there are in fact no "methods generally used to test new pesticides to see if they cause cancer or mutations," and there never have been. You will have recognized the old "safe dose" argument, discussed in an earlier chapter, and have noticed that Mrak ignored the cumulative effects of carcinogens.

The overall effect, of course, is obviously to suggest that the 2,4,5-T teratogenicity tests were far out of line, when in fact you don't even need to consider a normal 100-to-1 safety factor to be worried about them. Besides, at the time of Mrak's statement almost nothing had been banned anyway.

The story did point out that "Mrak has been a long-time defender of pesticides." But the story on page 6 of the same paper was a far better rebuttal. It was about the Lapland area of Sweden.

In the summer of 1969, it seems, the Swedish government used a mixture of 2,4,5-T and 2,4-D (similar to the DOD's "Orange") to attack some undergrowth in a pine forest near the Gulf of Bothnia. In the following winter a Lapp family with a herd of 600 reindeer moved into the forest, which no one told them had been sprayed. In April, 1970 (note the time lapse), a cold spell and heavy snowfall stopped the reindeer from reaching their normal fodder, and they started to eat the leaves of the forest undergrowth.

Within a few weeks 250 of the 600 animals had disappeared. By the time of the story in June, 1970, 100 corpses had been found. About 40 of the females aborted their young in April and May. As soon as regular fodder could be reached, the illnesses and deaths stopped.

A year after the spraying, the leaves contained 25 ppm of 2,4-D and 10 ppm of 2,4,5-T.

While we were pondering these stories, the Center for the Study of Responsive Law was pondering Hardin's denial of its petition. Three weeks after the denial, the Center and several others sued the Department.

On August 19, 1970, President Nixon asked the Senate to ratify the Geneva Protocol of 1925 against the use of chemical and biological warfare (the United States was a prime mover in the drafting of this protocol, but has never ratified it). In the message, however, the President made it clear that the Administration does not consider defoliants to be included in the protocol.

As mentioned earlier, practically everybody else in the world *does*. Even some Americans do; there was a bill in the Senate that month to ban defoliants in Vietnam, but it lost, 62-22.

The President also said that he had ordered the Department of Defense not to use defoliants any more—once the war in Indochina was over! In October a reporter finally got the military command in Saigon to admit what everyone there already knew: that they had gone on using 2,4,5-T for several months after Packard had ordered them not to. In December U.S. Ambassador Ellsworth Bunker and General Creighton Abrams "recommended" discontinuance of defoliants in Vietnam after present stocks were used up.

That last item is from a strange news story distributed by The *Times-Post* Service.[11] It misdescribes "Orange," but a lot of stories do that; it makes up for that by saying flatly that defoliants are used for the deliberate destruction of rice crops. What is strange is that the Bunker-Abrams recommendation said that until May, 1971 (when they estimated that all the defoliants would be used up), the agents should be used "only for the deforestation of uninhabited jungle areas." That would seem to come pretty close to wanton and useless destruction—if any faith could be placed in it.

What is also strange is that no explanation is offered as to why General Abrams, who is the military commander of the American mission in Vietnam, doesn't just order the defoliation stopped if he wants it stopped. Since the United States insists that it is never done for any but military purposes, it should be his decision. In fact, it probably would be—if the purposes were military.

Just to show that our civilian agencies keep up with the news, the United States Forest Service, in January, 1971, announced in response to a criticism of their use of 2,4,5-T and 2,4,-D (which continues) that no animal deformities or deaths could be caused by the combination.

In the first week of January, 1971—the legal action having been shifted to match Nixon's creation of the Environmental Protection Agency—Chief Judge Bazelon of the Appeals Court ordered the EPA to reconsider the use of 2,4,5-T. In refusing the Center for the Study of Responsive Law's petition to declare it an "imminent hazard," the court said, Secretary Hardin had "failed to assign sufficient importance to the risk of harm to human lives."

It is still not banned.

It probably will be banned, though—at home and in Vietnam (it is at this writing under study by EPA[12]). The environmentalists will cheer. We will shake hands all around and congratulate ourselves on a job well done. And we will thus fall subject to the greatest insecticide snow job of them all.

It is both unpleasant and time- and space-consuming to devote so much space, in a book primarily concerned with our domestic food supply, to "military" action in Indochina. But public affairs, too, are ecological—things that fit together cannot be understood in bits and pieces. The herbicide residues in our daily diet are, whether we like it or not, directly related to the conscious and deliberate decision to use the herbicides, in the way they are used, in Indochina.

Working from the studies of independent scientists who have visited Vietnam, Whiteside has done some calculating about the possible effects of herbicides in that country. A number of scientists, as he points out, have noticed that the "ideal" use of Orange and White (see Note 9) rarely takes place in the real world. The military claims that in the South it does not move against crops, but instance after instance have been observed in which that does happen, either through accident or because the people doing the spraying aren't told the truth about the herbicide's spray-drift characteristics.

And whatever happens, the herbicides get into the water, directly or by drainage. Whiteside writes:

> It has been calculated that, taking into account the average amount of 2,4,5-T in Agent Orange sprayed per acre in Vietnam by the military,

and assuming a one-inch rainfall (which is quite common in South Vietnam) after a spraying, a forty-kilo (about eighty-eight-pound) Vietnamese woman drinking two liters (about 1.8 quarts) of contaminated water a day could very well be absorbing into her system a hundred and twenty milligrams, or about one two-hundred-and-fiftieth of an ounce, of 2,4,5-T a day; that is, a daily oral dosage of three milligrams of 2,4,5-T per kilo of body weight. Thus, if a Vietnamese woman who was exposed to Agent Orange was pregnant, she might very well be absorbing into her system a percentage of 2,4,5-T only slightly less than the percentage that deformed one out of every three fetuses of the pregnant experimental rats.

Whiteside makes no mention of any 100-to-1 safety factor, nor does he note that the rats didn't get their dose every day. An advisory committee to the Secretary of the Interior, not long ago, made a recommendation (never implemented) as to the amount of 2,4,5-T that might be considered "safe" in drinking water. The Vietnamese woman in Whiteside's example (who is not unusually small by Vietnamese standards) would be getting 600 times that amount.

We are not dropping herbicides in Vietnam in order to deform Vietnamese babies; no one suggests that we are. We are, however (and to a much greater extent than we admit) doing so in order to destroy crops, because it is a part of our overall strategy in Indochina to move as many as possible of the South Vietnamese out of rural areas and into the cities by any means possible.

It has been concluded at high levels that this is the ultimate way to defeat the guerrilla tactics of the National Liberation Front and retain control of Southeast Asia. Without, of course, any mention of deliberate crop destruction, the strategy has quite clearly been set forth in public writings by Samuel Huntington and Henry Kissinger, both top Administration advisors.[13]

It must be remembered that the United States is destroying crops not only in Vietnam but in Laos and Cambodia. American scientists have already evaluated damage to crops in Cambodia as far too extensive and severe to be attributed to "spray drift." It is a strategy for Indochina, not for Vietnam. It was not undertaken with the knowledge that 2,4,5-T is teratogenic; but even without that knowledge, it is biological warfare, and by the standards of almost every other nation in the world it is a war crime. This is why the United States has consistently refused to pay claims in South Vietnam for crop destruction; if they were paid, it would be necessary to admit the extent of the crops destroyed. This is why the Bionetics report was suppressed literally for years, and why, at last, it had to be dealt with, not by a statement from a cabinet member, but by the top science advisor in the White House itself—and with what we can only describe as a smoothly delivered series of outright lies.

But even this is not the ultimate con job. The ultimate con job was to focus all of our attention (including that of the Nader group, that of the press, and even that of many, though not all, scientists) on 2,4,5-T. For with all the fuss, with all the statements by Secretary Hardin and Deputy Secretary Packard and Dr. DuBridge and even Chief Judge Bazelon, 2,4-D remains untouched. And that is no accident.

2,4-D is listed in the Bionetics report as "potentially dangerous" as a teratogen; it had teratogenic effects, but they were less dramatic than those with 2,4,5-T.

2,4-D is cheaper than 2,4,5-T, and the military likes it better.

2,4-D is more widely used in Indochina than 2,4,5-T. Both Orange and Purple are half 2,4-D; White, the other major defoliant, is mostly 2,4-D with a small amount of another ingredient. A fourth agent, called Blue—the names come from color codes on the drums in which the herbicides are shipped—is used in relatively small quantities and is not closely related chemically.

2,4-D is one of the six best-selling "pesticides" in America (the measurement includes both insecticides and herbicides), with annual sales of more than $25 million.

What happened seems, in retrospect, quite clear. Several reporters have discovered that government officials, faced with an impending scandal about the Bionetics report, expressed the opinion that there would be far more pressure from The Bugkillers' Conglomerate if action were taken against 2,4-D. Others have uncovered the fact that the military exerted extremely strong pressure against any ban on 2,4-D.

Obviously what was decided was to throw the relatively expensive, not so widely used 2,4,5-T to the wolves, please everybody—and hope that nobody notices that 2,4-D is still with us in all its horror.

For years, while both these substances were used on many of our foods, knowledge of their teratogenicity was deliberately suppressed, obviously because of the effect that knowledge would have on our master strategy to move the people out of the country into the cities of Indochina. When the suppression would no longer work, attention was diverted to the more expensive and less widely used of the two substances, with the hope that in the flush of victory, critics might, at least for a time, overlook the also teratogenic herbicide that the military really wanted to use.

This is a harsh judgment, harshly written. We do not like writing it any better than you enjoy reading it. But one fact stands out above all the others, confirmed and undeniable.

Regardless of one's opinion of the overall strategy (we find it inhumane, but at least some Presidential advisors obviously do not), regardless of the original acts of what most of the world regards as criminal biological warfare, regardless of whether one feels that almost any measure is justifiable in the name of confining Communism or that the United States is engaged in suppressing self-determination—regardless of individual opinions about any of these things, it remains true that the United States for four years knowingly continued the widespread and escalating use of chemical agents which it had learned were teratogenic, and which were distributed, with the loosest of controls, over civilian-populated areas.

And it remains true that for those same four years, the rest of us were forced to eat food containing residues of those teratogens in order that the nature of that crime could be concealed from the rest of the world.

CHAPTER FIFTEEN

The Nazis' Gift to Your Diet

Lear bade the wind to blow the earth into the sea. His creator could hardly have guessed what would go along with it by the middle of the twentieth century.

As we write, attention is focused on the sea's suddenly startling content of deadly mercury, put there for the most part by manufacturers who go on dumping it despite all knowledge of its danger until the threat of massive fines or worse forces them to change. The mercury turns up in our food (seafood, particularly), but it is a pollution contaminant, not a food additive even by our wider-than-legal definition.

When it is used as a fungicide, however, it is a food additive, and it is just as poisonous. It has been used to control rusts and fungi since 1929, applied to the seeds of wheat, barley, peas, rice, other grains, and cotton. It is still used in the United States, although its dangers have been understood for some time and it is banned elsewhere. Sweden banned mercury compounds from agricultural use early in 1966. Other countries, like Denmark, have never allowed their use.[1]

Sweden has published a stream of information on the effect of mercury as a fungicide (for the most part, the seeds were treated with Panogen, a product of Nor-Am Agricultural Products of Chicago). The data show clearly that treatment of seeds results not only in contaminated grain, but that the chain of contamination proceeds from the grain through hens to eggs to man, or through livestock to man.

A 1969 publication by Swedish scientist Goran Lofroth, for example, indicated that as long ago as 1965, he and his colleagues were convinced of the danger from alkylmercury compounds (methylmercury, most widely used in Sweden and here, is one kind of alkylmercury compound). They knew, for instance, that "the use of methylmercury in agriculture was responsible for the poisoning and drastic decrease of wild bird populations." When the mercury compounds were banned, eggs, which had averaged a mercury content of .029 ppm, dropped to .01 ppm in a year, a drop of two-thirds.

What mercury will do to humans who ingest it varies considerably with the amount, but there is no real indication (despite recently set FDA tolerances) that there is any "safe level." Apparently even the smallest amounts of mercury have some effect.[2] Some doctors started to pay close attention after fishermen began eating contaminated fish in Minamata, Japan, in 1958; the whole story of the then mysterious "Minamata disease" is told by Wesley Marx in *The Frail Ocean*.

About 10 percent of any mercury ingested will go to the brain, where it is certain to destroy at least a few brain cells. You have a lot of brain cells, and the damage may not be noticed for some time, but it is of course both cumulative and irreversible. When the damage does begin to make itself felt seriously enough to involve medical help, it is likely to be diagnosed as encephalitis or, among older people, senility.

Milder doses can and often do produce fatigue, headaches, irritability, anxiety, excessive self-consciousness, difficulty in concentrating, blushing, excessive perspiration, numbness, blurred vision, or impairment of hearing, speech, or motor coordination. In more serious cases the result is likely to be prolonged coma and involves irreversible damage to the liver, kidneys, and central nervous system.

To cap this, Swedish scientists also discovered that mercury causes severe damage to human chromosomes, and that mercury contamination in food led to an increase in spontaneous miscarriages and in abnormal births. We might repeat that any damage to the gene pool is an irreversible damage not only to the individual and his/her family but to the entire species.

All of this was known well in advance of December, 1969, when three children in the Huckleby family of Alamogordo, New Mexico, became ill—and ultimately attracted worldwide attention. They had eaten meat from a hog. The hog had previously been fed with the sweepings from a granary. The grain had been treated with Panogen, the fungicide that was the principal source of agricultural mercury in Sweden and banned there four-and-a-half years earlier.

At the end of 1970, one Huckleby child was still in a coma, one is blind and crippled (apparently permanently), one has severely spastic muscle responses. Although a *New York Times* reporter was able firmly to pin the blame on Panogen, the Pesticides Regulation Division of USDA did nothing about it until February, when NBC featured the Huckleby tragedy on a news program. The next day, Dr. Harry Hays, who was then the director of the division, decided that Panogen was an "imminent hazard" and ordered it off the market. Nor-Am promptly took the division to court.

Incredibly, Dr. Hays and the Pesticides Regulation Division presented no evidence from Sweden or Japan, nothing to indicate that the Huckleby incident was any more than an isolated or freakish accident. Hays says that the division went to the court so convinced that the court had no jurisdiction that he was not prepared to argue the case! To make it more incredible, Hays, a toxicologist, was certainly familiar with the Swedish data, and he had even canceled the registration of a mercury algaecide used in swimming pools (canceled, not suspended; the cancellation was appealed and the stuff is still being used).

The court reversed Hays' order, and Panogen went back on the market. Since then, the registration of mercury compounds for agricultural use has been canceled in the United States, but the cancellation is under appeal, and as of right now, all those mercury-coated seeds are still being used.

Your body, of course, isn't concerned with authors' definitions of food additives; it doesn't care whether mercury poisoning comes from the eggs of hens who have eaten treated grain or from fish that have absorbed the mercury from waste-discharge pollution. The FDA has never done much about this, although their office of compliance has a "mercury project officer." A story distributed by The *Times-Post* Service in July, 1970, said that the FDA was "uncomfortable" with its 0.5 ppm tolerance for mercury in fish. That tolerance is five times higher than the tolerances in those other nations which still allow mercury levels in fish. However, the tolerance still stands.[3] The tolerance of 0.5 ppm is itself relatively recent; until a couple of years ago there was no tolerance for mercury in American food. We don't mean a zero tolerance. We mean that any amount of mercury was legally okay.

What is interesting about the fish pollution is the way in which it came dramatically to public attention, and the attitude of the "regulatory" agencies toward doing something about it. Early in December, 1970, a private scientist who specializes in ocean pollution, Dr. Bruce McDuffie of the State University of New York at Binghamton, reported finding canned tuna with mercury levels well above the permitted tolerances.

On December 15, the FDA responded by pulling a bunch of canned tuna off the market. Following the usual practice of obfuscating the issue as much as possible, however, FDA Commissioner Charles Edwards called the tuna "contaminated but not a health hazard." The tolerance of 0.5 ppm, he said, is a purely artificial number used as a "guideline," and "offers a substantial margin for safety." Apparently he didn't read about his own scientists' concern, expressed to the press five months before, that the tolerance level was far too high.

Nor, of course, had that concern led to any action until Dr. McDuffie broke the story independently and forced the FDA to act. We would all be eating contaminated tuna for the rest of our lives without knowing it if we had left regulation to the FDA.

McDuffie, in fact, came right back at them. In trying to get some fresh tuna for analysis, he said, he had been sent, in error, some frozen swordfish imported from Japan. Not one to waste food, he tested the swordfish and found 1.3 ppm of mercury, nearly three times the tolerance for fish allowed in the United States.

McDuffie made the announcement a few days after the FDA had seized the tuna and Edwards had issued his reassuring statement. Asked about that action, McDuffie snorted, "It sounds like the typical don't-panic-the-public line that public food and drug officials put out." And of course it was.

Again, after the McDuffie finding, the FDA decided to test some swordfish, and wound up seizing most of what they tested. They also announced the *beginning* of a testing program for imported fish.[4] The announcement was timed to hit the newspapers on December 24, when a lot of Americans would be sure to read it.

When you're dealing with an industry which markets thousands of products, and in which one of those products may bring in as much as $25 million a year, the fast shuffles come at you with bewildering complexity. Any criticism of The Bugkillers' Conglomerate that might seem effective leads them to build not one but several lines of defense, none of which are genuinely concerned with alleviating problems (except balance-sheet problems), but some of which may bring results that are worse than the original problem was.

As protest against DDT has swelled, one line of defense quickly thrown up by the Conglomerate had to do with the existence of alternative chemical insecticides which, as they carefully pointed out, do not persist for long in food or soil, and against which far fewer strains of insects have developed resistance. Their names are Malathion, parathion, schradan, phosdrin, chlorthion, thimet, tetraethyl pyrophosphate (TEPP). Together they are called the organic phosphates, or organophosphates. The highly dangerous Shell No-Pest Strip is an organic phosphate.[5]

The organic phosphates come to us from Nazi Germany. The Nazi Mad Scientists were not looking for insecticides. They were looking for nerve gas. And that's what they got; the organophosphates are nerve poisons. If you put a drop of one of them in your eye, you will die.

They will kill you if you swallow them, even in tiny concentrations. They will kill you if you breathe them in. They will kill you if they get on your skin.

Not always. A tiny amount may only make you very, very sick. But the people who apply them on crops must dress very much as Neil Armstrong did when he first strolled on the airless moon. Human beings are not supposed to be allowed into the fields for specified periods of time—days, not hours—after they are applied.

Permitting their use as we do would be dangerous even if all the rules were followed. In practice, growers follow the rules for using organophosphates no better than they follow the rules for using any other of the dozens of compounds of various kinds with which they decorate our foods in the field. One result can be acute granulocytic leukemia. A more common result comes from the connection between organophosphate insecticides and nerve gases.

Some parts of the human nervous system are made up of what are called cholinergic nerves.[6] When a nerve impulse arrives at one nerve cell, a very tiny amount of acetylcholine (a combination of the vitamin, choline, with acetic acid) is formed. The acetylcholine reacts with a protein on the next cell, and the impulse thus moves along the nerve.

As soon as that happens, however, another chemical reaction must take place within a few millionths of a second. The hydrolysis of the acetylcholine ("hydrolysis" is just a word for any reaction in which water takes part) must happen, or the next impulse will not be able to move along the nerve.

That reaction requires the presence of an enzyme—in this case an enzyme called cholinesterase.[7] Without it, says Holum,

> the reaction is too sluggish; for all practical purposes it does not take place at all. The excess acetylcholine produces a prolonged, unwanted, and therefore dangerous stimulation of the nerves, and the glands and

muscles they control. Eyes, salivary glands, heart muscles, the digestive tract, and the adrenal glands, among other things, are affected.

Organic phosphates react with cholinesterase at the site where it does its work, and they deactivate the enzyme, just as nerve gases do. Only one of the thousands of tiny reactions in the body is thus directly inhibited. In a very few minutes, the result is death.

Of course, if the inhibition of cholesterase kills you, you don't have to worry about other possible effects of organophosphates. But it doesn't always. In tiny amounts it simply slows up your reactions (including your involuntary reactions—like heart action). Breathing becomes difficult, excessive fluids are secreted, vision suffers, headaches and nausea are common. The effect, if exposure to the organophosphates is at a low level but continuous, is a gradual onset of symptoms. External contact brings on eye irritations and skin rashes. All of these reactions are common among farm workers.

Near Visalia, California, in the heart of that part of California in which the giant corporations do their billions of dollars worth of factory-in-the-field farming, a handful of doctors have set up a tiny clinic, called Salud, to serve farm workers and their families (the name is Spanish because most of the workers are *chicanos*). In 1969 and 1970, in cooperation with a study being conducted by the state's Public Health Department, one Salud doctor offered to test the cholinesterase level of 58 children in a study he was conducting for other reasons.

He found that 27 of the 58 children showed some effects from toxic phosphates. Some had marked cholinesterase depression by adult standards. There are no children's standards. "It is tragic and absurd," said Dr. Lee Mizrahi, who did the study, "that such a study, by a rural doctor, should be the first ever done."

But it's not surprising.

If the low cholinesterase level doesn't get you, the aliesterase inhibition will. Aliesterases are enzymes (any chemical whose name ends in "ase" is an enzyme) which make possible the body's ability to get rid of small amounts of drugs and toxic chemicals. Many people puzzle over the fact that doctors use as medicines small amounts of compounds which, in larger amounts, are poisonous. This works because it is known (to some extent) how much of those compounds the body's detoxification system, involving the liver and the aliesterases, can handle, and how long it takes.

Succinylcholine, for example, is a muscle relaxant.[8] Too much of it will paralyze you. Procaine, a common local anaesthetic, becomes a toxic substance in larger amounts. The point is that the difference between aid and danger depends on the body's normal detoxification system. If, as with some headache remedies, a little will help your headache but a whole lot must be taken to hurt you, then the gap between what's therapeutic and what's toxic is pretty wide, and there is some margin for safety.

This is not true of either succinylcholine or procaine. They cannot be tolerated in even small overdoses. The organic phosphates inhibit the function of the aliesterases. In simple terms, this means that the body can't do so good a job of getting rid of potentially toxic substances. This means, in turn, that an otherwise normal dose of something like succinylcholine or procaine can become a dangerous dose for people whose systems have been exposed to organophosphates.

Do they get in the food? They do indeed. Some years ago, for example, samples of Greek olive oil were tested and found to contain up to 14 milligrams of parathion per kilogram. Tests on the people who lived in the area where the olives were grown showed low cholinesterase levels. Theoretically, this happens when farmers violate the rules about using the insecticides too close to picking time. In fact, this isn't always the case, because the organophosphates have the nasty habit of not always doing what the Mad Scientists say they're supposed to do.

One California case, during the 1963 peach harvest in Stanislaus County, resulted in one death, several hospitalizations, and 94 cases of clinically identifiable phosphate poisoning. A subsequent investigation showed that, depending on the orchards in which each person had worked, as many as 85 percent of the workers in each crew suffered from insecticide poisoning. But the investigation also showed that no grower had violated any rules for using the stuff.

There were other cases, later, in the same area, and the investigations went on. What the investigators ultimately discovered was that parathion—the organophosphate involved in these cases—is supposed to break down fairly quickly into a harmless residue, but that in fact it did not. It broke down into something more poisonous.

No one knows why it does not break down "correctly" sometimes when it does so at other times. No one can predict when it will do one and when the other. But obviously it is important, for it can determine how much of the highly poisonous substance may reach your dinner table. According to Bicknell, 0.1 percent of parathion will break down into an equally toxic substance called para-axon. Investigators looking for parathion residues in food won't find the derivative substance because they aren't looking for it.

Needless to say, none of this information, eight or more years old, has stopped anybody from using this highly profitable Instant Death. Whether Bicknell is right or not, the use of organophosphate insecticides is clearly another case of the Mad Scientist syndrome: use it now, make as much money as possible, and find out about its effects later. When you find out about it, keep selling it anyway. Sell it to anyone who will buy it and tell them, if they'll believe it, that it's a substitute for DDT because it won't persist.

That may be what was told Clarence Boyette of Pink Hill, North Carolina. He grows tobacco, and organophosphates were widely sold to tobacco farmers in the wake of the public fuss about DDT. As of mid-1970, Clarence Boyette couldn't harvest his crop. He couldn't even drink the water in his house. His 11-year-old son, Curtis, spent six days in the hospital because of phosphate poisoning.

And Daniel Boyette, aged seven, was dead. Two others died in that part of North Carolina at the same time.

The North Carolina Board of Health said that anyone who comes into contact with parathion (parathion killed Daniel Boyette) should take a bath, and that no one should go into fields where it has been used for five days afterward. But no one knows whether *that* parathion broke down into deadlier poison or did what the manufacturers said it would. Clarence Boyette certainly doesn't know. But he'd probably like to.

Even if we could be sure of the "disappearance" of organic phosphate residues before any food reached our table, we could not accept their use. None of us, who may be secure in urban or suburban surroundings, can genuinely be comfortable with our deliverance from the dangers of chlorinated hydrocarbon residues in our foods, if we know that that deliverance comes at the price of a startling rise in occupational disease among farm workers. We do not want our safety at the expense of the life of the worker who died in Stanislaus County in 1963, or the lives of the others who have died from the same cause, or the life of seven-year-old Daniel Boyette. Especially when it is unnecessary.

Since we are not, in fact, safe, there is all that much more reason to pay attention to the plight of farm workers. If most of our examples have been from California, it is not only because we live here; California is the only state that keeps reasonably accurate records of known occupational poisonings among farm workers. Be assured that things are as bad, or worse, in every state where today's giant agricultural corporations grow our food and fiber.

Ronald Taylor has written several articles on this subject; the one listed in the bibliography is a must for anyone who cares about the effects of organophosphates. Taylor can be specific—as in the story of the three-year-old girl playing in the field where both her mother and father were working. She poked her finger into a bottle of organophosphate insecticide on an idle tractor, stuck the finger to her mouth, and was dead within minutes.

Or he can look at the overall figures:

> This year [1970] 1,400 Californians will be poisoned, or injured, by pesticides or other agricultural chemicals. Nearly half will be disabled for a time; most will recover, nine or ten will die. These are predictable figures, based on state public health records that show *half the casualties are farm workers* [emphasis Taylor's].

Farm labor occupational disease rates are three times as high as the California average. The largest contributor to that startling statistic is insecticide poisoning. The largest contributors to that fact are the organic phosphates. The diseases and the deaths happen not only to the workers themselves (as with, for example, the "black lung" poisoning of coal miners) but to their children, and even, occasionally, to bystanders.

The organophosphates, as you've gathered by now, are very tricky materials. But the farm workers are not organic chemists or trained scientists. It is the unskilled and relatively uneducated who suffer. Not only the *chicano* and *Latino* and Filipino workers who struggle through the fields but, as Taylor also notes:

> Of the 156 accident reports filed by aircraft dusting operators in 1967, most of the poisoning occurred among unskilled men hired to load airplanes, mix chemicals or to flag the aircraft's path through the field.

Note the last seven words. Obviously *somebody* doesn't give a damn about the known effects of organophosphates.

The Office of Economic Opportunity has for some time financed a group of attorneys in California who are grouped together in an organization called California Rural Legal Assistance. In addition to using the courts to improve the lot of welfare recipients, CRLA has filed a number of personal injury suits on behalf of farm workers, attempting to fix responsibility for insecticide poisonings.

At this writing, CRLA has had its funds cut off, except for a temporary six-month reprieve, at the urging of Governor Ronald Reagan and the state government. The city newspapers have treated the fuss as a result of CRLA's activity on behalf of welfare recipients—but the realities of California politics are rarely reflected in the urban press. It is the corporate growers who run California, and they want no part of courts fixing responsibility for insecticide control.

The United Farm Workers Organizing Committee (UFWOC) headed by Cesar Chavez deserves national support if for no other reason than that scientists do not know how organophosphates break down and because there are residues in our foods. The organophosphates have FDA tolerances, so we know the residues must be there. The UFWOC alone is fighting the insecticide problem in the fields—not just to eliminate DDT so that we in the middle-class can be comfortable, but to eliminate the overuse of all insecticides so that everyone can be safe. Only those farms with UFWOC contracts have meaningful control over the insecticides used in the field. It is written into the contracts and enforced by joint grower-worker committees.

In principle, one's sympathies may be with the Teamsters (as in the lettuce struggle, current as this is written), or with UFWOC, or with nonunionized growers. But in practice, to support UFWOC—to seek out, for instance, the grapes and lettuce that bear their label—is to support our own health and the health of our children. And that question goes far beyond the political.

What, then, can be used to kill pests? To be honest, it is difficult to trust any chemical insecticide, herbicide, or fungicide. Take for example the fungicide known commercially as Captan, which has been in use for nearly 20 years. In 1958, after experiments with a number of animals showed no toxic effects, it was given a generous tolerance of 100 ppm residue on raw agricultural products (a tolerance which still exists). The generosity of that tolerance can be deduced from the fact that the tolerance for chlordane, a chlorinated hydrocarbon, is .03 ppm—only three ten-thousandths as much.

But Marvin Legator of the FDA, whose outstanding work has been mentioned earlier in this book, took a look at Captan in 1967, and found interference with the production of DNA, the basic genetic molecule, and the breakage of chromosomes. "Captan," he wrote, "is only one of a number of chemicals formerly considered safe that are now suspected of causing genetic damage."[9]

And he didn't just pick Captan out of the air. The Bionetics report listed Captan as a teratogen on the same level as 2,4-D. It is of course still marketed and used.

Another in the same group in the Bionetics report was ethyl carbamate, one of another group of chemicals being used as a "substitute" for the discredited chlorinated hydrocarbons. Urethane, another carbamate, is a known mutagen (and was so described by Rachel Carson in *Silent Spring*). Rachel Carson described how the carbamates (two were then used) "prevent[ed] sprouting of potatoes in storage—precisely because of their proven effect on stopping cell division." Sevin, one of the carbamate insecticides, breaks down in part into 1-naphthol, a highly toxic substance; the human intestine can deal in part, but not completely, with 1-napththol.[10]

"The governing principle for the development of chemical pesticides," Turner wrote, "seems to be selling pesticides rather than controlling insects." That is so obvious after we have investigated the subject that we find ourselves unjustly wondering why everybody doesn't know it already. Beyond that we can add that the governing principle for the *regulation* of chemical pesticides seems to be selling pesticides rather than controlling manufacturers or users.

Otherwise, how can we account for the fact that the FDA set tolerances for deadly aldrin on apricots that were two-and-a-half times the tolerances allowed on cherries? Or as the Rienows ask, "Why is Carbaryl, for example, twice as poisonous on beets (tolerance, 5 ppm) as on apples (tolerance, 10 ppm)?" Why the variations in DDT tolerances?

There can be only two answers, and in fact they undoubtedly work together. One is that the agencies refuse to apply zero tolerances to foods (like milk) in which a zero residue is impossible. That is, they set the tolerance to suit the fact, instead of trying to get the fact to fit the tolerance. A police department may know that in a given year 1 percent of the population will commit murder, but no one tries to make 1 percent of the murders legal. The other answer is that the tolerances are by and large what the manufacturers and the growers want them to be—for their own economic good, and for no other reason.

The fact is that, no matter how much propaganda is issued about how you demand pretty and pest-free foods, and about how chemical insecticides are necessary if the world is to be fed, it is the chemical industry, not the rest of us, that *needs* chemical insecticides. If one insecticide is discredited, it is easy enough (relatively) for Shell or Dow or Monsanto to switch to the manufacture of another. But if another method of pest control were to gain favor, they would have to fall back on gasoline, detergents, or napalm, and a chunk of their enormous profits would simply disappear.

The growers, too, need chemical insecticides. The family farmer who has managed to put a little money in the bank might be able to make the transition to other methods, but (unless the rest of us were willing to provide him partial support) he would have to take a loss during the transition period, and also make do without an absolute maximum crop for a year or two. The massive corporate "farmers" who in fact produce most of our food would as soon give up their Continental Mark IIIs, their boxes at the San Francisco Opera, or their bank directorships.

"I am convinced," said Emil Mrak in January, 1970, "that for many years to come we will need pesticides if we are to feed the world or even our own people." It is not insignificant that he was speaking to the National Canners Association.

We don't need insecticides at all—or it looks very much as though we don't. We certainly don't need them in the massive amounts in which we use them. We need agriculture; not even the most dedicated ecologist suggests that the human race go back to being hunters and gatherers, and of course we have multiplied to the point where there is not enough to be hunted and gathered. As a result, we may need some pest control, and we certainly need some in some places at some times.

But so long as The Bugkillers' Conglomerate controls both the propaganda and, in this area at least, the government, we will find it difficult to explore how that need might best be met.

The USDA even continues to insist that growers must be allowed not only to use chemical insecticides but to use all they want, lest the world go hungry. But the argument is false. Sherrill quotes a Swedish paper by Dr. Lofroth:

> In February, 1964, the Plant Protection Institute of Sweden recommended a reduction of the seed dressing formula [the use of fungicides on seeds] by 50 percent . . . and also recommended that only infected seed should be treated. These recommendations were made compulsory in October, 1965. In 1965 and earlier, about 80 percent of the Spring sowing was treated, whereas in 1967 only 12 percent was treated. *This drastic reduction has had no deleterious effects on the crop yield* [emphasis Sherrill's].

Dr. Chester Himel, an entomologist at the University of Georgia, has conducted studies which he insists demonstrate that 99 percent of the insecticide currently sprayed on crops is wasted! The sprayers used, he said, make droplets too big to be effective except in a very few cases.

The danger of insecticides to agriculture is being amply demonstrated, for all who want to watch, in Guatemala. In the tropics insect generations succeed each other more quickly than they do in more temperate regions. In that country, cotton has become (after coffee) the most important export, worth $14 million (at official rates of exchange) to the nation in 1968.[11] But despite the warnings of agronomists, the entire cotton crop is now in danger—not only for a year but, as far as we know, forever.

The problem is simple: the insecticides have led to resistant insects, which have led to more powerful insecticides, which have led to even more resistant insects. It is already quite clear that unless other methods are introduced quickly (perhaps at the cost of the growers' profits and the nation's foreign exchange for a year or two) the insects must inevitably win, and it will be impossible to grow cotton in Guatemala.

Only the tropical speed makes the situation in Guatemala different from that in the United States. In parts of Texas, the problem has been recognized. Worried by the increasing insect threat, a few growers combined earlier agricultural knowledge with new techniques; careful crop rotation, carefully timed planting, and careful tillage procedures designed to eliminate insect reproduction have gone far to wipe out the problem without chemical sprays. In California, attempts are under way to control the pink bollworm by introducing irradiated, and therefore sterile, males.

The sterile-male technique has worked quite well with cattle formerly threatened by screwworm flies. The fly used to cost stockmen $100 million a year in the United States alone. In a joint effort with the Mexican government, American scientists have been releasing a large number of sterile male flies at regular intervals, and it is already evident that the population of the fly is declining. Another similar effort is under way to deal with the Mexican fruit fly.

Another technique, already successful with some plants, is the breeding of resistant strains. The California citrus crop was once dramatically saved by this method,[12] wheat has been bred to resist the Hessian fly, resistance to the borer and earthworm are bred into corn, and alfalfa is being bred to resist leaf hoppers and aphids (which once threatened to wipe out alfalfa as an industry).

A third method of biological control, quite promising although without any dramatic successes as yet, is the breeding of insect parasites: bacteria and viruses that are specific to the insects to be controlled, or even wasps that lay their eggs in other insects. With this method, as indeed with the others, attention must be paid to making certain that balance is being restored rather than further destroyed; but it appears that they are being conducted with care, and at least they are likely to poison no one, either in the field or in the kitchen.

But "the biological control of pests," says Lester Brown, "is still in its infancy." And in 1969, Edward Knipling of the Department of Agriculture explained why, apparently without apology:

> [A] single insecticide that may with minor modification control a hundred different kinds of harmful insects can be developed at a far lower cost than the research needed to develop biological or selective chemical methods of insect control.

It depends on what you mean by cost. The Boyette family might have a different definition from Mr. Knipling's.

The companies presently set up to manufacture insecticides are not, as noted previously, set up to breed sterile male insects or resistant strains of apricot trees; consequently, they aren't about to do research on biological methods. As a spokesman for the National Agricultural Chemicals Association said, "They would research themselves right out of the market."

"Pesticide" sales for 1968 reached $1.7 billion, and the growth rate in dollars is 16 percent a year. Thus, although the agricultural research service of the Department of Agriculture does spend some money on researching biological methods of control, there is no real financial support for it. The government doesn't come up with much, either for its own agencies or for independent researchers at universities and elsewhere; nor do the universities themselves come up with the money, although they can find financing easily enough for research into chemical agents.

As a result, although there have been successes, even the most "successful" biological method, crop resistance, is not understood in principle, but has worked so far only on a trial-and-error basis. All that is known is that crop resistance works in one of three ways: through antibiosis, tolerance, or nonpreference.

Antibiosis means that something in the plant's system kills the insect (this is the most dangerous to fool around with artificially until we have a pretty good idea of what that same "something" will do to people). Tolerance means that the plant goes on growing no matter what the insect does. Nonpreference means that the plant doesn't particularly attract the insect any more.

With enough money, scientists could learn the plant chemistry that underlies these mechanisms; they could make crop resistance into a universal solution to crop pests. But they don't know now because of virtually no financial backing.

One university campus, the University of California at Riverside, has a distinguished faculty working on biological controls (you never know; the growers might *have* to fall back on them). But the former chairman, who helped to put that faculty together, has gone to the University of Toronto. Except for Riverside, he says, there are only scattered individuals here and there working on the subject and there is no government money to be had to support it.

"To be frank," says Dr. D.A. Chant, the scientist in question, "USDA does not have top-flight people and is going about the work in a superficial way." Despite the Department's own internal expenditures, Chant adds, "I don't think they really have much of a commitment to biological control."

On the other hand, a spokesman for Shell Chemical dismisses the idea with the statement that "no one but the federal government can afford this research." He didn't add—he didn't have to—that Shell of course has no motivation to afford it anyway. In the most delicate possible way, Robert Holcomb set forth the entire biological-control situation in *Science:*

> We have, therefore, knowledge of pest control methods that will solve many of the problems associated with the use of conventional pesticides and that could pay for themselves in the long run, but the structures that have evolved among the government, industry and farmers to implement the use of conventional pesticides appear to be unsuitable for the initiation of new control measures.

Clothe it however you will in the diplomatic passive voice ("the structures that have evolved") or the detached academic judgment ("appear to be unsuitable"), what that means is that there's too much money in poison for Dow, Shell, Monsanto and the rest of The Bugkillers' Conglomerate to let go of control of what happens in our fields.

The fact remains that the chemicals are poisons, and clearly poisonous to us as well as to insects. The fact remains that many of them are also mutagens and/or teratogens. And the fact remains that the chemical insecticides, herbicides, and fungicides do *not,* in fact, rid the crops of pests or disease; on the contrary, the problems get worse *because* of their use.

It is all a deadly game being run on America and increasingly against the rest of the world, for it is demonstrable to any investigator not buried under industry propaganda that healthy soil and healthy plants do not need all those insecticides at all, and are very likely better off without them. Nor, for that matter, is our nutrition helped by the other half of the chemical one-two punch of commercial agriculture: the chemical fertilizers, which also may do more harm than good.

The Mad Scientists do not like this point of view, and some of them are masters at sneering at it, mixing in what one semanticist used to call "purr-words" and "snarl-words" to make sure that you go along—as in this quotation from Paul Cannon:

> According to this point of view, animal manures and rotting composts presumably produce purer and more wholesome foods than do clean solutions of nitrogen, phosphorus and mixtures of trace elements.

Silly, isn't it? But of course it is with "animal manures and rotting composts" that our species has evolved to its current form—complete with mechanisms for using trace elements which we must have been able to get naturally from our food.

We are Antaean, whether we like it or not. When Heracles was able to lift Antaeus away from contact with the earth, he was able to conquer him. It may flatter the Paul Cannons among us to see themselves as Heracles—but when we lose our contact with the earth, we too will die, and the Mad Scientists with us.

Soil is not just a bunch of ground-up rock. It is, in its natural form, a community that teems unimaginably with life. Von Haller writes of experiments performed on a single gram of soil—no more than can be pinched between a child's fingers:

> It is estimated that [in a gram of soil] there are 600 million of bacteria, and in addition many millions of other vegetable living organisms, especially fungi and algae. In addition to this, we must mention the creatures of the soil, beginning with the microscopically tiny one-celled creatures, protozoa, on up to the earthworms. . . . All these animal and plant forms do not constitute just a senseless mixture. . . . The living community of the soil has its own metabolism, a continuous cycle of circulating substances, in which every link in the chain of its thousandfold life does its part.

Of course an earthworm doesn't fit in the pinch between the fingers, but the point is clear. It is this living community that makes soil "fertile," and thus able to grow healthy plants, resistant to disease and filled with the vitamins and minerals and trace elements we need.

It is possible, certainly, to find soil that lacks one of those trace elements; but few would propose that you grow all your own food in your backyard. Agriculture will go on being commercial, or at the very least collective; and to plant the right crop in the right soil is not the most difficult problem of agriculture.

Insecticides, herbicides, fungicides destroy the living community of the soil. The quality of our food, its "healthiness," its delicate relationship with our inner ecologies, suffers when that community is destroyed. And as this unhealthiness becomes evident in the plants themselves, which appear stunted and sparse, chemical fertilizers are added to "boost" growth. This they do for a while—but the plants are still not healthy, and just as you are when you haven't been eating right or are tired, they are more than usually susceptible to disease (or, in the case of plants, to attack by "pests"). This brings on more spraying, which kills more of the soil, which leads to more fertilizers—and so on.

It is not some nutty "food faddists" who have demonstrated the nutritional importance of healthy, natural soil to healthy, natural, disease- and insect-resistant plants. Sir Albert Howard was knighted by the British government for his many years of conclusive experiments on the subject—despite the fact that he was at first derided by the Mad Scientists of his time and for all of his life jeered at by the spokesmen for chemical companies, particularly those that manufactured fertilizers.

Sir Albert was convinced that in naturally fertile soil, there are elements that defied chemical analysis; he anticipated the more modern concept of "trace elements," but his argument went beyond that to a genuinely ecological perception that is still true: we don't know how all those things work together.

Natural growth conferred immunity to plant disease and insect attack, as he demonstrated in several experiments. He didn't know why, but he was convinced that the chain went all the way from the soil to man, through the plants and, if they intervened, through the animals that ate the plants.[13] To demonstrate it, he once grew oxen on nothing but organic food and then deliberately exposed them to others which had the highly contagious hoof-and-mouth disease. No infection took place.

The basic theory underlying Sir Albert's concept is now well enough established: the content of the soil determines the content of the plant. Soil deficient in copper grows grain deficient in copper, and animals which eat that grain will fall sick of a copper deficiency. In a dramatic study, Dr. Firman Bear, an agricultural chemist, once analyzed the mineral content of vegetables (beans, tomatoes, spinach, cabbage, and lettuce) grown all across America.

Take only the iron in spinach. Everybody knows that spinach has a lot of iron; the doctor tells us to feed it to our kids to prevent or to correct anemia. But in Dr. Bear's samples, the iron content of spinach varied, from the least to the most, in a ratio of 19 to 1,587! Closer analysis demonstrated beyond any doubt that the variation in the spinach matched variation in the soil.

Still, our children are taught in today's schools that such-and-such a food contains such-and-such a nutrient, and sometimes they are even told how much. This teaching is at best misleading and at worst simply wrong. One head of lettuce does not equal another.

As mentioned earlier, however, we don't simply take in this element and that, and put them together. At the top of a food chain, we organize into a higher form material that is already partly organized by lower forms—and we have to ingest it in that already organized form in order to use it. From our foods we must get amino acids and protein substances, properly organized by the plants and animals we eat.

When commercial agriculturalists are faced with problems like this, they often respond oddly. In the Midwest, a finally exhausted soil in one area refused to yield not only quality but quantity in the corn it mothered, no matter how much fertilizer was used. In response, a new hybrid variety of corn was bred, one that would provide a good yield per acre despite the exhaustion of the soil. The farmers sold the new corn and were happy.

But the corn itself contained 30 percent less protein than the older variety. Beyond that, because of a lack of essential amino acids, the protein in the new corn was of a markedly lower quality. One ear of corn does not equal another.

Whether or not the reason is to be found in the pervasive influence of the chemical companies, the Department of Agriculture does not like this kind of thinking. It is interested in yield, not quality. In its public statements about nutrition, at least, and in the "educational" material it assiduously distributes to schools and libraries, a bunch of spinach is a bunch of spinach, wherever grown; and the quality of the soil affects the *quantity* of food, but not the quality.

There is no doubt that their methods (the methods of extensive, continuing, escalating chemical tinkering) lead to food that is nutritionally less and less valuable. Added to the growers' insistence on yield per acre, we don't help matters with our insistence on "pretty" fruits and vegetables. As von Haller notes:

> If we are interested only in the size of the head of cabbage, then we will get cabbages of enormous dimensions. If we are concerned only with the good looks of the tomato or the apple, they will doubtless keep on getting more and more handsome. But if we put the emphasis on the health content of our fruits and vegetables, then the development will move in that direction.

As mentioned earlier, the growth of these nutritionally inferior products is maintained with chemical fertilizers, virtually all of which follow the NPK formula—that is, they are composed almost entirely of nitrogen, phosphorus, and potassium. They do not restore to the soil what has been killed in it; much less do they restore the soil as a living and balanced community. All they bring is growth.

There are, of course, other bad effects in the external ecology. The escalating use of phosphorus, which eventually winds up as a sediment at the bottom of the ocean, is rapidly using up the world's supply, and after that there isn't any more. Mining phosphate rock from which to make fertilizer (and organophosphate insecticides) is already a major industry wherever the deposits exist. It does not restore itself to its natural place in the planetary ecology nearly so fast as we are using it; we will either have to quit soon or find a way to mine the ocean bottom, or life will disappear from a lack of phosphorus.[14]

The nitrogen, in the meantime, wrenched out of the natural cycle that has evolved with the earth and artificially "fixed" by vast factories instead of tiny microbes, washes into streams and lakes and sinks into water tables. Excessive nitrate content in drinking water (it's already a problem of toxic proportions in parts of California) means a human disease called methemoglobinemia; the blood can't carry oxygen as it's supposed to. Children under five are particularly susceptible.

And of course, the fertilizer washed into the streams and lake would, if algae had emotions, delight the algae. They grow and grow and use up the available oxygen, so that the fish die (one species at a time). So, eventually, does the lake. The process is called eutrophication; you can see it working in Lake Erie.[15]

But most of all, fertilizers destroy the soil, and as noted above, lead to unhealthy plants which need insecticides which further damage the soil, and so on. You wind up not only with low-quality food but, eventually, with hardpan: dead soil. In effect, the farmer who falls into the insecticide-fertilizer cycle is creating a world of hardpan and resistant insects—and no farmers.

In 1957 Dr. William A. Albrecht, head of the school of Agriculture at the University of Missouri, demonstrated that with increasing exhaustion of the soil, the nutritional quality of wheat (measured by its ability to manufacture protein material in the berry) went steadily downward.

Dr. Albrecht testified before the Delaney Committee that his experiments had dramatically proven the need for healthy soil. In adjoining rows he grew plants that were 100 percent insect-free and plants that were *all* attacked by pests. Except for the treatment of the soil, the handling was identical. "Using spinach," he told the committee, "we had two rows of bugs, then no bugs, two rows of bugs, then no bugs. Not because we sprayed but because we got a different condition in the soil under it."[16]

He did the same thing with corn; solely by controlling the soil he eliminated attack by the corn borer in one crop, while another crop alongside was decimated. The two crops were so close that in some places they touched. The only damage to the "healthy" crop was where the borers apparently tried a few bites but couldn't take the "taste."

"Organic farming," as Sir Albert Howard used the term, doesn't mean merely abstaining from chemical "pesticides" and fertilizers. It is necessary to see that what is taken from the soil is returned to the soil, in the form of what Dr. Cannon correctly, if unattractively, called "rotting compost." What we now call garbage can be directly returned to the soil, and some of that "animal manure" is not at all a bad idea (alas, our own is now too unhealthy to use).[17]

Only by some such method can we be sure of the retention (the recycling, to be precise) of the elements we must have in our food. Dumping our garbage in the ocean (as many coastal communities do) is recycling it, all right, but it's doing it over the span of centuries, not crop cycles. Otherwise we must face the fact that the chemical-depletion cycle imposes on our foods a *double* burden.

First, the depletion of the soil reduces the amount of nutrients and essential substances in our food. Second, the residues from insecticides and other chemicals put a demand on our systems for more nutrients, because of the need to detoxify those residues. We wouldn't need to build so many enzymes to aid in the detoxifying process if we weren't taking in so many poisons; but the same farming process that is bringing us the poisons is bringing us fewer nutrients with which to combat them.

And that's a double burden that exists even on "fresh" foods—before any of the rest of those 3,000 additives are shoveled, shoehorned, or needled into what we eat.

It is of course possible to restore fertility to the soil; but it is not possible to do so and to make a fast buck at the same time. It is obviously stupid in the long run for agriculturalists to destroy the source of their income in the pursuit of the goals of their own shortsighted greed. But agriculturalists do not care about the long run any more; nor do the executives of chemical companies.

The men who control those corporations—the corporations that do the growing and the ones that make the chemicals—are interested in getting rich. When they are rich, they will leave their money to their children and grandchildren, who will be rich by inheritance and will doubtless invest their money in insurance companies or mutual funds or possibly even young actresses. They are not "inheriting farms," and there is no reason for the giants to pursue long-range wisdom on their own.

And no lawsuit by the Environmental Defense Fund can be brought against corporate growers to force them to return fertility to the soil. What can, perhaps, be done by that method is to channel a few resources so that, ultimately, we might score a few points against them.

If through the practice of environmental law we could force zero tolerances for all insecticides, and then assure the enforcement of those tolerances, it would be uneconomical to produce or to use many of them. If through such means we could force meaningful tests for total toxicity, cumulative effects, and synergistic effects, plus tests for carcinogenicity, mutagenicity, and teratogenicity, for all such chemicals (and, indeed, all additives), and then force the removal of any found suspect as imminent hazards, some incentive might appear for research on biological controls, and on "healthy soil" as well.

This might be a good time to try it, while the Environmental Protection Agency is new and appears to be at least a little more responsive. But the EPA has no money to grant for research. And we shouldn't kid ourselves: as long as the FDA is for all practical purposes controlled by the industry it's supposed to regulate, and

USDA is controlled by agribusiness, very little of this is likely to happen.

We certainly cannot rely, as even some of the otherwise sophisticated among us seem to do, on "trusting" the federal agencies. We can hardly expect much, for example, from an agency like the FDA, which was pushed into seizing $2 million worth of swordfish but wouldn't admit that there was anything dangerous about it. A United Press story early in 1971 made this point clear:

> A spokesman for the agency said the department has taken no stand on whether persons should purchase swordfish, terming it a personal decision.[18]

Of course it would be "a personal decision" if "our" agency were in the habit of giving us the facts about things like mercury contamination instead of patting us on our collective head and telling us that there's nothing to worry about. But until it is genuinely our agency, working for us and only for us, functioning as it was intended to function, we simply cannot trust it.

Taylor wrote of the insecticide situation among California farm workers:

> There are child-labor laws, pesticide-control laws, regulations and recommendations to make pesticide use safe. On paper, in California at least, these all look effective, and when state officials explain the enforcement procedures, it seems that the system should work.
>
> But it doesn't.

It does, now, in one small part of California agriculture, because the farm workers themselves took and used the power to make it work. "Power to the people" has become a slogan, a catch-phrase, and it calls up an automatic, semantic wall of resistance in mild, ordinary, middle-of-the-road Americans.

But it is the answer all the same.

CHAPTER SIXTEEN

The Consie Manifesto

In the 1950s, bleak days as they were of Joe McCarthy and the rampant House Committee on Un-American Activities,[1] worried intellectuals talked much of novels like Orwell's *1984* and Huxley's *Brave New World*. A science fiction magazine of the time, *Galaxy*, published a serial, Frederick Pohl's and C.M. Kornbluth's "The Gravy Planet," that might today be equally pertinent.[2] In their future world, U.S. senators were elected to represent not states but corporations (with the number of Senators determined by the size of the corporation). All, all was profit and exploitation. The educational system, of course, was merely a device for conditioning the young into not only an acceptance of this situation but an eager support of it.

And there was, of course, an enemy: a vicious and unscrupulous conspiracy, bound to overthrow all that was decent and good and right in the world of complete profit and total exploitation, and violently detested by the general public though few in the populace had ever seen a conspirator. But the conspirators were not called Commies.

They were called Consies.

Pohl and Kornbluth may have been the first writers—the first popular writers, in any case—to see that the genuine ecological concern they referred to as "conservation" is a subversive attitude, in a sense that goes beyond most of the political arguments with which most of us are familiar. In any case, their hero of course first hates and then joins the Consies, and ultimately they triumph.

There is only one difficulty. Pohl and Kornbluth set their story in a time of interplanetary travel, and the Consies "triumph" by leaving an exhausted planet to live on a new one, where ecological harmony will reign.

We cannot do that. We are stuck with the planet we have.

There is no escape, and there will be no new "input" to our planet's ecology. We have in our biosphere all the oxygen, all the cobalt, all the phosphorus we will ever have. Our planetary ecology is not something we *ought* to live with; it is something we have no choice but to live with.

We are also stuck, each of us, with the bodies we have. Our inner ecologies, too, are things about which we have no choice. We will not magically become, in a generation, a species that can exist without biotin, or can tolerate a mouthful of parathion.

Nor will we become static; we are processes—and limited processes at that. In each of our bodies, for example, there is one and only one liver. It is always working, and changing as it works. But it can detoxify just so much.

You cannot build a new sewage treatment plant or an effluent processing center to clean up your inner ecology. The one you have—the liver—is all you're ever going to get. Your body can handle just as much poison as your liver can detoxify. After that it either kills you whammo! or it kills you over a period of years.

As with ocean or river or bay, the solution is to stop dumping the stuff in there; at least it won't get any worse, and it might get better. Oceans and rivers and bays, of course, do not get cancer and do not have to worry about mutations; we know that some of the changes in us are irreversible.

Nonetheless, it is intolerable that we cannot make choices about our own bodies, that we cannot say a roaring "NO!" to the Mad Scientists who insist on poisoning us with one hand and making us more and more synthetic on the other. The answer to air pollution is not gas masks; the answer to poisoned bodies is not the development of a new, handy, easily insertable, guaranteed-for-50,000-miles plastic liver.

That, though, is the way in which the Mad Scientists now want to deal with the problems they have created: bigger and better fish ladders. Medicine, for instance, is not the few valiant doctors of Salud trying to help with the ravages of organophosphate poisoning and pushing for the abandonment of the substances that cause it. Medicine in America is *The Bold Ones*, all electronics and pacemakers and transplants (and did you ever watch that program and wonder who could afford to be in that hospital?).

It is intolerable that we cannot eat what we want to eat. But more than individual choice is involved—for we may choose to eat frozen rarebits or pizza rolls or coal-tar-dyed chocolate cake mixes or hydrogenated peanut butter, and let our inner ecologies go to hell. We do not, however, have the right to make these choices for our children (or to allow our children to make them, ignorantly, for themselves), any more than we have the right to feed them poison outright or to beat them senseless. What is truly intolerable is that we cannot obtain decent food for them. How can we bear the fact that an American woman cannot feed her child from her own breast without knowing that she is poisoning it?

A part of our frustration, too, is that it is intolerable that the Mad Scientists and their bosses should control not only our food but our information about it. We can know that we are being lied to, even that our children in school are being lied to—but where is the truth? By now, surely, it must be evident that you must not only question every word from a government agency, every word from an "expert," every word in a newspaper—but in wisdom you must question what you have read here as well. Where is the truth?

There may be no answer. Because of a previous book I wrote, I have been asked to speak on ecological subjects, and when in those talks I have outlined some ecological dilemmas, I have frequently been criticized because I describe the problems without offering a solution.

It is a common criticism but a bad one. We do not need to have answers in order to see and to describe problems. You are as intelligent as we are; we have shared our knowledge of the problems, but you can figure out the answers as well as we can.

But we can offer some observations. One of them is that we all have to learn to "break sets," as the psychologists put it—to find new ways of looking at the problems, or to look in new places for solutions, and to get out of thinking habits that keep us restricted to the same old approaches over and over again. We must look at some of the assumptions on which our ideas are built and see whether, in fact, those assumptions might be in error.

It appears to us that we must begin by joining, all of us, the Consie Conspiracy.

This is not at all easy for ordinary Americans to do, for we are not accustomed to thinking in the way that we shall call conspiratorial. Some of the educated middle-class young will have no problem; many have already done what we propose. But we are not suggesting long hair or rock music, after all. We are suggesting that "straighter" Americans break a set.

There are some things that we all know and that we do not talk about much. We know, for instance, that there is a We and there is a They. They want to get rich at our expense, and they're doing it—not by building a better mousetrap (we do not begrudge such people their reward), but by exploiting our bodies and our minds. They poison us and tell us that it's good for us, and underneath all of our I-don't-want-to-get-involved complacency, we know it.

We know that we have nothing much to say about what government agencies do, but They know this too. We know that we cannot afford to engage them in battle, even if we knew how. We know that electing one Congressman instead of another has been offered to us as a solution for years, but that things don't get any better.

Let us then return to the subject of food and, whatever your real attitude may be, let us ask you to join us in what you might even want to call a fantasy. Let us look at the whole subject of abuse against our bodies as though we were, in fact, members of The Consie Conspiracy, determined somehow (but peacefully if we can) to overthrow the Mad Scientists, their bosses, the PR men, and the government agencies which behave as though the food industry owned them.

The first fact we must face is that we are isolated, and that our enemy counts on that fact. The structure of Western life, based as it is on the nuclear family living in isolation, with one dwelling unit to each (except among the poor), helps to keep us isolated. We often do not know the people next door, or the people across the hall.

We cannot win if we do not break that isolation. It really does not matter whether your car has an American-flag decal and the car next door has a peace symbol or a clenched fist. You are both being poisoned anyway.

The second fact is that we live in different kinds of situations, urban and suburban and rural, some of us with available land to varying degrees, a lot of us without even a window box (but if you have one, you can at least grow your own herbs and spices with a minimum of care anywhere but in the soot that passes for air in Manhattan). We must adapt our conspiracy, of course, to our own situations. We will come back to that.

The third fact is that we must, most of us, learn to look beneath the surface of changes to see whether they really change anything.

The first concern is whether we are getting better food, more control over our bodies, and especially providing a better supply of natural energy to the inner ecologies of our children. The second concern is whether our neighbors, our fellow conspirators, are benefiting as well. Nothing else matters. Public statements by high officials, expressions of concern that don't bring any results, are useless.

The fourth fact is that conspirators must make sacrifices. We are going to have to give up at least a little ease. The children may have to give up the television-conditioned satisfaction of seeing a box of Sugar Frosted Junkies on the table in the morning (which means that we must give up taking that easy way out with their breakfast). Those of us who can afford it are going to have to spend a little more money for some things.

Now then: what can we do?

The first thing we can all do is stop doing some of the things we do now. We've talked about bread and cereals. Depending on time and how much we care about our families, we can extend the same thinking to cake mixes, and eventually to all kinds of things that are prepared for us through the medium of adding junk to healthy food.

We can avoid the unhealthy easy way. Don't use Crisco or Spry for anything when you can use a cheap vegetable oil, like soya oil (if you want to pay a lot more for a soya-cottonseed oil mixture because it's called Wesson and buys a lot of advertising, that's up to you). Go back to old-fashioned, unhydrogenated peanut butter; stirring it up is good for the kids' muscles.

We can also do without some other nonfood items that help to support the same people; our doing without will go toward giving them the message. Saran Wrap will do for an excellent example: it's nonbiodegradable anyway, it's a by-product of working with chlorinated hydrocarbons, most of its uses are either unnecessary or can be served just as well by other products, and it's made by one of the giants of The Bugkillers' Conglomerate who also makes napalm, dumps mercury in the water, and opposes the farm workers in California.

But you can make such individual decisions without our specific suggestions. Withhold your money from the enemy wherever possible, give it to your friends wherever possible, and keep looking for new ways to do it—including a serious consideration every once in a while of what you're using that you could really do without.

If you live where it's possible (and it doesn't have to be in a very big city any more), look over the local health food and natural food stores. If you're not used to it, remember that health food stores and natural food stores are not the same thing (though in some places one establishment may be both). The natural, or organic, food stores are usually just what they say they are; buy there whatever you can afford, after you've checked to be sure that the local supermarket doesn't sell the same thing for a nickel cheaper. Ask about bulk prices; you may want to know when your part in the conspiracy gets going.

Health food stores are something else. Some are only there to cash in on a fad; some are devoted to a particular mystic approach to diet; some are staffed by people who really know what they're talking about. But they all *think* they do. Enter with skepticism.

Look for what you're sure is healthy food that may not be available elsewhere. Things made with grains are usually good. Avoid the supplements, like Vitamin E pills or more exotic bottled goodies, unless you know exactly what you're doing. Skip the Upper Kashmir Yak's Blood Concentrate that will restore to you the vitality of a 20-year-old and cure your bunions. Watch the prices; if you can, compare prices in health food stores. Some are grotesquely high.

Of course, if you have some soil, and if it and the climate are any good, you can grow some food in it. There are plenty of sources for inexpensive material on organic gardening (it works for your flowers, too). You don't need any sprays on most gardens most of the time, no matter what the nice man at the seed (and spray) store tells you.[3] There is a man in Castro Valley, California, who has worked out a method of helping his neighbors achieve an ecologically valuable way of ridding their yards of excessive numbers of snails: he rents ducks for a dollar a week. That's what we mean by breaking sets.

You cannot, of course, grow everything. This is one of the places at which you start talking to the people next door. Depending on your respective time and ability, one of you grow the lettuce and the other the tomatoes—or whatever. Don't worry about money unless there is a marked difference; then share the cost. But never, never let the human relationship of the conspiracy get hung up on money; you are fighting for your life, and doing it together.

Now get the people across the street to come up with some carrots. You're on your way.

As fellow conspirators, you naturally won't each buy all the books or pamphlets or whatever. If there's something you need to know, you can find it out together; it's more fun anyway. You will be astonished to discover how much you know as a group, while each of you may feel relatively ignorant individually.

Don't, of course, restrict it to just the women, or just the men. For one thing, you cut off half your combined knowledge, and for another, you all eat, don't you? Get the kids in on it; when they understand that it's something you believe in and that it's important, even if they don't understand why, they'll come up with a lot of help and they'll learn along the way why the conspiracy is necessary. They'll get in the way sometimes but they do that anyway, don't they?

Since we are a conspiracy, in opposition to the way things are, we must break our mental sets about who does what. To accomplish what we must accomplish, we cannot afford the luxury of "men's roles" and "women's roles." We must look at what we all do, and divide it up according to time and ability and an equal share in the dull parts. It makes the group function better as a group too. So does sharing the children among households.

This is only the beginning of The Consie Conspiracy (and we haven't yet turned to the apartment dwellers), but already your mind is probably running ahead. You can see, for instance, that this kind of combination of cooperation and concern can lead to other things: things related to schools, street lights, and external environment. The details we can safely leave to your particular conspiratorial cell; we will only comment that you will notice the difference between a conspiracy based on a continuing endeavor (and one that is designed to overcome an enemy) and mere cooperation for a particular short-range neighborhood goal.

If there is a Co-op where you live, or near where you live, join it (look in the phone book under "Co-op" or "Consumers Cooperative"; we're sure of several in the San Francisco Bay Area, several in or around Los Angeles, and one in Manhattan, but there are a lot of others). The Co-ops are not part of The Consie Conspiracy, although they once were. They do, however, have a lot of relatively inexpensive good-quality food, more or less enlightened label practices, and in some areas they carry lines of food products that will enable you to get away from those 3,000 additives. For instance, they are likely to sell Cornell bread and in some places a little organically grown produce.

That is not, however, the principal reason for joining one. The principal reason for joining one (get the whole neighborhood conspiracy to join it, or organize the conspiracy from among the members) is that they are member-owned and observe genuine democratic forms. You can demand that they stock what you want, and if enough of you want it, they will. You can't always do that so well with Safeway or the A&P, for instance.

In fact, not buying *anything* at a Safeway is one of the best things The Consie Conspiracy can do to subvert the enemy. Despite any good things they may sell, Safeway is a part of the industry that is trying to get rich without any concern for whether it kills you along the way. They were the last major retailer to hold out, for example, against union-grown grapes and to sell table grapes with residues of aldrin, one of the worst of the chlorinated hydrocarbons. Among the people who own Safeway are the people who grow the grapes.

In Berkeley, as you might expect, there are several stores which sell organically grown foods. One of them is run by a cooperative, and one of the members of the cooperative set forth in a recent interview the attitude that The Consie Conspiracy simply must have if we are ever to win:

> I see [natural food] as a way of understanding oppressed people and the political implications of food. Eating natural food is a way of subverting Safeway. The food industry is a heavy political thing.[4]

If the language puts you off, then it is necessary for us to argue that the language is irrelevant. The poisoning of your body and those of your children is all that is relevant.

Apartment or suburban dweller, there are a number of things your friendly neighborhood Consie Conspiracy can do. High on the list is getting together to deal with the neighborhood supermarket.

You can think of it, depending on your relative mildness or belligerence, in either of two ways. You can bring organized consumer pressure on the supermarket: stock this good stuff or we won't shop here. Or you can offer a friendly incentive: stock this good stuff and we will promise you a market for it.[5] In either case nothing will happen if you hit the store as an individual and hope that the family across the hall or down the block has read the same book at the same time. Nor will anything happen if you wait for the family across the hall or down the block to read the book and organize *you*.

Remember that you don't need an organization with a president and a secretary and the minutes of the last meeting. In fact, that's the last thing you need. All you need is a friend in the next house or apartment, and a friend in the house beyond that or the apartment below, and the realization that together you can do what you can't do alone.

For instance: maybe one family doesn't have any kids and is not pushed for time and can bake bread for two or three houses (it isn't that much more work, and you have the potential for getting occasional help). The others can do something else in return (help vacuum the baker's house, or cook a meal for that evening). Bread can even be baked in a New York apartment. You can see how the "too much work" argument begins to dwindle away.

Next step (in apartment or suburb): keep enlarging the conspiracy. Once you've got a couple of good things going (it doesn't matter which things), you'll think of a dozen others that you could all do if you got the people in the next block involved.

Here's one: over in Jersey or out on Long Island or down by Peoria or over the hill in Contra Costa County or someplace out by Riverside (suit the phrase to your locality) there's a farmer growing a few things on a little land and making a modest living at it. You get your cell together and decide what all of you will use in the way of vegetables on a regular basis. Then you go and find that farmer out on the Island or wherever, and you tell him that you'll guarantee to buy (at a negotiated price) so much lettuce, so many carrots, and so on if he'll agree to grow them without the use of chemical fakery. Maybe you can even loan him a book.

This is an important ploy, because there is not enough food organically grown in America to feed everyone who might want it; a big part of The Consie Conspiracy is to get more people to grow more of it. The farmer we're talking about isn't going to quit using chemicals because he is concerned about the inner ecologies of a bunch of city types he's never seen, unless he's a very unusual farmer. But he will quit if you promise him that he can sell his produce. Tell him you don't give a damn whether it's pretty.

Back to the supermarket: if you use much lettuce and/or grapes, you can do this either as an individual or as a group, but it will have much more effect if you do it as a group. Ask to see the box in which the lettuce and/or the grapes came. If the people in the store won't show it to you, buy your lettuce and grapes somewhere else (do not trust anything they tell you unless they are old and trusted friends). If they do show it to you, look for the symbol of the United Farm Workers; you'll know it when you see it. It's a stylized, Indian-looking symbol in the shape of a black bird, in a circle on a black flag. If it's not there, buy your lettuce and grapes somewhere else (a Teamsters Union label will *not* do; there are insecticide residues on that lettuce). In fact, buy *everything* somewhere else, and tell the store that you'll go on doing so until they have UFWOC's pesticide-free produce.

Through your cell, or indirectly through the clout that your group now has with the neighborhood store, you can put pressure on a nearby dairy to produce milk from cows whose forage is not sprayed with DDT or anything else. You do the same thing: promise a market for the milk.

If your group is composed of people who don't have among you enough time to pursue some of these ideas (especially the trip to Jersey or Peoria or wherever) or if you're living in such an isolated place that the whole thing is hopeless, or even if you'd just like to get a little variety into your conspiracy, here's a suggestion that will cost the whole cell (to begin with) just five bucks.

Send the five dollars, in the name of one member of your local conspiracy, to The Ecological Food Society, 114 East 40th St., New York City 10016 (they don't know us, there's no rakeoff, and they will be as surprised to read this as you are). Tell them you want to become a member. You don't have to do anything else if you don't want to. They'll send you a newsletter and a catalogue, and the five bucks will go toward bringing more land under organic cultivation.

The Society is a sort of Consie Conspiracy by mail. It does some of the things we've suggested: it guarantees a market for organic foods, but it then distributes them by mail. They cost a little more that way, but of course it's worth it if you're not squeezing every penny, and the cost is not as much higher as you might think (more members would bring the cost down, too). After you see the catalogue, the rest of the cell can decide whether to join; or you could, we suppose, cheat on the membership and just order a whole lot in the member's name from the catalogue. That wouldn't be nice, but it will work.

The Society has a lot of interesting things, all of them guaranteed to be organically grown and with no additives: fruit, vegetables, all kinds of meat (including fish and fowl), dairy products, cereals—almost all the staples. It also tells you how to get some perishable kinds of food near where you live. And there are a few very far-out items for fun, like organic toothpaste. The catalogue alone is worth the five bucks.[6]

Also, the catalogue will give you a few ideas for using your own home-grown food conspiracy to cut your food costs. There is a food conspiracy in Berkeley, for example (we don't belong to it, but we know about it), whose members insure themselves a supply of good-quality cheese by buying it (through a cooperating local cheese store) in full wheels. Conspiracy members take turns doing things like picking up the cheese from the distributor, cutting it up into individual sections and wrapping it, and delivering it to members.

The more such activities you can get your group into, the more of them can be combined, so that the time factor is cut down to manageable proportions. If there's a butcher in your neighborhood, you've really got it made.

The more of us who buy natural food, by mail or in person, and who insist that stores stock it, the more the prices will drop to near the cost of the mass-produced, chemical-infested, nutrition-poor produce most of us eat. That's good not only for us; poorer families who can't join our conspiracy yet, for a lack of both time and money, can get some benefit from it all the same.

It seems to us that if, at this point, you are with us—if you are willing seriously to consider trying to get together with some of your neighbors to improve your and your children's inner ecologies—you will by now have begun to wing it a little on your own. Ideas will be coming faster than you can read.

If, on the other hand, you have read the book and are concerned, but you find yourself reading about The Consie Conspiracy as though it is all a little utopian and pretty far-out, and what can *real* people do?—then please permit us another plea.

There is *nothing* else that real people can do. These are the hurdles that we have to overcome, because it is precisely our own isolation as individuals that makes the exploitation of our bodies, and the bodies of our children, possible.

Of course you can follow certain shopping, and eating, practices on your own, and benefit thereby. There is a great deal, too, that can be accomplished through more formal organizations, as we will see in a minute. But the Mad Scientists and their employers expect—they are perfectly confident—that you will react as you are reacting, that the worst that will happen to them is that a few of you will act on your own, as individuals, unorganized, and that you will not break the set in which you are something called a "consumer" and therefore someone with relatively little power.

Whether you're talking about an election, a new way to buy cheese, or an old revolution: put enough of us together and we have all the power we want to take. Those first Americans knew what they were saying when they made that flag with the snake chopped into pieces.

One of the most important things that ordinary people can do when they're together, instead of isolated, is that they can begin to regain some control over the most important of all commodities in our culture: information. As it is, however concerned about our food we may be, we are forced to rely on books like this one (about which we cannot know whether we should trust them or not, for there is always another book that says the opposite), on magazine articles that may appear in publications supported by advertising bought by the food industry, and on oversimplified newspaper columns. We may read a reporter's distorted version of a federal official's statement that may have been a lie in the first place. We may see on television a respectable-looking businessman saying that Cesar Chavez is a Communist. We are at the mercy of fragmentary information when we are alone. Together we are not.

Ten families, together, will include someone who is willing and able to take the time to spend some evenings with a chemistry book like Holum's, and then to explain what seems important in it. Ten families, together, is a large enough group to induce some not-yet-captured graduate student from a nearby university or college to spend an evening answering elementary questions. Ten families, together, is a group of enough size so that someone in it will be skeptical enough of any particular item of information to be sure that it's checked out. And the graduate student may bring her husband (or his wife) and make it 11 families.

Then, when your conspiracy does have some idea of what it's about, when you are convinced that most of what you hear about food is wrong, that you are being poisoned, that the whole chemical cycle in agriculture is folly—then you can begin to exert a force for better information.

You can, for instance, find out what the schools in your neighborhood are teaching, and what materials they are using. You will make your own decisions, then. Here are some of our ideas.

Cooking classes in the schools are ridiculous. Anyone can learn to cook; I started learning at 36, and because I enjoy it so much my son is learning in his teens (we have that horrible male stereotype to break). But everyone, male and female, should be learning, certainly before they're out of junior high school, at least the kind of nutritional science and organic chemistry that we have talked about in this book. We *all* need to know about how vitamins work as parts of enzyme systems, about how the liver goes about its vital process of detoxification, about what the hell an intestine is for.

We all need to know the link between evolution and food, to gain the sense of ourselves not as machines but as functioning, growing ecosystems, as continually evolving processes. We all need to know how food functions—and when we do, we will learn to cook in such a way that will not destroy those functions.

It goes without saying that every piece of industry propaganda in the school should be meticulously examined by your neighborhood conspiracy. We will leave you to do your own screaming after you see it.

If there is no teacher in the school who can teach about food in this way, make them get one. There aren't enough to go around, of course; so create a demand and there will be more of them. The same thinking applies to doctors who are willing to provide nutrition-oriented diagnosis and treatment instead of relying on concepts that are a generation out of date. You can't pressure the medical schools directly, of course; but you can change medicine. Don't get the idea that every headache is a coal-tar dye or every sleepless night an insecticide residue; there are viruses and microbes and infections. But if you have read this far, it is extremely likely that you know as much about the effect of food, good and bad, on health as your doctor does.[7]

Beyond that, it is with food as with anything else: the more you know, the harder it is for somebody else to lie to you. Your existence as part of a group, instead of merely as a person who lives alone or is a member of a single isolated nuclear family, means that all the information that one of you has becomes known to all of you; and the group can learn by leaps and bounds what one person would have to devote long and grueling weeks to finding out.

In the meantime, the food problem is national, and it is urgent. All the things we have suggested should be done; but in a sense the answer to "What should I do?" is "Everything."

By that we mean that it is necessary to organize locally and to do the things we have mentioned. It is necessary also to keep pressure on legislators about the things they can do. It is necessary in addition to bring the environmental lawsuits and to seek other new ways of bringing about change. For we must do both the immediate and the long-range at once, or perish.

If you have the money, seek out the Environmental Defense Fund,

162 Old Town Rd., East Setauket, N.Y., and help to support it, for environmental law is one more way to attack what we must learn to see as the enemy: the engineering mentality, the Mad Scientist syndrome, the corporate conglomerate control of our bodies and our lives.

The EDF and the other organizations which have joined it in the courts have established an extremely important legal precedent in America. For the first time, courts have agreed that the public has a direct legal interest, has "standing" to sue for its own protection and to force government agencies to do what they're supposed to do.

It used to be held that citizens only had standing if they could show that they were personally injured in some way. The new rulings make clear that the courts will recognize a general public interest. This is an enormous step forward in legal thinking, and it cannot be allowed to rest there.

It is barely possible that by the time you read this DDT and 2,4,5-T will have disappeared from the market directly because of the environmental-law approach. But there are 3,000 substances behind them that can yet be the subjects of suits; and there are attorneys ready to push some new legal concepts.

One possibility which the EDF is trying to explore is an interpretation of the Ninth Amendment to the Constitution, which says that just because some rights are enumerated in the Constitution, that doesn't mean that there are not other rights—and if there are, they are retained by the people. In an air pollution case brought by the EDF, they argued that the right to breathe clean air is a legal right under the Ninth Amendment, and that pollution is a "nonnegotiable hazard" because the people can't negotiate away their rights.

This is the kind of advanced legal thinking that usually causes conservative lawyers to snort in derision but often turns out to be the crucial Supreme Court decision of five or ten years later. At any rate it is being explored; and obviously the right to eat healthy food would be a simple application of the idea.

Another interesting area of the law that is being explored involves something called the "trust doctrine." This is actually a very old theory in law, and holds that all land was once held in trust for the people by the government. The crucial point is that the government cannot entirely give up responsibility for the way in which the land is used, even if it's now completely privately owned.

The doctrine is routinely recognized in narrow areas, as in the law about submerged lands, but has never been applied generally. There is some hope, though, that it might be. There was, for example, a case in Massachusetts in which the highway department was stopped by the court from using a public marshland as a right-of-way, even though state law seemed to permit it. EDF attorneys hope to be able to expand this area of law also.

In addition, courts have begun within the last five years to regard the environment, and ecological values, as valid considerations to be given weight in decisions, where not long ago they would have been considered irrelevant. One reason for this is the increasing willingness of scientists to be a part of organizations like the EDF. Where the private interests have always had highly paid experts, they are now likely to find themselves opposed by men who are equally expert and not suspect because of their position as employees; the "public's experts" are often appearing for nothing, or only for expenses, as their contribution to the general good. We all owe them a great deal.

All of this activity can be applied to questions about food, and particularly about the safety of food additives. The EDF should be supported, but any group can bring a suit if it can raise the money. There have even been cases where apparently disparate groups have joined for such a purpose; in one such case the local NAACP joined an ecology-oriented group to file a suit against a public development; one group thought it was ugly and the other thought it was discriminatory. No doubt the members of each group learned from each other in the course of the action, and there will be more help for each from the other in the future.

Certainly there are enough scientists willing and able, for expenses alone, to tackle the wearying and dangerous dominance of the Cannons, the Stares and the Mraks. Certainly an environmental-law drive to take the junk out of our food will bring out the more knowledgeable authorities who can challenge the industry's propaganda brigade of run-of-the-mill, small-time MDs and out-of-date "home economists."

We can imagine, for instance, a suit to force the FDA to a broader definition of a carcinogen, perhaps using one of the coal-tar dyes as a vehicle. We can imagine a suit to make them define the word "ingest" to include injections when they're interpreting the Delaney Amendment.

We can imagine a suit to make it impossible for manufacturers to keep selling us junk, or putting it into our food, while questions of its safety are being decided. We can imagine a test of the validity of the GRAS list. And of course we can imagine a series of tests dealing with the safety of particular additives of all types; maybe we could finally get rid of sodium benzoate after all these years.

More than that, we can imagine a suit whose purpose is to require that additives, those in use as well as those newly to be proposed, be tested *at the manufacturer's expense* for mutagenicity and teratogenicity. The manufacturer is supposed to show that they're safe; let him do so.

A suit like that could have several purposes and several effects. It could, if successful, result in the withdrawal of a number of additives, because it would be cheaper to quit making them than to test them. And it could be a deterrent to the constant flow of new and dangerous goodies from the Mad Scientists, who might not be willing to spend the money or to take the chance.

All of these actions, the private or small-group actions of our Consie Conspiracy and the public actions of the environmental lawyers, have effects of a social importance we should never forget: they benefit all Americans, including those too poor or isolated or uneducated to fight for themselves, and including the unfortunate children of those who think we're silly, paranoid, or quixotic. If a coal-tar dye is banned, it is banned for everybody, rich and poor, concerned and complacent, alike. If we force the store to carry UFWOC lettuce, none of the customers gets lettuce with insecticide residues.

Among those who are seeking out new legal routes, of course, are the Center for the Study of Responsive Law and its now famous founder, Ralph Nader. We all owe to Ralph Nader, and to those who have worked with him, an enormous debt for his constant, persistent efforts to bring responsibility to our regulatory agencies, for his continuing attempts to find new ways to influence private corporations, for his tireless efforts to make us all better informed.

Nader has accomplished some specific results that have made all of our lives just a little bit healthier. Besides that, he has made a great many of us much more aware than we were of what James Ridgeway calls "the horrors of the multi-national corporation." And in a crisis situation (for both the condition of our inner ecologies and that of the external environment are certainly at crisis stages) all approaches must be taken at once, and Nader has taken his approach and done a hell of a job.

But we are a little concerned about the adulation he gets, particularly among the young; we fear that they may come to see his approach as *the* approach, and we believe that there is in the long run a flaw in the Nader approach as there is a flaw in the approach of environmental law.

Ridgeway has pointed out that the small local groups that follow Nader are more populist than anything else, and that there is nothing about Nader's approach that keeps him from being embraced by followers of George Wallace. He also points out that the entire "Nader movement" apparently does not pose too much of a threat to the powers that be, since ambitious politicians from all directions are moving in and taking over its rhetoric.

One reason may be (and this is not to criticize Nader personally) that the movement often seems to ignore two overriding American problems: racism and poverty. Problems of pollution—of air, water, or food—are very much problems of the poor, who are often black and often urban. The coalition of minority organizations with ecology-minded organizations makes total sense in urban surroundings (who is more victimized by the supermarket than the ghetto resident?), but the Naderites do not seem to push it much.

Certainly Nader's movement is more populist than radical, more libertarian than liberal. But the flaw seems to us less complex than that. It was perhaps best summed up by Nader himself, as a guest on *The Dick Cavett Show*, when Cavett asked him what he thought Americans should do. "Organize yourselves into groups," Nader said, "and put pressure on your Congressmen."

Putting pressure on your Congressman is simply not going to do it. It never has. There is no use pretending that it ever will. You may indeed change his vote once or twice; you will not change the economic facts about the United States.

We have argued (amiably) with biologist Charles Wurster, who is one of the mainsprings of the EDF, about the extent to which it is useful to work within the limits of the governmental system. The record of the EDF is perhaps the best argument for his side: if DDT is banned, then it is banned and that's that.

But after DDT there will be something else, and something else again—from the same people, and in the same way. The kind of reform that Nader urges on the federal bureaucracy has been done; Tugwell was given *carte blanche* by Franklin Roosevelt to reform the FDA, and he did his best to turn it into a genuine consumer-protection agency and to obtain for it the legal muscle it needed.

Within a few years it was back firmly under the thumbs of the people it was supposed to regulate. The strong, courageous approach of environmental law and that of Naderesque reform are useful, even essential, in bringing about particular changes—the passage of a Delaney Amendment, the banning of a particular poison. But that is *all* that the approach does, or can do.

Of Nader and the EDF we must argue that there is danger even in their successes: the danger that we be dazzled into mistaking the particular for the general. The crucial distinction is that those successes do not bring about changes in the ways in which, for example, we and our children are poisoned.

That can happen only when *we* change, and the first and most important change that must take place within us is the realization that, in fact, the "system" does not work and cannot. Multi-billion-dollar organizations will always be able to control the agencies that are supposed to control them, except for a flurry of action perhaps every 20 years or so. Politics in any capitalist country will always be finally decided by money.

We all know that, really—as much as some of us may abhor "revolutionaries" in the streets or militants in cellars. To know that the system cannot work as advertised, however, is not necessarily to prepare for guerrilla warfare. It is only to take the next step, to admit that there is nothing we can do about it on our own, and to recognize the need for what we must see, genuinely and not in fantasy, as a kind of conspiracy.

The Consie Conspiracy cannot become something that can be taken over and used by a presidential aspirant. If it does, we're doing something wrong. There *is* a They, and there *is* a We, and we must always keep them separate. It has to be *our* Consie Conspiracy.

Our conspiracy, our revolution, doesn't have to drive us into the streets. All that is necessary is that it drive us into each other's lives.

Notes

Chapter Two: . . . and Half a Bag of Peanuts

1. Quoted by the Associated Press, in *The San Francisco Chronicle,* June 24, 1970.
2. *The San Francisco Chronicle,* July 5, 1970.
3. For a complete rundown on vitamin research, see Longgood and von Haller.
4. Quoted in Rorty and Norman.

Chapter Three: Look, Ma, No Cavities!

1. October 26, 1949.
2. In June, 1971, anthropologists discovered a "lost tribe" called the Tasaday in an isolated mountain rain-forest on Mindanao, in the Philippines. None had ever tasted sugar or salt. Dr. Norton J. Winters, an American dentist who has practiced for 20 years in Manila, examined their teeth. "Not one of the men," he said, "had a single cavity." So of course the first thing one of the anthropologists did was to offer one of the Tasaday some sugar. Associated Press story in *The San Francisco Examiner,* July 25, 1971.

Interlude: In Mournful Numbers

1. Except where noted, figures used are the latest for which reliable data are available. They come from all the standard sources: the United Nations agencies, the U.S. Public Health Service, life insurance companies, etc. Unrecorded births and deaths don't show up, of course. It should also be noted that some figures are affected by accidental deaths, which were just over 6 percent of the deaths in America in 1967. This not only may effect conclusions about health (though not too much: people of all ages die in accidents), but certainly affects comparisons with other countries, since America's accident death rate is the highest in the world except for West Germany's.
2. *The New York State Medical Journal,* September 15, 1955.
3. We hasten to note that contrary to popular impression, the U.S. motor vehicle accident rate per car-mile is low relative to that of most other mechanized countries (New Zealand is horrendous), and has steadily *de*creased, not *in*creased, over the last 40 years. In world terms Americans are superb drivers. But car statistics is another interlude for another book. Maybe someday.
4. We are skipping the semantic examination of just what constitutes "heart disease," but if you ever meet a heart specialist, it's a fascinating subject. The definition just isn't all that clear in a number of cases, and then there are special subjects like fright or shock which further complicate the definitions. Since most cases are relatively clear, we'll just stay with the published figures.

Chapter Four: Ethylenediaminetetracetic Acid and Its 2,999 Friends

1. James Noonan, "Color Additives in Food," in Furia.
2. As mentioned earlier in the text, a very few chemicals are listed twice when they are used in two different and unrelated ways. The list is Publication 1274, NRC-NAS.
3. They are azo dyes, triphenylmethane dyes, fluorescein type, and sulfonated indigo. Furia has chemical diagrams. We are also ignoring the technical distinction between "dyes" and "lakes."
4. A fairly recent rundown on irradiation, FDA action, and the results of conflicting studies is offered by Edward Gross in *Science News*, March 22, 1969.
5. Bicknell quotes a British authority, R.S. Hannan, as follows: "[I]onizing radiations are potentially carcinogenic when applied to living animals, even at very low dose levels. There is no positive indication that this effect is produced through the medium of a stable carcinogen or that such a carcinogen, if formed, would be transmissible to a consumer of the irradiated tissues, but the possibility cannot be discounted unless exhaustive testing has been carried out."

   Testing for this result, and the many other potentially dangerous results, from irradiation of all the foods for which the process has been suggested is probably more than even America's laboratories could handle. But it is in just such situations that the industry likes to start talking about "unnecessary" testing.

Chapter Five: Progress Is Our Most Important Murphy

1. Mahoney quotations from J.F. Mahoney, "Industry's Needs for Food Additives," in *The Safety of Foods*, edited by H.D. Graham. Mr. Mahoney's title is more candid than his article.
2. *Life*, March 6, 1970. Stare said the additives are safer than the natural foods because natural foods are sometimes prepared badly or eaten unwisely. He also says manufacturers should put salts into baby foods, rather than have mothers do so individually "in an uncontrolled manner."
3. It is Northridge brand "wheat berry" bread, distributed in the San Francisco Bay Area by Oroweat. There is also a Northridge white bread, among other varieties; it also keeps well, and likewise has no preservatives. It *is* well wrapped.
4. Ingle and Mahoney are in Graham.

Chapter Six: The Webs of Anacharsis

1. Turner has written an extremely valuable book about the FDA, its inner machinations, and its almost constant obeisance to the food industry. Our last chapter, as will be seen, disagrees strongly with his, and Ralph Nader's, analysis of what it all means and what must be done about it. But Turner's facts are devastating and detailed, and if you haven't read his book you should. Some of this chapter and the next is drawn from Turner.
2. House Report No. 2284, Eighty-Fifth Congress, Second Session, 1958.
3. Start at 10 CFR 21 and keep going through 53 CFR 21 if you'd like to know what some of the standards are.
4. Catsup is 53.10, and tomato paste is 53.30, in CFR 21.
5. This quotation from the *CRC Handbook* is from the Introduction, written—in

terms so proindustry as to be almost unreadable—by T.D. Luckey, a University of Missouri biochemist.

6. 1.3 CFR 21.
7. *Food Processing*, March, 1961.
8. *The San Francisco Chronicle*, January 24, 1970.
9. *The Federal Register*, July 28, 1970.
10. A story finally moved on the wire service of *The Chicago Daily News* in late August, appearing in *The San Francisco Examiner*, August 30, 1970.
11. 8.12 and 8.36 CFR 21. 8.13 and 8.14 are procedurally related. 8.36 is the one that can fool you if it's quoted by itself.
12. The standards are in 19.765 CFR 21 and 19.750 CFR 21, respectively. In either case, incidentally, the emulsifier can constitute as much as 3 percent of the solids in the product.
13. For technical reasons, the number "17" is a little arbitrary. There are other ways you could count them that would make the number higher.
14. Turner describes at some length the industry pressures that put "the Coca-Cola Amendment" into the regulation.

Chapter Seven: "Is She Not Pure Gold, My Mistress?"

1. It is irrelevant, but interesting, that we found a reference to the adulteration of coffee with "oats, peas or beans, or any other adulterant" in a royal decree issued by Louis XIV of France in 1692. Merchants don't change much.
2. The Division of Chemistry had been upgraded to a Bureau during Wiley's tenure. Responsibility for enforcement of the laws we're talking about passed to a new Food, Drug, and Insecticide Administration, within the Department of Agriculture, in 1927. It became the Food and Drug Administration in 1930, was transferred out of the Department of Agriculture to the Federal Security Agency in 1940, and was placed under HEW when that department was created in 1953.
3. The Dunbar-Wiley controversy is reported in detail by Anderson.
4. Some producers and retailers, including Co-op outlets like ours in Berkeley, practice grade-labeling voluntarily on most products. It should be noted that a lower grade may be just as good nutritionally. Size, appearance, and other factors are also involved.
5. He did ask Washington to find out a couple of things for us, and was otherwise helpful.
6. Quoted by Turner.
7. Including the times when it harasses some small but pesky competitor to the big boys, as it has done with tiny, nonprofit Pacifica Radio for years.
8. See my *America the Raped*, Chapter VI.
9. *Science*, January 16, 1970; article by Andrew Hamilton.
10. *Chemical and Engineering News*, October 10, 1966.
11. 3.41 CFR 21.
12. Breakfast food: Kellogg, General Foods, General Mills, Quaker Oats. Cheese: National Dairy (Kraft) and Borden. Salad dressing: National Dairy and Corn Products (Best Foods). An aside on Campbell: they now sell both their familiar "condensed" soups and some ready-to-eat soups in the same flavors. The ready-to-eat soups cost a great deal more per pound. The principal difference, however, is that Campbell has added the water!

13. Maximum presence of mineral oil in yeast is given in the FPC's list as 0.015 percent, in the *CRC Handbook* as 150 parts per million.
14. On the FDA private-eye bit, there are numerous sources. See for instance the hearings on invasion of privacy by government agencies, Subcommittee on Administrative Practices and Procedures of the Committee on the Judiciary, United States Senate, Eighty-Ninth Congress, 1965.
15. William Turner, *Hoover's FBI* (Los Angeles: Sherbourne Press, 1970). Bill Turner, an ex-FBI man, is also a former colleague. A lot of publishers were afraid of his book; but if your bookstore doesn't have one, they can order it, and you can find out why. And should.
16. *The San Francisco Chronicle*, September 1, 1970.

Chapter Eight: The Hopeful Ostrich and the Hypersusceptible Mouse

1. A fuller treatment of the political—as opposed to the scientific—furor surrounding cyclamates can be found in Turner's first chapter.
2. There were fewer than 100 rats in the FDA tests (which is not necessarily too few), and six such rare tumors. The FPC, in a 1959 pamphlet, says that in 75 animals, a 5.3 percent incidence of tumors would be statistically significant, and in 100 animals, 4 percent would be enough. The pamphlet (*Problems in the Evaluation,* etc.; see bibliography) is specifically concerned with carcinogenic hazards; all FPC quotations in this chapter are from the same source unless otherwise noted.
3. For rundowns on the scientific data used in this section, see *Science News* for December 7, 1968, and October 25, 1969, and *Consumer Bulletin* for November and December, 1969.
4. The Legator quotation, and the story of the memo, are from Turner. There has been some disagreement about who changed the memo, so we have omitted the name.
5. The Long Island laboratory in question does between $1 and $3 million a year worth of business for customers in the food and drug industry. Its president is Dr. Bernard L. Oser, also a member of the Food Protection Committee.
6. Quoted by Jean Carper, a former Congressional committee staff member whose article is quite clear on animal experiments.
7. Quoted by Jean Carper.
8. One paragraph in the AP story (see *The New York Times,* June 24, 1970) cited "an agency source," but obviously the reporter already had the story and simply got official confirmation for one fact.
9. The products of course are still marketed, but they don't taste the same.
10. It is interesting, and not irrelevant to this book, that once the political argument over nuclear testing was settled, the "experts" who thought there was a "lower threshold" for genetic damage all suddenly disappeared, and it is now completely accepted that *any* amount of radiation, however small, causes some genetic damage if it has a chance.
11. It was, to be strictly accurate, the company's Naugatuck Chemical Division, perhaps most famous for Naugahyde.
12. Quoted by Longgood.
13. This was discovered by accident. Some animals were kept in wooden cases that had been treated with creosote, and it was only a researcher's zeal at finding *all* the variables in an experiment that led to the discovery.

14. There is also a very recent study by Dr. Steven Martin, involving a mutation of Raus sarcoma virus in chickens, in which cells rendered cancerous by the virus at one temperature appear to return to normal at another. This is a tentative finding and its meaning is not well understood at this writing; it has been reported in *Nature* (Fall, 1970).
15. We are leaving out of the discussion the fact that repeated injection itself, at the same site, may cause subcutaneous cancers in some animals regardless of the substance injected.
16. Cannon and Mrak both contributed essays to Ayres *et al.*
17. A chemical formula doesn't necessarily identify. It depends on how the atoms are joined in the molecule. C_2H_6O, for instance, may be ethyl alcohol or methyl ether, compounds with markedly different properties.
18. The single and double lines in the diagram represent single and double covalent bonds. Obviously we can't explain organic chemistry here; but we would like to emphasize, for those who are frightened by its formulas, diagrams, and strange words, that basic organic chemistry is a fairly simple subject. You need no prior training (except the ability to read English) to sit down with a book like Holum's for an hour a night, starting at the beginning and doing all the exercises; you'll be astonishing yourself in two weeks and incidentally learn a lot that can help in understanding arguments about food. If two people work through the book together it's more fun than dirty-word Scrabble.
19. Don't be misled. Some aromatic hydrocarbons which are *not* polycyclic, like benzene itself, are nonetheless carcinogens.
20. Fifteen of the 26 FDA-cleared flavorings are also listed as GRAS by FEMA. The three on the FEMA list for which we can't locate an FDA approval are beta-naphthyl ethyl ether, beta-naphthyl anthranilate, and 6-methylcoumarin. The first is also known as 2-ethoxynaphthalene, or nerolin. Source for the structures is the FPC's *Chemicals Used in Food Processing;* it also lists a thirtieth polycyclic aromatic hydrocarbon used as a flavoring, beta-naphthyl methyl ether, but that one is not listed in the *CRC Handbook* and we can't find an authorization for its use.
21. The booklet says that these chemicals are "the most investigated group of carcinogens," which is true. But it also describes in some detail how limited the tests are and how few are the compounds tested. See also the booklet's source list for this section.
22. This is a distorted pseudoexperiment in several ways—it makes no mention of the amount ingested, for just one instance—but the principle involved in the argument is sound despite these oversimplifications. It is not an attempt to describe a real experiment, but only to show, in a gross way, a kind of mathematical relationship.
23. Quoted by Jean Carper. It should be emphasized that Dr. Saffiotti was and is in the department that was headed by Finch, and that Saffiotti's evaluation of the cyclamate experiment was available to Finch before the Secretary made his various misleading statements.
24. If the coefficient of risk (P) is 0.025, if there are *no* effects in the control group, and if all 200 animals live until the experiment is complete. The surer you want to be of "significance," the more animals you need to make sure that your 4 percent is meaningful. That decision is in a sense arbitrary, but $P = 0.025$ is often used in tumorigen tests on animals.

25. P = 0.05. In testing for carcinogenic activity, the statistical requirement might be more stringent—which is good for avoiding error, but bad if it means that a substance may be called "safe" because stronger "proof" of carcinogenicity isn't present.
26. More complete data are in the FPC booklet (see note 2).
27. See *The New York Times*, March 20, 1970, and *The San Francisco Chronicle*, July 23, 1970. Neither *Times* reporter Richard D. Lyons nor the unnamed *Washington Post* reporter whose story was reprinted in *The Chronicle* (nor their editors, copy desks, etc.) appear to have been aware of earlier studies on saccharin and cancer.
28. The report is quoted by Turner. It is interesting, though probably not significant, that the association of saccharin is with *stomach* cancers, and that since the widespread use of cyclamates began, there has been a decided drop in the number of stomach cancers in America.
29. It should be noted that saccharin, besides tasting awful, doesn't provide in soft drinks the "body" that sugar does, and it's likely to feel strange in your mouth. "Diet" soft drink manufacturers therefore include other additives to provide body and what they charmingly call "mouthfeel." The combinations are known (see Salant in the *CRC Handbook*), but of course no synergistic tests have ever been made.
30. *Science*, November 7, 1969. Abelson emphasized that he was not making an official statement for the American Association for the Advancement of Science, which publishes the magazine.

Chapter Nine: Dirty Pool

1. J.B.S. Haldane traced the family connections given here; see the chart in Papazian. Edward, Duke of Kent, is not the only possible mutant, but the most likely. Hemophilia is a sex-linked recessive characteristic, which means in effect that women carry it as a single gene but only sons (half of them on the average) will be hemophiliacs. A woman can be hemophiliac only if both her parents carry the gene and she happens to get them both—a chance of one in four. The present British royal family is descended from Edward VII, a son of Queen Victoria who did not receive the hemophilia gene, and is therefore fortunately free of the trait.
2. Rather than continue to write apologetic footnotes, let us say right now that biologists and geneticists may wince a little at some of our formulations in this chapter. Of necessity we have to omit some possible qualifications and oversimplify some processes; the alternative would be to write a book on genetics. As an example, we later use figures which are guesses at best, though they are guesses by experts. All we can do is ask forgiveness of the expert; the lay reader who would like to learn more can do so, without any prior knowledge, from Papazian or Moody.
3. Senator McGovern is quoted from an Associated Press story in *The San Francisco Chronicle* for October 25, 1969, the same story which reports the action of the Baby Food Kings. The same story includes the sentence, "There is no proof it [MSG] is harmful to human babies," another example of misunderstanding the meanings of "proof," discussed in the previous chapter.
4. So was I, once. See Marine, "Molecules and Mental Illness," *The Nation*, April 20, 1957 (condensed as "New Clues to Mental Illness," *Science Digest*,

July, 1957). I should perhaps note that the excitement of spending some time, during the preparation of that article, with Dr. Linus Pauling—surely one of the great men of our age—sparked an interest in biochemistry which has never gone away and which has a lot to do with our being able to write this chapter.

5. MSG is, of course, the sodium salt of glutamic acid.
6. When it was first discovered that glutamic acid is used by the brain, experiments were conducted in which extra glutamic acid was fed to children in the hope of increasing their intelligence. It sounds a little like Dachau now, but it didn't bother anybody at the time. Nobody knows what may have happened to the children's endocrine systems.
7. At about the same time, people began to hear about something called the "Chinese restaurant syndrome"—so called because monosodium glutamate, originally developed in Japan (by someone who wondered why seaweed tastes so good), is widely used in Oriental cooking. The syndrome involves symptoms similar to those of food poisoning, and is experienced by some—not all—people when they eat rather large doses of MSG with their food. It may be an allergy rather than food poisoning. If so, it is another example of how people who suffer from allergies are unable to avoid allergens in their food, since MSG does not have to appear on the labels of all products, and no one can tell when it's used in a restaurant.
8. Huntington's chorea is a hereditary disease that appears in late adulthood. Presumably the "defect" is present in some form at birth, and therefore technically "congenital," but the word usually means that you can see the defect in the newborn baby.
9. Acheiropody—unlike hemophilia—is not sex-linked. *Both* parents must carry an acheiropody gene before the condition can show up. Like all such conditions ("autosomal recessives"), it is relatively rare except in situations involving much inbreeding; in fact, it has been observed so far only in Brazil.
10. There is one other possible situation: teratogenic damage to the *germ* cells of an unborn child could turn out to be mutagenic. The result would show up in the child's child, if "dominant."
11. 1966 figure from Furia; current figure an estimate from *Consumer Bulletin*, March, 1970. In 1966, more than 220 million pounds were produced worldwide, much of it in Asia; the Asian commercial product is believed to be somewhat different from ours.
12. It should be noted that the *Consumer Bulletin* story of March, 1970, erroneously describes the Olney studies as reported in *Science*. Scientific errors, sometimes minor and sometimes not, crop up in other *Consumer Bulletin* stories on food additives as well; readers would do well to check original sources. Note that we are *not* talking about *Consumer Reports*, an unrelated publication.
13. The relationship between hypertension and high salt consumption is not established, but is widely accepted, based on a series of experiments by Dr. Dahl at Brookhaven National Laboratory. The quotation is from *The Washington Star*, October 29, 1969.
14. It should be clear that there is *no* analogy except chemically. It has nothing to do with the value or danger of any salt or acid as a food. It's only an illustration of how an innocuous-sounding phrase may conceal content for many people.

15. As an indication of absurdity, LSD provides a superb example. Not long ago, widespread publicity was given to a report that LSD might cause broken chromosomes (explained later in this chapter). Since few critics of LSD have been able to find too much wrong with it, this rumor is still widely used in support of the illegality of LSD. But in *Science* for August 7, 1970, three French scientists, Roux, Dupuis and Aubry, reported on an exhaustive teratogenic test of LSD—1,713 fetuses of three kinds of animals. The tests "failed to prove any abortifacient, teratogenic, or growth-depressing effects."

    That doesn't, of course, "prove LSD safe"—which can't be done (and we hasten to add that the impure stuff clandestinely circulated—a consequence of illegality—is apt to be extremely dangerous for other reasons). The point is that if such a test were done on MSG, any suggestion that it be banned as a teratogen would be met, properly, with howls of derision. LSD is illegal because it is part of the youth culture, not because of demonstrated harmful effects. MSG is legal, despite strong scientific evidence of potential danger, because it's big business. Maybe if people under 30 could get high on MSG, they'd ban it.
16. Spuhler's figure and a discussion based 20 years later are in Moody.
17. If you spend the time with Holum suggested in Note 18 for the previous chapter, you will learn along the way much more detail about genes, chromosomes, and reproduction than you will here. If you then want *more* detail, Moody, Papazian, McKusick, and Stahl are all in language clear to the layman without previous training. They take a little concentration, of course.
18. Just in case you should run into the terms somewhere, *genetic* mutations are sometimes called *germinal* mutations, and *gene* mutations are sometimes called *point* mutations.
19. A somatic mutation can take place in one of the cells of an embryo or a fetus after that cell has taken on specialized functions. This might cause the baby to abort, to be born dead, or to be born with something wrong with it, but it would have nothing to do with heredity. A substance or event which caused a somatic mutation in an embryo or fetus would thus be teratogenic in effect, not mutagenic, as we are using those terms.
20. To be strictly accurate, we aren't sure whether it's chromosome No. 21 or No. 22 that's involved. But it seems to be the same one that's damaged in myeloid leukemia cases. As we mentioned, mongoloids have a higher incidence of leukemia than is "normal."
21. Most authorities believe now that if there are broken chromosomes in the somatic cells, there are probably also broken chromosomes in the germ cells. There are tests for finding broken chromosomes in humans. As you can gather from the text, this is an area of much current research, and it is impossible not to report many uncertainties.
22. Moody explains all of this in clear detail, with excellent diagrams. Papazian diagrams and explains some variations of "crossing over."
23. Moody goes into the eye-color question in detail. The general idea is that blue eyes and brown eyes are not really different characteristics at all, but simply involve relative amounts of pigment. Combinations of genes work in such a way that eye color often *appears* to follow a Mendelian single-gene, dominant-recessive pattern, but in fact it does not—so that the exceptions are not really exceptions (and therefore not socially suspicious). Green and hazel eyes, and flecks, are accounted for as well.

24. These chemicals have all been shown to be mutagenic in some creatures—not necessarily in humans. As we hope is becoming clearer, there is no easy way to test for mutagenic effects in humans. This is the importance of the fact, shown a bit later, that the same chemical, DNA, is the hereditary substance in fruit flies, elephants, bacteria, coyotes—and men.
25. There is other protein material associated with the DNA in a chromosome, and probably more than one single molecule of DNA. The role of the protein material is not completely clear, and the explanation here should not be taken as *the* explanation of heredity. Learning is progressing more rapidly in this subject than in any other area of science; it is literally impossible, even for an expert in the field, to write an up-to-date book. Things happen too fast.
26. Technically the two run in opposite directions—i.e., their polarity is opposed—but that's not crucial here.
27. This does *not* mean that the old racist myth about a white person, marrying a person who is partly black but has a light skin, having a "coal-black baby" is true. That would involve such astronomical odds that it would be wiser to bet on a touch-football team of ten-year-olds beating the Green Bay Packers.
28. The number of chromosomes, and the total *amount* of DNA, are of course different. But DNA is DNA.
29. "Temperature rising" *means* that molecules move faster. One doesn't "cause" the other: they're the same thing.

    We have not mentioned all types of mutational activity. See especially Moody.
30. "Proteins" aren't magic either. It's just the name for any chemical with a certain kind of structure. Holum makes it clear. Papazian has an excellent chapter on enzymes, what they are and just how they work; no special prior knowledge is necessary.
31. If brothers and sisters mate with other, the odds are one out of four that their offspring will be homozygous for any particular gene. With first cousins the odds are lower, but still much higher than in the general population with random mating. Hence, where there is "inbreeding," recessive characteristics turn up more frequently. This is easy to observe in breeding animals, once the dominance patterns are known.

    "Sex-linked" recessives, mentioned in Note 1 for this chapter, are exceptions to the rule in the text. Cf. also Note 9.

Chapter Ten: No Other Gods

1. Children in sizable cities with ice cream specialty outlets may be exceptions; but sometimes the only difference they encounter is in price.
2. We have only just begun to find out about pesticide residues; but then, "equally dangerous" is a silly concept. You can only be killed so dead.
3. Although the opposite is not true. "Uncertified" colors may be synthetic, as noted later in the text.
4. From the beginning, ads quoted are for Green and Green, Inc., Stange Company, Ac'cent International, and H. Kohnstamm and Company, Inc. The others are identified in the text.
5. Formally, diacetyl is 2,3-butanedione. It's GRAS.

6. 21 CFR 8.301 and 8.306. The regulations provide for supplementation with "xanthophyll and associated carotenoids," which may incidentally provide some Vitamin A—but that isn't why they're there. They're there to help fake the color, as the regulations explicitly make clear. Other additives are included in the same regulations.
7. Mystery fans will find a novel fictional use of cochineal in Alan Caillou's recent super-intrigue novel, *Assault on Loveless.*
8. Honest. See 21 CFR 31.1, paragraphs (b) and (c)(2). The FPC list gives a lot of chemicals used to provide grape flavor, including some, like 3-phenylpropyl cinnamate, which are polycyclic aromatic hydrocarbons. See Chapter Eight.
9. 21 CFR 8.201. The chemical name is 1-(2,5-dimethoxyphenylazo)-2-naphthol. Small amounts of other substances may be added.
10. There are an increasing number of what might be called "overlaps" in the food industry. The largest producer of frozen orange juice, Minute Maid, is now owned by Coca-Cola. Some readers may recall that their treatment of agricultural workers in Florida was featured on a 1970 television documentary that caused something of a flap.
11. For example: octyl acetate, manufactured for wide use as a fruit flavoring in beverages, ice cream, candy, and baked goods, occurs naturally in the oil of green tea laves.
12. Indigotine is called indigo carmine in England. The chemical names of the dyes still in use are in Part 9 of 21 CFR. We will not use them here, for reasons that may be evident if we tell you that FD&C Yellow No. 5 (tartrazine) is technically the trisodium salt of 5-hydroxy-1-*p*-sulfophenyl-4-*p*-sulfophenylazopyrazole-3-carboxylic acid. If you think that's bad you should see Green No. 3.
13. Quoted by Longgood.
14. 21 CFR 8.34.
15. Also provisionally listed are the "lakes" of those nine dyes. You combine the dye with alumina hydrate or aluminum hydroxide, and you get a "lake"—in effect an insoluble form of the dye, that colors by dispersion instead of by solution. Beyond that we cannot go without getting highly technical. Lakes must be made from certified dyes, and must then be separately certified after manufacture. The dye content may vary: 21 CFR 9.100.
16. We don't mean to give the impression that the more benzene rings, the more dangerous the product. It doesn't work that way as far as is known. It's just that two or more indicate a strong carcinogenic potential.
17. 21 CFR 8.12, 8.13, and 8.14. There are a couple of other fast shuffles in the regulations. Part 8.12 says that anyone who is adversely affected by the cancer clause can *ask* for an advisory committee, but has to post a fee in advance (which turns out to be at least $2,500). Part 8.50, however, gives the FDA Commissioner the right to "waive or refund" such fees any time he wants to. The advisory committee, incidentally, is bound to silence about the information made available to it until a final decision is made and a regulation issued. And Part 8.36, read by itself, sounds as though the full Delaney Amendment applies; the other Parts, however, modify it in fact—though it's handy for quoting.
18. Using a blue dye doesn't mean that the food comes out blue. A small amount

of blue might be used, for instance, to counter an undesirable tinge of yellow. Combinations, too, are common.

19. After this section was written, a United Press International story in *The San Francisco Chronicle* for January 4, 1971, reported a letter from Ralph Nader to the FDA suggesting carcinogenic danger from FD&C Violet No. 1. The news story, very short as printed and accordingly undetailed, mentions a Canadian study in 1964 showing tumors in rats after ingestion of Violet 1 (but doesn't say what is meant by ingestion). Also mentioned by Nader was information regarding possible cancer danger from Citrus Red No. 2, the orange dye; fears about this dye have been voiced before.
20. "At present," Bicknell writes, "it seems probable that any dye is capable of producing cancer if it is broken down by the body with the formation of an aromatic amine with the position *para* to the amino-group blocked by a large substituent, so that subsequently the cancer-producing *ortho*-hydroxamine group is formed. Such a breakdown can be seen to be probable with the common hydroxyazo dyes . . ."
21. Not all surfactants, technically, are emulsifiers. Some are antifoaming agents, or catalytic agents to enable other emulsifiers to work. But emulsifiers are by far the largest class of surfactants, in terms of volume in use. The possible specific dangers from the others, aside from the general danger of introducing *any* strange chemical into our food, are simply beyond this book, which necessarily has to stick to those additives which are most common, or most dangerous, or both.
22. Where it becomes a concern of those who worry about the *outer* ecology—but not nearly so important a concern as the residual detergent itself. The grease is usually biodegradable.
23. A common kitchen emulsion (besides dishwater) is mayonnaise: normally olive oil or corn oil won't mix with water, but in the presence of egg yolk—the emulsifier—they blend pretty well. What actually happens is that the protein molecules of the egg yolk "surround" the oil droplets so that they won't coalesce, and thus they stay uniformly suspended in the water. Put differently, you can say that the egg yolk takes big globs of oil and turns them into little globs so that they are easier to suspend in water—and incidentally easier to digest. Detergent emulsifiers work somewhat differently, as described in the text, but the principle is the same.
24. One of those pieces of intelligence that doesn't fit anywhere else in the book is that the familiar plastic polyethylene (used in making a lot of toys, for instance) is an approved food additive (GRAS) as a "plasticizer" in chewing gum.
25. Mayonnaise may be emulsified only with an egg-yolk product. Other common emulsifiers that don't fit neatly into the two categories under discussion are lecithin and carboxymethylcellulose. The latter is a carcinogen by most non-FDA definitions, and is mentioned in Chapter Eight.
26. Large libraries may have the FPC's *Chemicals Used in Food Processing*, and you can look on your own for the multiple benzene rings. But it won't help when a food label says only "artificial flavoring." Almost all of those we spotted are used in beverages, candy, ice cream, and baked goods, among other uses. "Baked goods" as used here includes mixes.
27. Mr. Salant, incidentally, also talks about the problems involved in making "dietetic" or "low-calorie" salad dressings taste like the real thing. Along

the way we learn that the low-calorie dressings "differ from their high-calorie counterparts as a result of the reduction or complete elimination of the oil content. . . . In the case of the French-type dressing there is no vegetable oil used at all." It should be pointed out, for those who watch their weight, that the federal standards for salad dressing *do not apply* to the low-calorie product.

Chapter Eleven: Black Is Beautiful

1. "Niacin" is used in this chapter interchangeably with "niacinamide," though they are not precisely the same thing (one is an acid, the other an amide). There is no evidence that one functions differently from the other, and we will treat them accordingly, though it should be emphasized that knowledge of the functioning of the B-complex is still a lot more sketchy than some people would like you to think. Niacin is also called nicotinic acid (the word "niacin" is used partly because the Mad Scientists are afraid you wouldn't like the look of "nicotinic acid" as an additive), and niacinamide is also called nicotinamide.
2. *The San Francisco Chronicle*, September 1, 1970.
3. There are long and learned essays on why Melville chose an opposite symbolism for *Moby Dick*, but that is one of the very few exceptions.
4. Genesis XVIII:6.
5. If you learned it as "Coventry Cross," don't feel bad; that was probably the original version.
6. Tortuous as the process was, it was more efficient by today's standards than was the milling of flour. Emperor's Sugar had *all* the vitamins and minerals refined out of it, just as white sugar has today.
7. When wholemeal bread was the wartime standard in England, Bicknell refused to believe that the good fortune of Englishmen could continue: "For feeding animals, the germ has such high value, that it is unlikely that it will be retained in English bread in peace times, since the feeding of pigs is more profitable than the nourishing of human beings."
8. The work of McCance and Widdowson, which was perfectly respectable as far as it went, was largely responsible for the phytic acid bugaboo.
9. 15.1 CFR 21. Even "whiter" flour, ground finer, and with nothing added to it, is called farina. The same fineness of grind, using durum wheat, is called semolina. Neither is worth a damn nutritionally. 15.130 and 15.150 CFR 21.
10. Ascorbic acid may be used as a "dough conditioner," and must be shown on the label if so used. The standard does not provide an extraction percentage; it describes in effect the size of a hole through which 98 percent of the flour must pass. It's too small for anything good to get through.
11. As much as we would wish for everyone to understand nutrition and metabolism, there just isn't room for the complexities of the fatty acid cycle and the attendant citric acid cycle. It's all in Holum.
12. 17.1 CFR 21. All references to the standard for bread are to this Part. "Enriched bread" is 17.2, "whole wheat bread" is 17.5.
13. 15.80 CFR 21.
14. This started in 1925 and went on for many years afterward. AMA spokesmen regularly argued that anyone who claimed superior nutritional value for whole wheat flour was a "food faddist" and a "quack"—despite the

fact that all serious scientific evidence was (and is) to the contrary. The campaign still echoes, as in Dr. Stare's nonsensical references to "key vitamins."

15. Which is why so many basic food discoveries (canning, and the use of beets as a source of sugar, are examples) were made under Napoleon in anticipation of a blockaded France.
16. *International Clinics*, December, 1938.
17. This is technically inaccurate. Enzymes are almost always proteins, but not *entirely* proteins. The protein portion of an enzyme is called the apoenzyme. The nonprotein portion may be a metal ion, in which case it's called an activator, or it may be an organic substance, in which case it's called a coenzyme. Thus an apoenzyme plus a coenzyme equals an enzyme. It just seemed to us a little easier to follow if put the other way.
18. The works of the nutritionist Adelle Davis are highly controversial, and some of them should at least be read with caution (*e.g.*, in *Let's Get Well*, she is simply wrong about the bacteria in yogurt, unless *we're* misreading something). The book *Let's Eat Right to Keep Fit*, however, has an excellent description of vitamin balance, and on the whole we'd recommend it.
19. The standards exist both for "enriched white bread" and "enriched white flour," so it may be the miller who does the adding. The calcium option was put in partly because of the phytic acid fear described earlier in the chapter.
20. At the Berkeley Co-op. There are Co-ops in many parts of the country, and they may sell it. The one in Manhattan is on Sixth Avenue approximately opposite Waverly Place; I mention it because a surprising number of New Yorkers don't know that it's there.
21. Quoted by Brown.
22. So what about rice in your kitchen? Brown rice is best. Don't in any case use nonenriched white rice. If it's a major part of your diet, you risk beri-beri; if it's a steady part of your diet, you risk a "subclinical deficiency" as described in the text. If you keep rice around, you should know that it can lose up to 30 percent of its Vitamin B in room-temperature storage (and the oils in brown rice can turn rancid). Either use it quickly or keep it in a tight container in the refrigerator. Properly stored, tight and very cool, rice has kept its vitamins intact for 100 years. Cook it in a double boiler or otherwise away from the heat, by the way; the closer it gets to the heat, the more its B-vitamin content drops.
23. Rice loses about 80 percent of its thiamine in polishing.
24. To avoid any possible confusion, this might be a good place to note that vitamins, as the word is now used, aren't even necessarily precise substances. Just as niacin (an acid) and niacinamide are considered interchangeable for most purposes, so there are at least two forms of Vitamin A, three of Vitamin D (Vitamin D_2 is also called calciferol), and two of Vitamin K.
25. And polio, which they said was epidemic among Texas migrant workers despite the availability of vaccine. Associated Press report in *The San Francisco Chronicle*, July 18, 1970.

Chapter Twelve: Who Dies in the Desert?

1. This is an extremely difficult thought habit to break. See for example the "explanations" of the different vitamins in *The Columbia Viking Desk Encyclopedia*.

2. Those more familiar with the scientific terminology will get the whole point from Schrödinger's classic sentence: "Life feeds on negative entropy."
3. Some people have used this concept to lead to the idea that artists go beyond physical science, by "organizing" otherwise chaotic images or sounds into orderly forms. This and similar concepts are fun to play with and may be important, but the spinning of philosophies and aesthetic theories from scientific processes is an intellectually dangerous game. Look what they did to poor Darwin.
4. There are more serious symptoms for serious overdoses. He also says—in case you're wondering what not to eat—that "the acute form of Vitamin A intoxication has been noted in persons ingesting . . . large quantities of polar bear liver." For a thorough discussion of hypervitaminosis A and a lot of other things related to this section, see Tatkon. He believes that added vitamins (as in pills or in "fortified" or "enriched" foods) are almost never justifiable and almost always dangerous. He makes a strong argument.
5. On Vitamin E and wheat germ: Vitamin E is destroyed by rancid oil. In wheat germ, the oils can go rancid. By the time you notice it, the Vitamin E is already gone. If you buy wheat germ separately, it should be kept tightly covered in the refrigerator—or used up in a few days. The same is true of whole wheat flour. And even if you do use it quickly, don't keep it near the stove in the meantime.
6. *Let's Get Well.*
7. See *Consumer Bulletin,* December, 1969.
8. It can be argued—I've argued it at symposia and whatnot—that pollution or population pressure or even nuclear warfare, all the concerns of the "ecologically aware" today, may wipe out a lot of us and all of what we call civilization, but they will not render the *species* extinct. Changes in food intake, spreading to wherever there are people, very well may. Of course we *can* do it with things like DDT or by rendering the air ultimately unbreathable (and maybe already have), but there is not the same kind of logic about it. But then many readers may not give a damn whether the species survives merely to do it all over again. Personally, I'd rather the leopards and dolphins make it and we don't.
9. Some cereals found in some health food stores are excellent. They just aren't that readily available to most Americans—and we have no space for evaluating them in detail. If you can't find out how they're made, nor of what, don't spend your money.
10. *Not* the individual packets over which you pour hot water.
11. No relation to the West Coast health food cereal, Crunchy Granola, made in Paradise, California. Crunchy Granola, besides tasting good, is nutritionally as good as *good* bread, though the sugar in it doesn't do much.
12. Moskowitz notes that Kellogg's profits fell behind the total profits of Ralston-Purina only because of the Salada acquisition.
13. In the real world, these chains are very long, and a completely saturated vegetable oil would be brittle, like tallow. Commercial hydrogenation leaves a few double bonds (exactly where they are along the chain doesn't matter). As generally used, then, the terms "saturated" and "unsaturated" are relative, not absolute.
14. Products high in polyunsaturates are sometimes described as "linoleates." You can figure out the percentage of unsaturated fatty acids in any given

oil from its melting point; the lower the melting point, the more unsaturates. From the least saturated to the most, the order of some common ones is linseed, soybean, cottonseed, corn, peanut, and olive. When you think of olive oil as "too rich" for everyday use in cooking, that's what you mean. Most everyday salad or cooking oils are soy and/or cottonseed oil; if you just buy a cheap-label bottle of soya oil, you'll save a lot of money. Keep it tight and cool.

15. 3.41 CFR 21.
16. The column in question was in *The San Francisco Chronicle* on December 16, 1970.

Chapter Thirteen: "What If We Do?"

1. On very technical grounds it can be said that DDT is more than one insecticide, depending on chemical configuration, but that makes no difference to anything in this book and we have no reason to go into it. We also include in statements about DDT its close chemical relatives such as DDD and DDE, unless there is a reason to distinguish. "DDT residues" may be other, related compounds, but ordinary-language usage is good enough for our purposes.
2. Ants aerate the soil and prey on harmful root organisms. Earwigs—often killed by householders only because they look "ugly"—keep down the population of other insects. Pillbugs help to keep the soil nourished. Aphids are an important part of the diet of many helpful insects, including ladybird beetles. And so on. The garden trick is to eliminate "pests" but to maintain balance.
3. See for instance McWilliams.
4. In America this is particularly true; less so in the rest of the Western World. Mad Scientists exist, of course, wherever there is anything for them to play with. In the technological rush following World War II, and with the need to restore food production in a devastated country, Russia too turned them loose. Since these scientists thrive on bureaucracy and bureaucratic ignorance, they are still there. For an excellent nonfood example, consider the SST. The difference is that in Russia the Mad Scientists do not dominate the agencies they work for, so that knowledge is better able to catch up with folly and might even yet do so. Folly, furthermore, is not the same as cynical and callous greed. See Marshall Goldman's illuminating and carefully balanced essay. In China, where the linguistic base is not Indo-European and the traditions are entirely different, Chairman Mao seems able to combine some genuine ecological values with revolution, though of course things are far from perfect. Orleans and Suttmeier have presented a really intriguing summary of developments there.
5. Much of the agricultural research at the university is supported by the Giannini Foundation, set up and funded by the family that founded the Bank of America. Billions of the bank's dollars are involved in California agriculture, and it is itself one of the state's major landowners, through subsidiaries. It is the bank euphemized as "The Bank of the West" in Steinbeck's *The Grapes of Wrath.*
6. This is aside from his being himself a director of a food processing firm which also manufactures additives.
7. Farm workers represented by the Teamsters Union have no such protection.

8. Heptachlor cannot now be used on foods at all.
9. In the copy of the book which we borrowed from the Berkeley Public Library, someone has penciled in at this point, "You must be retarded!"
10. Unless the area of the Redwoods National Park in northern California is enlarged, in a particular direction, very soon, there will be no redwood ecosystem left intact on the planet. Scott Thurber of *The San Francisco Chronicle*, a redwoods expert, has often written on this subject.
11. *Science News-Letter*, February 26, 1966.
12. *The New York Times*, April 30, 1969. The figure is lower than that of 1960, largely because of the substitution of other insecticides—*not* because of any diminution of their overall use. On the contrary. Exports of DDT have continued to increase.
13. See for example a *Los Angeles Times* story distributed by The *Times-Post* Service and published in *The San Francisco Chronicle* for December 31, 1970. Not the day on which most people read their newspapers most carefully.
14. The extinction of species is not merely poignant. If it is arbitrary as in these cases, it leaves ecological niches abandoned, with consequences that cannot be foreseen. It also wipes out stores of genetic information whose future usefulness is equally unforeseeable. There is not room to treat these subjects here; see my *America the Raped* and the many sources therein and, on genetic information, my "The Trees on the Back Lot."
15. Quoted by Winter.
16. Cirrhosis is a degenerative disease marked by the excessive growth of connective tissue in the liver and a consequent contraction of that organ. The Public Health Service lists cirrhosis of the liver as the eleventh cause of death in 1967; it accounted for 1.5 percent of total deaths by their measurement, a total of 27,816 in that year.
17. *Medical World News*, August, 23, 1968.
18. Insulin, a hormone, is a factor in keeping the blood sugar level even; remember the hyperglycaemia in the monkeys. We can never overstress that when we talk about the body we are talking about a single ecosystem. Dr. Martin is quoted by Winter.
19. Deichmann, Radomski, and Rey in *Industrial Medicine and Surgery*, March, 1968.
20. Compare Note 4 to this Chapter. If you really want to feel pro-Russian it might help to know that it is also illegal for a Russian, at sea or otherwise, to kill a dolphin.
21. They also filed a lawsuit, with other coplaintiffs, having to do with the Alaska pipeline. That doesn't concern us here. See *Science* for June 12, 1970, which includes an extensive discussion of the Appeals Court rulings on all three cases.
22. Among the totally banned (Hickel's successor has not rescinded the order—at least publicly—as of the fall of 1971) are aldrin, dieldrin, endrin, DDD, 2,4,5-T, some mercury compounds, and a few others.
23. An Associated Press story that moved on January 15, 1971, called 2,4,5-T a "weedkiller." It will come as news to the Vietnamese that it kills only weeds. Semantic fast shuffles never end.
24. To nobody's surprise, the EPA has since announced that neither DDT nor 2,4,5-T is an "imminent hazard." Cancellation proceedings (not suspension

proceedings) are still going on, and may go on forever unless either Senator Nelson or the EDF suddenly gets very lucky.

25. The chairman of the House Committee on Agriculture is a Texan.

Chapter Fourteen: Birth Defects at Home and Abroad

1. These things have such long and fascinating names that we feel compelled to say that the names are really rather simple. Hydrocarbons involve chains of carbon atoms comparable to those diagrammed in the section on fats. "2,4,5-trichloro—," as a prefix, simply means that three chlorine atoms are attached, respectively, to the second, fourth, and fifth carbon atoms in the chain. Take away the third one and the prefix becomes "2,4-dichloro—."
2. We want to stress that we are not talking about a My Lai incident, in which each participant must answer for his individual responsibility. There is no doubt that many individuals, on lower levels, who are engaged in defoliation are innocent of any intent to starve civilians and even ostensible allies. Much pseudoscience is talked by middle-level officers in Vietnam; some of them may even believe it, and certainly most individuals under them do. Nor can a soldier usually see an overall pattern, much less a strategy. But if nations could not con their own soldiers, there would be no wars at all.
3. Anita Johnson was a researcher for Turner—but Turner tells none of this story in *The Chemical Feast.*
4. There is no reason to assume that DOD knew of the teratogenic effects of 2,4,5-T when they began using it in Vietnam; that was plain dumb engineering-mentality behavior. But they obviously kept using it, and escalating its use, after they knew.
5. Richard Homan of *The Washington Post* gets the credit for this detail.
6. The Pentagon version (from a letter to a reporter for *The Yale Daily News*) is that "2,4,5-T, when fed in large amounts to highly inbred and susceptible mice and rats, gave a higher incidence of birth defects than was normal for these animals." Except for the "large amounts," that's true—but how cleverly they put it! The animals are "inbred" (all laboratory rodents are "inbred"—that's meaningless) and "susceptible" (anyone who suffers an effect is susceptible to it by definition). The dosages in the experiments varied considerably. The equivalents to a 150-pound human of the rat dosages would range from about three-tenths of a gram to about three grams. There are about 28⅓ grams to an ounce.
7. If all abstainers had voted with the United States, the resolution would still have carried, 58 to 38. The two stalwart allies of the United States were Australia and Portugal. Portugal uses defoliation and crop destruction (with American chemicals) in Angola.
8. Or maybe some tiny trace of dioxin always and inevitably remains. The effect is the same.
9. Orange is 50 percent the *n*-butyl ester of 2,4,5-T and 50 percent the *n*-butyl ester of 2,4-D. Purple is the same thing except that the *iso*-butyl ester of 2,4-D makes up 20 percent of the overall total. White is mostly a salt of 2,4-D, and some picloram. Blue is sodium cacodylate and a little cacodylic acid in salt water.
10. Distributed by The *Times-Post* Service; *The San Francisco Chronicle,* July 13, 1970.

11. *The San Francisco Chronicle*, December 17, 1970.
12. See Note 24 for Chapter Thirteen.
13. For a critical, but considered, examination of the strategy in detail, see the article by Banning Garrett.

Chapter Fifteen: The Nazis' Gift to Your Diet

1. Sherrill provides some comparative figures for pork chops, beef and pig liver, contrasting Sweden, before the 1966 ban, with Denmark, and then giving 1968 figures for Sweden. The effect of the mercury ban was dramatic.
2. Information is from Swedish sources, the Public Health Service, the National Communicable Diseases Center, and a private report to Senator Philip A. Hart of Michigan. Bud Boyd, former fish and game columnist for *The San Francisco Chronicle*, did extensive research on mercury after contaminated pheasants were found in California in the winter of 1969–70. See his columns for July 21 and August 13, 1970. Also a *Times-Post* Service story in *The San Francisco Chronicle*, July 30, 1970.
3. The newspaper story appeared in *The San Francisco Chronicle* for July 31, 1970. Britain's tolerance is one-fifth of ours; Denmark's is one-tenth. Germany has a zero tolerance, which follows the WHO recommendation. The California pheasants mentioned in the above Note were tested at from 1.6 to 4.7 ppm.
4. *The San Francisco Chronicle*, December 18 and 24, 1970, from The *Times-Post* Service and Associated Press respectively.
5. O,O-Dimethyl-2,2-dichlorovinyl phosphate, called DDVP for short.
6. Because their energy (*ergon* means "work" in Greek) is derived from choline. Choline, you'll remember, is a B vitamin. Another interconnection.
7. An "esterase" is an enzyme that acts on an ester. An ester is a chemical derived from combining an alcohol with an acid. Choline is chemically an alcohol, so acetylcholine in an ester. Cholinesterase is a *specific* enzyme: it acts on a particular substance. There are nonspecific esterases, which act on several different substances. One class is called aliesterases, mentioned later in the text.
8. You'll notice the presence, again, of the B vitamin choline. This time an ester is formed by combination with succinic acid.
9. *Science News-Letter*, June 17, 1967.
10. Consider that statement tentative. See Pekas and Paulson.
11. Figures are from *The Statesman's Year Book, 1970–71*.
12. This story is in my "The Trees on the Back Lot," and demonstrates the idea of saving genetic information, mentioned earlier. A genetic characteristic that existed only in two useless trees—products of an earlier, unsuccessful experiment—proved to be the one that saved California's oranges.
13. Several decades later, researchers "mined" the soil and isolated the substances we now call antibiotics. They are substances produced specifically by soil microbes (which can be killed by chemical sprays). That stuff isn't called *terra*mycin for nothing; the microbe that "manufactures" it came from Indiana soil.
14. Life will not, of course, literally disappear, though some species may. Humans will fight over what phosphorus is left, and some will presumably win and the others die. It's like the population problem—not everybody will die as the resources dwindle. Unless, of course, there's fallout. Etc.

15. There is a fuller explanation of this phenomenon in my *America the Raped.*
16. Associated with Albrecht in the experiments were Dr. Francis Pottenger, Jr. (whose later work is discussed in Chapter Eight) and Dr. Ira Allison.
17. There are many organic gardening manuals that provide detailed instructions for sanitary composting. It should be noted that it violates the letter of sanitation ordinances in some locations. There are also many manuals on civil disobedience.
18. *The San Francisco Chronicle*, January 13, 1971.

Chapter Sixteen: The Consie Manifesto

1. Which we should not forget still exists under another name.
2. Ballantine Books still lists the novel in print under the title *The Space Merchants,* in paperback, but it's hard to find a copy.
3. Send a quarter and a couple of eight-cent stamps to Consumers Cooperative of Berkeley, Inc., 1414 University Ave., Berkeley, Calif. 94702. Ask for a copy of "How NOT to Use Toxic Pesticides in the Home and Garden." It is a superb small book produced by a group of consumer types and some scientists, working together as a part of The Consie Conspiracy. The four-book bibliography is valuable.
4. Larry Miller, interviewed by Lucy Oakley in *The Daily Californian,* February 8, 1971.
5. The Berkeley Co-op has now put in an entire organic food department as a result of member pressure. They would not have done it on their own, no matter what anybody tells you about co-ops.
6. They advertise a DDT-less apple. They mean that no DDT is used on the tree or in the soil.
7. That is a statistical statement. Younger and more progressive doctors may be better; your 80-year-old doctor may be the best. The point is to remind you that doctors are not by definition nutritional experts at all.

Bibliography

NOTE: Citation of newspaper and some magazine sources is in the notes, and not repeated here. The magazines *Consumer Reports* and *Consumer Bulletin* were used generally and are cited in the notes only when quoted.

What follows is a list of materials used in preparing the book, NOT a recommended reading list. We have added comments, including some recommendations, after some of the titles, to help those who may wish to pursue the subject further.

Anderson, Oscar E., Jr., *The Health of a Nation: Harvey W. Wiley and the Fight for Pure Food* (Chicago: The University of Chicago Press, 1958).

Ayres, J.C., A.A. Kraft, H.E. Snyder, and H.W. Walker, eds., *Chemical and Biological Hazards in Food* (Ames, Iowa: The Iowa State University Press, 1962). Mad Scientists rampant. Essays by Mrak and Cannon are quoted in the text. John C. Ayres is a member of the FPC along with Mrak and Cannon.

Bicknell, Franklin, *Chemicals in Food and in Farm Produce: Their Harmful Effects* (London: Faber and Faber, 1960). A standard work by an outstanding physician.

Bitman, Joel, Helene C. Cecil, and George F. Fries, "DDT-Induced Inhibition of Avian Shell Gland Carbonic Anhydrase: A Mechanism for Thin Eggshells," *Science*, May 1, 1970. See comments under Peakall.

Boffey, Philip M., "Herbicides in Vietnam: AAAS Study Runs into a Military Roadblock," *Science*, October 2, 1970.

———, "Nader's Raiders on the FDA: Science and Scientists 'Misused,'" *Science*, April 17, 1970.

Braun, Armin C., *The Cancer Problem: A Critical Analysis and Modern Synthesis* (New York: The Columbia University Press, 1969).

Brown, Lester R., "Human Food Production as a Process in the Biosphere," *Scientific American*, September, 1970. The entire issue is a splendid guide to the ecology of the planet.

Campbell, Persia, *Consumer Representation in the New Deal* (New York: The Columbia University Press, 1940).

Cannon, Paul R.: see Ayres *et al.*

Carper, Jean, "Secretary Finch Is Not Alarmed," *The Nation*, March 9, 1970. A good report on the fuss about cyclamates.

Carson, Rachel, *Silent Spring* (Boston: Houghton Mifflin Co., 1962; Fawcett Crest Books, 1964).

Carter, Luther J., "Conservation Law I: Seeking a Breakthrough in the Courts," *Science,* December 19, 1969.

———, "Conservation Law II: Scientists Play a Key Role in Court Suits," *Science,* December 26, 1969.

———, "Environmental Law: Courts Demand DDT Action, Block Pipeline Road," *Science,* June 12, 1970.

Code of Federal Regulations, Title 21, Volume 1 (1970).

Courtney, K. Diane, D.W. Gaylor, M.D. Hogan, H.L. Falk, R.R. Bates, and I. Mitchell, "Teratogenic Evaluation of 2,4,5-T," *Science,* May 15, 1970.

Cox, James L., "DDT Residues in Marine Phytoplankton: Increase from 1955 to 1961," *Science,* October 2, 1970. The study reported here took place in Monterey Bay, California.

CRC Handbook: see Furia.

Crow, James, "Do Chemicals Sow the Seeds of Genetic Change?" *Medical World News,* April 26, 1968. Anything that Dr. Crow says about genetics is worth listening to.

Cummings, Richard Osborn, *The American and His Food* (Chicago: The University of Chicago Press, 1940).

Darlington, Jeanie, *Grow Your Own* (Berkeley: The Bookworks, 1970). Includes a good bibliography on organic gardening.

Davis, Adelle, *Let's Eat Right to Keep Fit* (New York: Harcourt, Brace and World, 1954). Recommended for information, though we have serious reservations about so many pills so freely used.

———, *Let's Get Well* (New York: Harcourt, Brace and World, 1965). Has an error or two but the charts in the back are worth the price, if you remember that spinach grown in one place is not the same as spinach grown in another.

Delwiche, C.C., "The Nitrogen Cycle," *Scientific American,* September, 1970.

Department of Agriculture, *Food for Us All: The Yearbook of Agriculture, 1969* (Washington: The U.S. Government Printing Office, 1969). A mixture of solid information and best-fed-people-in-the-world propaganda.

Deutsch, Ronald M., *The Nuts among the Berries* (New York: Ballantine Books, 1961; revised edition, 1967). Straight AMA propaganda. Stare wrote the introduction and found the book marvelous.

Epstein, Samuel S., Willa Bass, Elsie Arnold, and Yvonne Bishop, "Mutagenicity of Trimethylphosphate in Mice," *Science,* May 1, 1970. TMP is a gasoline additive, not a food additive, but the paper contains an excellent description of genetic and mutagenic mechanisms, and test methods, in general.

Epstein, Samuel S., Alexander Hollaender, Joshua Lederberg, Marvin Legator, Howard Richardson, and Arthur H. Wolff, "Wisdom of Cyclamate Ban" (letter), *Science,* December 26, 1969.

Food and Agriculture Organization and World Health Organization,

Joint FAO/WHO Expert Committee on Food Additives, *Evaluation of the Carcinogenic Hazards of Food Additives* (5th report, FAO Nutrition Meetings Report Ser. No. 29, Rome, 1961). Other FAO, WHO, and joint reports go on and on about food additives; it's impossible to list them all, but the WHO contributions especially are an excellent balance to American propaganda.

Food and Drug Administration, "Nutrition, Nonsense and Sense," May, 1967. The "fact sheet" quoted in the text.

Food and Drug Administration, *What Consumers Should Know about Food Additives* (Leaflet No. 10, Department of Health, Education and Welfare, Washington, D.C., 1959).

Food and Nutrition Board, National Academy of Sciences-National Research Council, *Recommended Dietary Allowances* (Publication No. 1146, 6th Revised Edition, Washington, D.C., 1964).

Food and Nutrition Service, Department of Agriculture, *Food Donation Programs* (Publication No. PA-667, Washington, D.C., July, 1970).

Food Protection Committee, Food and Nutrition Board, National Academy of Sciences-National Research Council, *Chemicals Used in Food Processing* (Publication No. 1274, Washington, D.C., 1965). This is the "FPC list" referred to frequently in the text; it lists chemical additives according to functional categories, and gives chemical formulas for synthetic flavorings.

———, *Food Packaging Materials—Their Composition and Uses* (Publication No. 645, Washington, D.C., November, 1958).

———, *Principles and Procedures for Evaluating the Safety of Food Additives* (Publication No. 750, Washington, D.C., December, 1959).

———, *Problems in the Evaluation of Carcinogenic Hazard from Use of Food Additives* (Publication No. 749, Washington, D.C., December, 1959). This is not quite about how cancer is good for you, but it comes close.

———, *The Relation of Surface Activity to the Safety of Surfactants in Foods* (Publication No. 463, Washington, D.C., October, 1956).

———, *The Safety of Artificial Sweeteners for Use in Foods* (Publication No. 386, Washington, D.C., August 1955; reprinted May, 1959).

———, *The Safety of Mono- and Diglycerides for Use as International Additives in Foods* (Publication No. 251, Washington, D.C., December, 1952).

———, *The Safety of Polyoxyethylene (8) Stearate for Use in Foods* (Publication No. 646, Washington, D.C., December, 1958).

———, *The Safety of Polyoxyethylene Stearates for Use as Intentional Additives in Foods* (Publication No. 280, Washington, D.C., July, 1953).

———, *Safe Use of Pesticides in Food Production* (Publication No. 470, Washington, D.C., November, 1956).

———, *The Use of Chemical Additives in Food Processing* (Publication

No. 398, Washington, D.C., 1956). An earlier version of Publication No. 1274, the "FPC list."

———, *The Use of Chemicals in Food Production, Processing, Storage and Distribution* (Publication No. 887, Washington, D.C., 1961).

———, *Use of Human Subjects in Safety Evaluation of Food Chemicals* (Publication No. 1491, Washington, D.C., 1967).

Food Technology, Journal of the Institute of Food Technologists, July through December, 1969.

Furia, Thomas E., ed., *Handbook of Food Additives* (Akron, Ohio: Chemical Rubber Co., 1968).

Garrett, Banning, "Vietnam: How Nixon Plans to Win the War," *Ramparts,* February, 1971.

Goldman, Marshall I., "The Convergence of Environmental Disruption," *Science,* October 2, 1970.

Graham, H.D., *et al.*, eds., *The Safety of Foods:* An International Symposium on the Safety and Importance of Foods in the Western Hemisphere, University of Puerto Rico, Mayagüez, P.R. (Westport, Conn.: The Avi Publishing Co., Inc., 1968).

Grant, Doris, *Your Bread and Your Life* (London: Faber and Faber, 1961).

Gross, Edward, "Nuclear Energy and Edibility," *Science News,* March 22, 1969.

Gruchow, Nancy, "Curbs on 2,4,5-T Imposed," *Science,* April 24, 1970.

von Haller, Albert, *The Vitamin Hunters,* translated by Hella Freud Bernays (Philadelphia: Chilton Co., 1962; orig. Dusseldorf: Econ Verlag GMBH, 1959).

Hamilton, Andrew, "FDA: New Pressures, Old Habits Bring a Change at the Top," *Science,* January 16, 1970.

Holcomb, Robert W., "Insect Control: Alternatives to the Use of Conventional Pesticides," *Science,* April 24, 1970.

Holum, John R., *Elements of General and Biological Chemistry* (New York: John Wiley and Sons, 1962).

Hueper, W.C., and W.D. Conway, *Chemical Carcinogens and Cancers* (Springfield, Ill.: C.C. Thomas, 1964).

Hunter, Beatrice Trum, *Gardening without Poisons* (Boston: Houghton Mifflin, 1964; New York: Berkley Medallion Books, 1971). A good overall guide to organic gardening.

Huntington, Samuel P., "Bases of Accommodation," *Foreign Affairs,* July, 1968 (reprinted in *Current,* October, 1968, as "Ending the War in Vietnam").

Igel, Howard J., Robert J. Huebner, Horace C. Turner, Paul Kotin and Hans L. Falk, "Mouse Leukemia Virus Activation by Chemical Carcinogens," *Science,* December 26, 1969. A paper suggesting that ingestion of chemicals may lead to a virus leukemia.

Irving, George W., and Sam R. Hoover, eds., *Food Quality* (Washington:

American Academy for the Advancement of Science, 1965). Proceedings of a December, 1962, AAAS symposium on the effects of processing on the quality of food.

Johnson, Eric V., "DDT and Safer Substitutes" (letter), *Science,* October 2, 1970.

Johnson, Paul E., "Health Aspects of Food Additives," *American Journal of Public Health,* June, 1966.

Johnson, Wendell, *People in Quandaries* (New York: Harper and Brothers, 1946).

Kirk-Othmer Encyclopedia of Chemical Technology, 2nd edition, 1963.

Kissinger, Henry A., "Viet Nam Negotiations," *Foreign Affairs,* January, 1969.

Knipling, Edward F., "Alternative Methods of Controlling Insect Pests," *FDA Papers,* February, 1969. Prochemical propaganda.

Kramer, Joel R., "Pesticide Research: Industry, USDA Pursue Different Paths," *Science,* December 12, 1969.

Landis, James M., *Report on Regulatory Agencies to the President-Elect* (Washington, D.C., 1960).

Latham, M.C., *Human Nutrition in Tropical Africa* (Rome: Food and Agriculture Organization, United Nations, 1965).

Legator, M.S., K.A. Palmer, S. Green, and K.W. Petersen, "Cytogenic Studies in Rats of Cyclohexylamine, a Metabolite of Cyclamate," *Science,* September 12, 1969.

Lindsay, Dale R., "Food Safety," *FDA Papers,* July–August, 1970.

Longgood, William, *The Poisons in Your Food* (New York: Simon and Schuster, 1960).

Loyd, Harold J., and Harold F. Breimyer, *Distribution of Food to Lower Income Families in Harrison County* (Columbia, Mo.: University of Missouri Agricultural Experiment Station, 1969).

Manufacturing Chemists' Association, Inc., *Food Additives—What They Are/How They Are Used,* 1961.

Marine, Gene, *America the Raped* (New York: Simon and Schuster, 1969; Avon Books, 1970).

———, "The Trees on the Back Lot," *National Parks and Conservation Magazine,* August, 1970.

Marx, Wesley, *The Frail Ocean* (New York: Coward-McCann, Inc., 1967; Ballantine Books, 1970).

McCance, R.A., and E.M. Widdowson, *Breads White and Brown: Their Place in Thought and Social History* (Philadelphia, J.B. Lippincott Co., n.d.). A fascinating mine of information, but its conclusions are in error and have since been disproven.

McCarrison, Sir Robert, *Studies in Deficiency Disease* (New York: Oxford University Press, 1921; Milwaukee: Lee Foundation for Nutritional Research, 1945).

McGraw-Hill Encyclopedia of Science and Technology, 2nd edition, 1966.

McKusick, Victor A., *Human Genetics* (Englewood Cliffs, N.J.: Prentice-Hall, Inc., 1964).
McWilliams, Carey, *Ill Fares the Land* (Boston: Little, Brown and Co., 1942).
Mead, Margaret, *Food Habits Research: Problems of the 1960s* (Washington: National Research Council-National Academy of Sciences Publication No. 1225, 1964).
Moody, Paul Amos, *Genetics of Man* (New York: W.W. Norton and Co., 1967).
Mooney, Booth, *The Hidden Assassins* (Chicago: Follett Publishing Co., 1966).
Mrak, Emil M.: see Ayres *et al.*
Mueller, Marti, "Nader: From Auto Safety to a Permanent Crusade," *Science,* November 21, 1969.
Naeye, R.L., M.M. Diener, and W.S. Dellinger, "Urban Poverty: Effects on Prenatal Nutrition," *Science,* November 21, 1969. Demonstrates among other things that babies born to the urban poor are 15 percent smaller than those born to others. Much additional information.
National Research Council-National Academy of Sciences, *Chemicals in Modern Food and Fiber Production* (Publication No. 1033, Washington, D.C., 1962). Symposium conducted by The American Chemical Society and The Agricultural Research Institute.
———, *Enrichment of Flour and Bread: A History of the Movement* (Bulletin No. 110, Washington, D.C., November, 1944).
Neel, James V., "Lessons from a 'Primitive' People," *Science,* November 20, 1970.
Nelson, Bryce, "Herbicides: Order on 2,4,5-T Issued at Unusually High Level," *Science,* November 21, 1969.
Olkowski, Helga, ed., *How NOT to Use "Toxic" Pesticides in the Home and Garden* (Berkeley: Consumers Cooperative of Berkeley, Inc., 1970).
Olney, J.W., "Brain Lesions, Obesity and Other Disturbances in Mice Treated with Monosodium Glutamate," *Science,* May 9, 1969.
———, and L.G. Sharpe, "Brain Lesions in an Infant Rhesus Monkey Treated with Monosodium Glutamate," *Science,* October 17, 1969.
Organic Gardening and Farming Magazine, eds., *The Organic Way to Plant Protection* (New York: World Publishers, 1969).
Orgeron, J.D., *et al.*, "Methemoglobinemia from Eating Meat with High Nitrite Content," *Public Health Reports,* March, 1957.
Orians, Gordon H., and E.W. Pfeiffer, "Ecological Effects of the War in Vietnam," *Science,* May 1, 1970.
Orleans, L.A., and R.P. Suttmeier, "The Mao Ethic and Environmental Quality," *Science,* December 11, 1970.
Papazian, Haig P., *Modern Genetics* (New York: W.W. Norton and Co., 1967).

Peakall, David B., "*p, p'*-DDT: Effect on Calcium Metabolism and Concentration of Estradiol in the Blood," *Science*, May 1, 1970. This paper and the one by Bitman *et al.* describe experiments backing up the idea that DDT causes the thin-shell phenomenon in birds. The footnotes to the two papers give references to much of the earlier evidence.

Pekas, J.C., and G.D. Paulson, "Intestinal Hydrolysis and Conjugation of a Pesticidal Carbamate in vitro," *Science*, October 2, 1970.

Present Knowledge in Nutrition, prepared from articles written for *Nutrition Reviews* (New York: The Nutrition Foundation, Inc., 1967).

President's Science Advisory Committee, *Report of the Panel on Food Additives*, May, 1960; printed as a public service by The Nutrition Foundation, Inc., New York City.

Price, J.M., C.G. Biava, B.L. Oser, E.E. Vogin, J. Steinfeld, and H.L. Ley, "Bladder Tumors in Rats Fed Cyclohexylamine or High Doses of a Mixture of Cyclamates and Saccharin," *Science*, February 20, 1970. *Not* the whole story, and makes an unjustifiable implication of safety. These are the Abbott Labs and FDA types.

Price, Weston A., *Nutrition and Physical Degeneracy* (New York: Paul B. Hoeber, Inc., 1939; Los Angeles: American Academy of Nutrition, 1950).

Reid, Mary E. "Interrelation of Calcium and Ascorbic Acid," *Physiological Review*, XXIII, 1943.

Ridgeway, James, untitled column, *The San Francisco Bay Guardian*, December 23, 1970.

Rienow, Robert, and Leona Train Rienow, *Moment in the Sun* (New York: The Dial Press, 1967; Ballantine Books, 1970).

Rorty, James, and N. Philip Norman, *Tomorrow's Food: The Coming Revolution in Nutrition* (New York: Prentice-Hall, Inc., 1947; revised 1956).

"Salting the Pure Food Mine" (no author given), *Science News*, October 4, 1969.

Sanders, Howard J., "Food Additives," *Chemical and Engineering News*, October 10 and 17, 1966. This special supplement is pretty naked about the profit motive behind all the PR.

Schrödinger, Erwin, *What Is Life? and Other Scientific Essays* (Garden City: Doubleday Anchor Books, 1956).

Sherman, Henry C., *The Nutritional Improvement of Life* (New York: The Columbia University Press, 1950).

Sherrill, Robert, "The Real Villains," *The Nation*, September 14, 1970.

Shute, E.V. and W.E., *Alpha-Tocopherol in Cardiovascular Disease* (Toronto: Ryerson Press, 1954).

Southwick, Thomas P., "Faster FDA Action Asked in Lawsuit," *Science*, June 26, 1970.

Stahl, Franklin W., *The Mechanics of Inheritance* (Englewood Cliffs, N.J.: Prentice-Hall, Inc., 1964).

Stare, Frederick, "The Balance Is All in Their Favor," *Life,* March 6, 1970. As Charles Schulz' famous beagle says, "Bleagh!"

Sternsher, Bernard, *Rexford Tugwell and the New Deal* (New Brunswick, N.J.: The Rutgers University Press, 1964).

Swan, Lester A., *Beneficial Insects* (New York: Harper and Row, 1964).

Swenerton, Helene, Ruth Shrader, and Lucille S. Hurley, "Zinc-Deficient Embryos: Reduced Thymidine Incorporation," *Science,* November 21, 1969. A paper suggesting a mechanism whereby a tiny mineral deficiency in diet may lead to gross congenital malformations. It's already known that zinc deficiency *does* lead to such fetal malformations in rats.

Tatkon, Daniel, *The Great Vitamin Hoax* (New York: The Macmillan Company, 1968).

Taylor, Ronald B., "Nerve Gas in the Orchards," *The Nation,* June 22, 1970.

Turner, James S., *The Chemical Feast* (New York: Grossman, 1970).

Welch, Henry, and Felix Marti-Ibañez, eds., *The Impact of The Food and Drug Administration on Our Society: A 50th Anniversary Panorama* (New York: MD Publications, Inc., 1956). It says everybody's wonderful.

West, Irma, "Pesticides as Contaminants," *Archives of Environmental Health,* November, 1964.

Whiteside, Thomas, *Defoliation* (New York: Ballantine Books, 1970).

Wickenden, Leonard, *Our Daily Poison: The Effects of DDT, Fluorides, Hormones and Other Chemicals on Modern Man* (New York: Devin-Adair, 1956).

Winter, Ruth, *Poisons in Your Food* (New York: Crown Publishers, Inc., 1969).

Woodwell, G.M., "Effects of Pollution on the Structure and Physiology of Ecosystems," *Science,* April 24, 1970.

Yannacone vs. Suffolk County Mosquito Control District et al., proceedings of court hearings, April, 1966, to April, 1967. As good a one-volume source as there is on what was known about DDT up to that time, though not easily available. The National Audubon Society used to have copies at $10 each; there may be some left.

Zimmerman, O.T., and Irvin Lavine, *DDT—Killer of Killers* (Dover, N.H.: Industrial Research Service, 1946).

Index